GOOD
TIDINGS

GOOD
TIDINGS

*The Belief in Progress from
Darwin to Marcuse*

W. WARREN WAGAR

Indiana University Press
BLOOMINGTON / LONDON

"The Kallyope Yell" is reprinted from Collected Poems by Vachel
Lindsay, copyright 1914 by The Macmillan Company. The quotation
from W. H. Auden appears in The Age of Anxiety, copyright 1947 by
Random House, Inc.

Published in Canada by Fitzhenry & Whiteside Limited, Don Mills, Ontario

Library of Congress catalog card number: 73-180484
ISBN: 0-253-32590-0

Manufactured in the United States of America

This book is for

JOHN

BRUCE

STEVEN

JENNY

CONTENTS

Contents

Part Four / The Survival of Hope, 1914–1970

ACKNOWLEDGMENTS

MY FIRST DEBT IS TO MY STUDENTS, at Wellesley College and later at the University of New Mexico, who participated in several seminars and upper-division courses during the 1960s on the history of the belief in progress. The experience of sharing my research with them has meant a great deal to me, and to this book. I also profited from an evening of "shop talk" with my colleagues in the Wellesley College Shop Club, and from the wise counsel of Professors G. Wylie Sypher and Franklin Le Van Baumer, who read the manuscript for Indiana University Press.

I am grateful to the American Council of Learned Societies and to Wellesley College for research grants that gave me the time to start *Good Tidings* during a happy year of study and writing in London in 1963–64.

Finally, I owe a curious debt to a rather priggish, difficult young man of twelve, from whom I have been separated for more than a quarter of a century. While sensible boys were camping in the mountains or playing baseball or stealing hubcaps, he spent much of the summer of 1944 writing a book called "The Philosophy of Progress." A second and much larger edition "appeared" in the summer of 1945. Leafing through these yellowing masses of juvenilia today, I am moved to two observations: they are incredibly bad; but without the questions they put in my mind, *Good Tidings* might never have been written.

W. WARREN WAGAR

May 1971

GOOD
TIDINGS

The Belief in Progress

Yet, sometimes glimpses on my sight,
Through present wrong, the eternal right;
And, step by step, since time began,
I see the steady gain of man;

That all of good the past hath had
Remains to make our own time glad,
Our common daily life divine,
And every land a Palestine.

Through the harsh noises of our day
A low, sweet prelude finds its way;
Through clouds of doubt, and creeds of fear,
A light is breaking, calm and clear.

That song of Love, now low and far,
Erelong shall swell from star to star!
That light, the breaking day, which tips
The golden-spired Apocalypse!

Henceforth my heart shall sigh no more
For olden time and holier shore;
God's love and blessing, then and there,
Are now and here and everywhere.

—*John Greenleaf Whittier* (*1851*)

I

Definitions

SINCE THE EIGHTEENTH CENTURY, nothing illuminates the spiritual character of Western civilization so well as the response of thinkers to the question of human progress. Their belief (or disbelief) in progress and their explanations of the mechanism of progress reflect the whole course of modern history, and sum up the hopes and fears of Western man.

Great ideas are inevitably difficult to define, if only because their meaning tends to change from generation to generation. Even at a particular moment they have many uses and operate on different levels of complexity. But the belief in progress poses special problems. Thinkers in all fields have filled it with every imaginable kind of substantive content. In itself the idea of progress is a thought-form, or in Arthur O. Lovejoy's terminology a "unit-idea," [1] rather than a doctrine with a prescribed ideological thrust. Depending on one's values, a belief in progress may be conservative or radical, religious or anti-religious, rational or irrational. Everything hinges on the thinker's definition of the good, by which he measures what is "better" or "worse" in the stream of history.

The word "progress" carries with it no necessary axiological implications. A Roman who *progressus fecit in studiis* advanced in his studies. The tour taken by a prince or prelate on official business was once known as a "progress." Even today, "progress" may denote nothing more than simple forward motion or development. "The army made little progress in its attempt to reach the sea." "The doctors were alarmed by the rapid progress of his disease."

During the eighteenth century "progress" acquired a further meaning. It became synonymous with the "perfection" or "betterment" or "improvement" of mankind, and theories of progress were advanced that explained how development toward continually higher temporal realizations of the good had occurred in the past, and might or would occur in the future. The best short definition of the idea of progress is still J. B. Bury's: the belief "that civilisation has moved, is moving, and will move in a desirable direction."[2] Morris Ginsberg retains it in his recent essay on progress, and much the same definition appears in John Baillie's *The Belief in Progress*.[3]

In each instance, however, the writer has found it necessary to fortify his definition with a variety of "implications" and "inferences." Bury attempts to rule out any confusion with the idea of providence by insisting that progress must be seen as "the necessary outcome of the psychical and social nature of man,"[4] a stipulation that both Ginsberg and Baillie implicitly reject. For Ginsberg, the criteria of what is desirable must be "rational," but for Baillie, they are "beyond the reach of the discursive understanding."[5] Bury describes the goal of progress as general happiness; Ginsberg as justice and equality; Baillie as wisdom, happiness, and goodness.[6] Bury and Ginsberg minimize the historical connections between Christian faith and the idea of progress, whereas Baillie, like most contemporary theologians, finds these connections of the utmost importance: the Christian hope, he explains, lies at the very "core" of the belief in progress.[7] These distinctions are not so much flat differences of opinion as a natural result of the scholar's struggle to keep bravely afloat in an ocean of possible meanings.

Constructing a comprehensive definition of the idea of progress is by no means easy. The scholar feels a temptation to insinuate his own values and to exclude from his definition any concept of betterment that he himself finds disagreeable. But, inevitably, thinkers committed to a belief in progress have reached widely divergent views on a great number of significant issues.

Consider, for example, the problem of the content of progress. Along what lines may progress be said to occur, what constitutes improvement in each of them, and how much improvement along how many lines is necessary before one may speak of progress in general? Humanity may imaginably achieve progress in technics, knowledge, methods of reasoning, or rationality; in comfort, health, material wealth, or happiness; in social relations, economic organiza-

4

tion, government, law, or justice; in the fine arts, language, literature, or aesthetic sensitivity; in freedom, goodness, spirituality, or godlikeness. Each of these, in turn, may be defined in various ways, some of them mutually contradictory.

Another important issue is the question of the subject of progress, i.e., who or what is considered to be progressing. Is the whole cosmos in process of meliorative change, or just life on earth, or mankind, or the "leading part" of mankind only, or a single race or nation only? Does progress involve improvement in nature or in nurture, in the germ plasm or in culture, in the inner life or in conduct, in the individual or in society? Or in both? Again, there are many different answers. Closely related is the question of the agency of progress, the instrument through which progress is alleged to occur, whether divine providence, laws of history or nature, organic destiny, chance, collective human effort, great men, social institutions, or some combination of these.

Then one may ask whether progress must be inevitable or conditional, and if conditional, is one speaking of a bare possibility, a likelihood, or something between the two? Also, what is the timeline of progress? Must progress be rectilinear, or could it also be spiraliform, discontinuous, or in some other way irregular? Can there be an idea of progress that sees progress only in the past or only in the future? How far back must progress be traced, or how far ahead must it be expected to continue? Can one speak of progress if life is improving in certain respects while at the same time in certain other respects it is not, or may even be deteriorating? The accounting problems involved in arriving at a belief in progress are often quite complex, even if the believer is not fully conscious of them. Most thinkers do not, in any event, keep careful accounts.*

Depending on one's definition, then, "progress" can mean almost anything at all. In the long history of the belief in progress, examples can be found for every plausible alternative. Indeed, with

* See Charles Van Doren, *The Idea of Progress* (New York, 1967). Van Doren's approach is analytical rather than historical, but his definition of the idea of progress resembles mine. For him it is the belief that there exists a known and generally irreversible pattern of change for the better in history. On the controversy concerning the nature or properties of progress, he identifies three main issues (Is progress necessary or contingent? Will progress in the future continue until the end of time? Does progress consist only in an improvement in man's products and institutions, or does it also involve improvement in human nature itself?) and five subordinate issues with respect to the types of progress (in knowledge, technology, wealth, social and political institutions, and morality). Van Doren, pp. 3–16.

only a few twists and shifts of emphasis in the conventional definitions, the idea of progress is easily followed into the sixteenth and seventeenth centuries, the Middle Ages, and even classical thought. At the other extreme, one can limit the idea of progress to the *philosophes* of the Enlightenment, or argue that progress as a historical concept dates only from the founding of modern philosophy of history by Kant, Herder, and Fichte, or of sociology by Comte and Marx.

In the present inquiry a belief in progress will be taken to mean a view of history or evolution that traces general improvement, by whatever idea of the good and by whatever agency, in the temporal life of mankind, extending from the past into the predictable or at least possible future. This is a somewhat broader definition of progress than Bury's, although on first inspection it may not seem to differ in any crucial respect.

So defined, a belief in progress is, first, a "view of history," a way of discerning meaning in the historical process (or, more broadly, in the evolutionary process). Utopism, often confused with progressivism because the two mingle intimately in many eighteenth- and nineteenth-century minds, belongs to a rather different and perhaps older tradition. Improvement is "general," suggesting the conviction that progress has occurred and can occur along a variety of lines, so that on balance humanity has achieved gains greater than its losses. Humanity has "turned out well," when all things are taken into account, and may be on its way to a still higher destiny.

But exactly what constitutes a "good" result depends strictly on the judgment of the individual believer in progress. Any attempt by the historian to define what is good inevitably deflects him from his proper calling. Two thinkers can observe the same amount of what both agree to be progress and still disagree on the larger issue of whether there has been net improvement. Two thinkers can observe the same facts and disagree profoundly on the issue of whether there has been any improvement at all. The historian, as historian, should not presume to sit in judgment on the evaluations reached by those whose thought he is studying. It is their idea of the good, and not his, that matters. *De gustibus* (or *de virtutibus*) *non est disputandum.*

The only exception to this rule envisaged in the definition relates to the sphere of human life in which improvement is supposedly traced. Our concern here is only with "temporal" existence, with the life of man in terrestrial space and time, and not with any real or imagined life hereafter. By this limitation, the orthodox Christian view of history stands outside our field of inquiry whenever it points to origins or destinies beyond the span of natural time. Also, "mankind" should be taken to mean either the whole human race or those special and superior elements in the human race who at any time are believed chosen to carry on the struggle toward a higher life.

Finally, the course of improvement "extends from the past into the predictable or at least possible future." Either past or future may be stressed, but neither may be absolutely excluded. The questions of how far progress is projected into the past or into the future, how the believer thinks he knows, and whether such progress is regarded as inevitable or not, are deliberately left open.

In actual historical experience, the idea of progress dissolves, therefore, into many separate ideas of progress, which in turn are statements of belief founded in a variety of value-impregnated worldviews. Every belief in progress, whether "scientific" or "intuitive," springs from what H. G. Wells once called the "synthetic motive," the impulse to invest life with a unifying purpose or meaning beyond the urgencies of everyday existence. The believer contrives to bring history into conformity with his faith, and looks to the future for the fulfillment of his highest hopes. As Bury pointed out, "The Progress of humanity belongs to the same order of ideas as Providence or personal immortality. It is true or it is false, and like them it cannot be proved either true or false. Belief in it is an act of faith." [8] Ideas of progress are theodicies, justifications of the ways of God or history or life, assuring us that we may take heart in the essential rightness of the scheme of things. The belief in progress is a great moral adventure. It sets itself boldly against the storm. Nothing in contemporary analytical philosophy gives us reason to believe that it has any cognitive content whatever.

But a definition of the progressivist faith need not include the observation that it is, after all, a faith. All speculative ideas of history are faiths, even those that have as their object the acknowledgment of human defeat or frustration in historical time. They are founded on belief, once again in Baillie's phrase, "beyond the reach of the discursive understanding." Man gives history whatever meaning it can have for him, just as he creates his own gods.

7

The present study aims at something much less than a complete history of the gospel of progress. Several substantial books, in French and in English, have examined at length the development of the belief in progress in its "classical" period, the eighteenth and early nineteenth centuries, from J. Delvaille's *Essay on the History of the Idea of Progress to the End of the 18th Century* and Bury's *The Idea of Progress,* to such recent studies as R. V. Sampson's *Progress in the Age of Reason* and Frank E. Manuel's *The Prophets of Paris.*[9] But little systematic attention has been given to the period since about 1880, to the generation that followed Marx, Spencer, and Darwin, and to the generation of the world wars.

In these years progress passed into general currency as the faith of the newly educated and enfranchised masses. Through Marxism it became the official hope of a world revolutionary movement (and, today, of fourteen socialist republics). It pervaded the speculative thought of scientists and metaphysicians, of theologians and sociologists, of political thinkers, economists, and not a few men of letters. In some respects, the thirty or forty years just before the First World War can be seen as the culminating period in the history of the progressivist faith. At the same time, counter-tendencies in thought, originating, for the most part, long before 1914, brought the belief in progress under attack during this same period, an attack so vigorous and so forcefully supported by the sanguinary logic of twentieth-century political history that the position of the believer in progress in the Western countries in recent decades calls to mind the position of the follower of Pascal or Bossuet in the high noon of the French Enlightenment. We who survive are like children who have caught their fathers in a lie that cannot be forgiven. We are the victims of a vast deceit, and for some of us historical hope is dead.

But this is not to say that the idea of progress has no history after 1914. Despite the holocaust of the First World War, many thinkers of the older generation continued to preach the gospel of progress. New prophets have emerged, especially since 1960, to carry on the tradition; and in some of the leading minds in the anti-progressivist camp, it is not difficult to detect lingering traces of a terrestrial hope that could have no other source than the very faith they abjure. Major fragments of the belief in progress have also remained cogent for the general public, as distinguished from the intellectuals. The idea of progress lies deeply embedded in the rhetoric of every politician, advertising writer, and commissar in the world.

8

Good Tidings is divided into four parts. The remaining chapter of Part One explores the history of the belief in progress down to the latter part of the nineteenth century. Part Two studies in detail those European and North American prophets of progress whose principal work appeared between 1880 and the First World War. The third and fourth parts discuss the attacks mounted on the belief in progress since the nineteenth century and the replies of some of its contemporary defenders in the Western world. In the epilogue, I remove my cap and gown, and offer some thoughts of my own on the problem of progress. But in all the intervening pages, I have tried scrupulously to present a value-free analytical survey of the belief in progress from Darwin to Marcuse. The historian's task, in my view, is empathy, not judgment; understanding, not censure.

At the same time, I anticipate that *Good Tidings* may disappoint some readers, not because it strives for value-neutrality, but because it fails to advance the kind of boldly overarching interpretative hypothesis that would disclose the "pattern" or "shape" of the recent history of Western man's belief in progress. What does it all "add up to"? In the next chapter, which recapitulates the history of the progressivist faith down to the 1880s, order seems to prevail over chaos. But in the second, third, and fourth parts of the book, our attention shifts to individual thinkers and to relatively specialized fields of thought. Except for the obvious point that the belief in progress reached its apogee just before 1914 and has had to fight a desperate battle to stay alive ever since, little is said about the nature and dynamics of the belief in progress itself, as opposed to the systems of individual thinkers.

I can only suggest, by way of defense, that boldly overarching interpretative hypotheses do not work well for the intellectual history of the late nineteenth and twentieth centuries. It is not difficult to see the "pattern" of the history of the belief in progress before 1880 because Western thought itself conformed to a certain broad pattern of development, from the neoclassicism of the Renaissance to the science of the seventeenth century, to the Enlightenment of the eighteenth, to the romantic-idealist revolution of the early nineteenth, to the resurgent positivism of the mid-nineteenth. All these movements no doubt have far less unity and structure than textbooks assign to them, but it is difficult to resist believing in their existence.

In the last hundred years or so, the only discernible common theme in Western intellectual and cultural history is disintegration.

Schools of thought, such as vitalism, Leninism, existentialism, surrealism, logical positivism, or psychoanalysis appear, but none dominates the age in the way that mathematical physics dominated the mind of the late seventeenth century, or Christian theology dominated the mind of the Reformation era. We may speak, and shall later do so, of a rebellion against positivism, of an age of longing and anxiety, of the discovery of the relativity of values. But these are all fundamentally negative concepts, which only underscore what must remain the central spiritual fact of the last hundred years: the ever-accelerating disintegration of Western civilization. Everything fragments. A steadily rising number of educated people are engaged in producing a steadily rising number of competing and essentially private systems of belief or escape from belief.

It follows that if we wish to understand the nature and dynamics of the belief in progress in recent years, we must consult the various individual thinkers who have published theories of progress or anti-progress, and the various rival movements of thought to which they belong. We shall find all sorts of tendencies and patterns, but none on the grand scale of earlier periods in the history of ideas. What should we have expected? If even before the 1880s several quite different versions of the belief in progress arose, it was inevitable that in our own time there should be a much greater number. As we have already noted, the idea of progress is a thought-form, not a doctrine with a specific ideological content. Like the chameleon, it takes on the coloration of its environment. In a historical milieu such as ours, which displays all the colors of the rainbow in dazzling confusion, the variety of ideas of progress and anti-progress that appear on the scene is inexhaustible.

2

Origins

E VEN A PROFESSIONAL geometer would be hard pressed to imagine all the possible configurations of speculative history. Solid and broken lines, circles, spirals, waves, pendulums, and angles appear regularly in its arcane symbology, although Frank E. Manuel, in his *Shapes of Philosophical History,* manages quite well with only two "archetypal shapes," the straight line and the circle.[1] But for a study of the belief in progress, geometry—simple or complex—is less helpful than the homely image of the balance sheet. In the final reckoning, the only relevant question is whether the movement of history results in net gain for mankind. Even circles may spiral upward to higher life; even straight lines may lead straight to hell.

Seen in this light, the belief in progress is primarily a late modern Western faith. But it was not created *ex nihilo* by the philosophers of the Enlightenment. We must begin our search for its spiritual and intellectual sources in the thought of antiquity, a search made difficult by the reluctance of ancient thinkers to philosophize systematically about history. A "philosophy of history" would have seemed a contradiction in terms to the ancient mind, both in the West and in the East, since most schools of religion and philosophy insisted that truth was timeless.

Yet thoughts about the meaning of history can be discovered in every ancient civilization. In perhaps the most representative view, especially in the settled agricultural societies, historical time was equated with natural time. The rhythms of birth, growth, and death, and the cycles of the heavens and the seasons furnished the ancient mind with a profound sense of the circularity of life. Empires rose and fell according to the same fatality. Thucydides hoped that his account of the Peloponnesian wars would be judged useful "by those who want to understand clearly the events which happened in the past and which (human nature being what it is) will, at some time or other and in much the same ways, be repeated in the future." [2]

Thucydides felt little awe for the remote past, but in an alternative ancient view of history, derived from sacred mythology, mankind had fallen from an age of primitive glory (the Golden Age of Hesiod, the *Ta T'ung* of the Confucian *Book of Rites,* the Eden of *Genesis*) to its present sinfulness and misery. Although such a view might be harmonized with the notion of cycles of political history, in either case a belief in progress could not arise. Both the cyclical and the primitivist concepts of history have survived, and continue to enlist adherents today.

The question of whether an authentic idea of progress emerged before the modern era is much more difficult to answer. Bury, with his arbitrary insistence that a faith in progress is incompatible with a faith in "providence," had little choice but to argue that "the notion of Progress . . . is of comparatively recent origin." [3] But revisionist studies of the Judeo-Christian idea of history by Christian scholars, of Greek thought by Ludwig Edelstein, and of Chinese thought by Joseph Needham, show that all the essential ingredients of a belief in progress were present in antiquity, and may occasionally have come together in formal doctrines of historical progress, depending on how one reads the evidence and defines his terms. [4]

What does seem clear is that the mythology of circular time, golden ages, and extra-temporal ideal worlds did not prevent many Zoroastrian, Jewish, Christian, Hellenic, Chinese, and Muslim thinkers from seeing hope for the historical future. Accounts of the gradual progress of the arts and sciences from an unromanticized age of primeval barbarism appear in the literature of every ancient civilization. A familiar example may be found in the fifth book of the exposition by Lucretius of Epicurean philosophy, *On the Nature of Things.*

Still more significant are the anticipations of the future progress or perfection of society embedded in ancient thought. The almost Rousseauian tradition in revolutionary Taoism of an approaching *T'ai P'ing* ("Great Peace") and the Confucian imperative to bring society into conformity with the *Tao* or Way of Heaven both have a powerful futurist orientation. The same is true of the prescriptions for good government of Plato, Aristotle, and the philosophers of the Middle Stoa. Among Roman writers, Virgil prophesied a new golden age, inaugurated in his own time by the reign of Augustus; and fourth-century Christians such as Eusebius and Prudentius viewed the Constantinian imperium as the earthly counterpart of the kingdom of heaven.[5] Of special interest is the expectation in prophetic Judaism of a New Jerusalem, carried into Christianity in the form of a millennial rule of the saints. Millennialism remained vigorous all through late antiquity and the Middle Ages, achieving its most ecstatic expression in the thought of Joachim of Floris in the twelfth century.

Nowhere do we stumble upon full-fledged theories of progress that might satisfy Bury. Those who described past cultural progress, like the Epicureans, may have harbored little hope for the future; and the heralds of golden ages did not necessarily trace an orderly course of human progress in earlier times. Thinkers who saw both past and future as progressive often qualified their views by adhering to a large-scale theory of cultural cycles. But it remains clear that not all ancient thinking about history is steeped in otherworldly despair or Stoic indifference to earthly fortunes. Ancient and medieval thought disciplined the Western mind to look for meaning in history, and supplied it with myths and symbols that later generations could exploit in articulating their faith in human perfectibility.

But I am not prepared to accept the judgment of Christian apologists that the modern belief in progress is nothing more than a "Christian heresy" or a "bastard child" of the Christian hope.[6] Nor am I convinced that it could have arisen only within the "horizon of the future" established by Jewish messianism,[7] or, for that matter, only in a civilization nourished by Hellenic and Latin humanism. The various ideas of progress that first appeared in the West during the Enlightenment responded, above all, to the historical situation of Western man in the modern age. They are modern ideas, which might well have grown in the rich topsoil of modern culture without any direct inspiration from Judeo-Christian and classical sources.[8]

All this becomes clearer if we turn to the modern period itself, and identify the more immediate sources of the progressivist faith. What happened in the intellectual history of Europe between the late fifteenth and the early nineteenth centuries, I suggest, was not only a revival of classical taste, or a revolution in science, but the growth of a distinctively modern religion. It was a religion of the mind, which built no churches and inspired no rites or creeds. Yet it ministered to the same needs as Christianity, and for many intellectuals it took the place of Christianity. This new religion was a rational and liberal humanism, a celebration of the dignity of man through the cultivation of reason. It originated in the Italian Renaissance, among men such as the Florentine neo-Platonists, and reached its culmination in the philosophies of Condorcet, Comte, and the German idealists. It borrowed much of its substance from the classical, Jewish, and Christian heritage, but it was fundamentally neither classical, nor Jewish, nor Christian. The rational and liberal humanism of the modern era has always been, at its roots, a religion of man, a faith in man and human possibility, which ultimately evolved into a faith in history, or, what amounts to the same thing, a faith in progress.

I prefer not to call this new religion a "secular" faith, or worse yet a "secularized" form of Christianity, because such terms evoke the traditional Christian antinomy of the sacred and the profane. Modern humanism was suffused by a deep sense of the sanctity of human existence, and by a metaphysical passion to ground human existence in its natural and even divine milieu. The same sense of the sacred recurs in most late nineteenth-century and twentieth-century survivals of liberal humanism, and in all those heretical fragments of the new religion, such as socialism and nationalism, which have sought to carry on in its stead.

Let us not overlook the obvious. Whatever other forces helped to shape modern liberal humanism, it was above all a product of the confidence awakened by the political, economic, and cultural achievements of modern Western civilization. These achievements may or may not have constituted real progress. No matter—they were perceived as progressive by the modern Western mind. Revolutions in commerce and agriculture; the voyages of discovery; the conquest of the New World and parts of Asia and Africa; the rise of the modern state, with its great armies and navies and efficient bureaucracies; the new science of Galileo and Newton; and the genius of modern art and literature, from Michelangelo to Goethe, produced a mental climate unlike any other in history. The rapid pace of change, as well as the scope and quality of change, encouraged thinkers to invest their faith in the creative power of man and man's reason.

In time, they also came to expound theories of general human progress, but the idea of progress did not arrive with anything like suddenness. Delvaille, Bury, Ginsberg, and others have traced its evolution between the Renaissance and the Enlightenment in detail. Confidence in human power was its prime mover, yet it could not have appeared in any formal sense until certain mental barriers had crumbled, and until certain basic assumptions had become widely distributed through the thinking population of Europe.

One of these barriers was the otherworldliness of the Christian tradition, which exalted the vocation of the monk and the mystic, and attached little value to temporal affairs. This Plotinist strain in Christian thought came under strenuous attack from within the church itself during the Reformation, so that one of the results of reform was a deepening of modern man's sense of the spiritual worth of action in everyday life. But otherworldiness could not have long survived in any case, in the mental climate generated by the spectacularly visible and tangible accomplishments of "secular" civilization.

Another barrier that had to fall was the reverential attitude of medieval civilization toward antiquity, which was carried still further by many of the scholars of the early Renaissance. The assumption that nature had exhausted herself and that modern men could not compare with Greeks and Romans was questioned vigorously by thinkers such as Jean Bodin in the sixteenth century and all but demolished by the uniformitarian physics of the seventeenth-century Scientific Revolution. "Nature has in hand a certain paste which is always the same," wrote Bernard le Bovier de Fontenelle in 1688, "which she turns this way and that, unceasingly, in a thousand different ways, and from which she forms men, animals, and plants; and certainly she has not formed Plato, Demosthenes, or Homer from a clay finer or better prepared than our philosophers, orators, and poets of today." [9]

But the new science erected a barrier of its own making to the belief in general progress. The same uniformitarianism that disproved the natural superiority of the ancients also disposed seventeenth-century scientists and philosophers to turn a blind eye to the problems of history and natural history. The world was now—they assumed—precisely as it had been since the Creation, obeying the same mathematical laws; men were born with the same faculties as always, and remained prey to the same vices and follies. Viscount Bolingbroke's definition of history as "philosophy teaching by examples," although borrowed from Dionysius of Halicarnassus, and by Dionysius from Thucydides, harmonized in every way with the scientific world-view. Throughout the eighteenth-century Enlightenment, which owed most of its optimism to the teachings and methods of seventeenth-century science, relatively few thinkers brought themselves to accept the idea that any significant change, much less progress, could occur as the result of long processes of historical development. A stubbornly anti-historical bias pervades the thought of nearly every *philosophe,* not only of men like d'Alembert, Voltaire, and Rousseau, but even of the Marquis de Condorcet himself, the high priest of progress, who took little satisfaction from large portions of the historical record, and had little feeling or reverence for history as such.

Eventually the Western mind overcame its reluctance to see history as a creative process. The rise of the romantic sensibility had its part, as did the persistence of the Christian view of the meaningfulness of history, especially in the more conservative thought-world of academic Germany. After 1789 the growth of acute interest in national cultural and political origins supplied further stimulus to historical study. Science itself helped, by turning its attention at last to natural history. When scientists, from the time of Buffon, could seriously address themselves to such problems as the origins of the solar system, the history of the earth's crust, and the evolution of life, it became much easier for other thinkers—and sometimes the same ones—to develop a philosophical interest in the processes of history. Such early prophets of progress as Kant and Schelling published major works in the field of natural history, which still enjoy a prominent place in the history of scientific ideas.

As the obstacles to the appearance of a full-fledged concept of general human progress gradually fell away, other developments in European intellectual history helped in a positive way to prepare the ground for the belief in progress. The new *Naturgeschichte* of the eighteenth and nineteenth centuries is one of these. Two other developments essential to the rise of the faith in progress were the discovery, in the scientific method itself, of a plausible mechanism

of progress, and the unfolding of an idea of human unity independent of all cultural or creedal tests.

As R. V. Sampson notes in his *Progress in the Age of Reason,*
seventeenth-century scientific philosophy proposed two antithetical
definitions of scientific method.[10] In one, the Cartesian, science
started Socratically from a clean slate and produced fresh, indubi-
table, and final truth by a process of systematic scepticism and de-
duction. In the other, defended by Pascal, science could never claim
absolute validity for its research, but held itself open to correction
by later and better ratiocination. Pascal in effect proclaimed the relativity of all scientific knowledge to the method of inquiry selected
by the scientist. In a similar vein, Lockean empiricism asserted the
tentativeness of scientific investigation, since all knowledge originated in the senses, and new sensory data could at any time compel
the restructuring of scientific thought.

Whether one followed the teachings of Descartes, Pascal, or
Locke, it now became possible to envisage vast future progress
through a scientific understanding of the laws of nature and human
nature, and through the application of this understanding to the
solution of man's greatest problems. From the earliest years of the
Scientific Revolution, thinkers came forward with schemes for applying the scientific method to the relief of human needs: an un-
broken line runs from Francis Bacon's New Atlantis to the humanitarian projects of Leibniz and the abbé de Saint-Pierre to the scientific utopism of Condorcet's Tenth Epoch and Comte's Western
Republic. Although the idea of the advancement of learning did
not immediately issue in a theory of general progress, it furnished
much of the oxygen of hope for such theories, and showed men how
general progress might be achieved.

The Scientific Revolution contributed in one other vital respect,
less obvious, but not less important, to the emergence of the belief
in progress. It gave rise to a new idea of mankind as a universal and
natural community, without which a belief in general progress
could have developed only with the greatest difficulty. As popularized by the *philosophes* of the Enlightenment, the new idea of mankind undeniably drew on the spiritual capital of antiquity as well:
on the ecumenicism and cosmopolitanism of the Stoics, and on the
idea of the fatherhood of God and the brotherhood of man in the
Christian tradition. But the idea of the unity of man in the new

science and philosophy was in and of itself neither Stoic nor Christian, and largely independent of earlier traditions of thought.

At its heart lay the concept of man as a being whose scientifically observed human nature inclined him always to seek the same pleasures and avoid the same pains, and who was everywhere, when not corrupted, a rational being capable of ethical behavior in accordance with natural law. The natural man was at home in all countries and all ages. By the uniformitarian principle, his nature wholly transcended the limits of geography and time, just as the laws of gravity applied on Venus and Mars no less than on Earth. Even the revealed religions could be dismissed as mere supplements, not indispensable in the final analysis, to the natural religion of reason common to all men at all times. Finally, the empiricist psychology founded by Locke regarded all men as naturally equal, born with the same natural freedom and the same rational faculties capable of like development in like environments. From these concepts of liberty and equality sprang the great revolutionary principle of fraternity. Just as the belief in a universal common nature led the Cynic and Stoic philosophers of antiquity to the idea of *Kosmopolis,* the world-city of man, so in the eighteenth-century Enlightenment, the doctrine of human equality created a sense of world community and world citizenship from which few *philosophes* dissented in any significant respect.

The stage was set, then, during the first half of the eighteenth century, for the appearance of an idea of general progress, holding out the promise of perfectibility and embracing all or most of humanity. The *philosophes* had only to assemble into a coherent, emotionally attractive credo what was already implicit in the advanced thought of the seventeenth century.

The history of the belief in general progress from the early eighteenth century down to the 1880s is, on the whole, a history of successively broader and more far-ranging conceptions of the progress of mankind, from the enthusiastic tracts of the abbé de Saint-Pierre to the heavy volumes of the *Synthetic Philosophy* of Herbert Spencer. As Peter Gay reminds us, the French Enlightenment abounded in programs for future perfection, but produced relatively few theories of universal progress.[11] With the partial exception of Turgot's Sorbonne lectures of 1750, French theories of progress throughout the eighteenth century clung tenaciously to the inspiration of the Scientific Revolution, and went no further. Some progress, they agreed, had taken place in the past; much more was scheduled for the future; but always as a result—in Condorcet's phrase—of "the progress of the human mind." Deep distrust of every other process at work in history permeated the writings of even the most optimistic *philosophes*.

Nevertheless, I am prepared to follow Bury in classifying Saint-Pierre, Voltaire, Turgot, Chastellux, d'Holbach, Volney, and Condorcet as exponents of a doctrine of progress in some form. One might add Rousseau, if he belonged to the Enlightenment at all. Jean-Jacques described modern civilization as a disaster, and traced through history a pattern of moral degeneration. But he also argued that the transition from primeval savagery to the first civil societies represented true progress, and he hoped for a resumption of progress in the future through democracy and education.[12]

The one ingredient most lacking in eighteenth-century French ideas of progress—an organic sense of history as a process of continuous development—was incorporated into progressivist theory late in the eighteenth century and throughout the nineteenth by thinkers from every national tradition. Three principal strategies may be distinguished, each working to the same broad ends: the approach of German historical idealism, which blended a radically immanentized version of the Christian doctrine of divine providence with an ingenious amalgam of rationalist and romantic thought; the approach of the new sociology, which took as its project the discovery of the scientific laws of progress; and the approach of the evolutionists, which accepted the positivism of sociology, but sought to root the idea of human progress in a larger scheme of biological or cosmic progress.

It is only in these more ambitious theories that the full implications of the belief in progress become clear, and man is seen as a creature who improves by the very nature of things. The movement in German thought from Lessing, Herder, and Kant to Fichte, Schelling, and Hegel invested progress with a spiritual dimension, a revelation of spiritual freedom, absent in the more narrowly eudaemonist theories of the *philosophes*. The sociological approach of Saint-Simon, Comte, Marx, and J. S. Mill, foreshadowed by the work of Vico, Montesquieu, and the early political economists, explored the dynamics of social interaction more perceptively than could the *philosophes*. The evolutionism of Lamarck and Darwin, translated into an elaborate doctrine of progress by Herbert Spencer, united almost all the tendencies of progressivist thought in a grand synthesis that rivals the *Summa Theologica* of St. Thomas.

By 1880, therefore, the modern religion of liberal humanism had reached full fruition. Its rationalism had already come under fire from the romantic movement. Its cosmopolitanism and liberalism faced the challenge of the tough-minded, exclusivist nationalism of the post-1848 era. But to anyone looking for spiritual and intellectual orientation in the nineteenth century, liberal humanism and its faith in progress offered a deeply founded hope. The source of its hope was mankind, conceived as a natural and moral organism, progressing through time to a realization of its fullest potentialities. Its true believers did not worship existential man, or human power, or progress itself, but rather a fundamentally spiritual idea of mankind seen as a transcendental reality—transcendental to the life of the present—advancing toward ever-higher levels of being. When Kant looked forward to the establishment of a perfect world polity in which all of man's powers would be fully realized; when Diderot reported that posterity had replaced heaven as the hope of the philosopher; when Condorcet discovered in the progress of the human mind the strongest motives for believing that nature had set no limits to human possibility; when Comte foresaw the triumph of the positive method in every department of human culture; when Marx and Engels proclaimed the inevitable leap of mankind from the kingdom of necessity to the kingdom of freedom; when Spencer noted that evolution can end only in the establishment of the greatest perfection and the most complete happiness—these were not the ravings of prideful lunatics, but the rational hopes of the prophets of a distinctively modern religion. It is a religion that grew from, and helped to sustain, modern Western civilization in its centuries of world hegemony. Despite widespread disenchantment, it has struggled on, in a variety of forms and combinations, since 1880; and it may survive, in ways that we cannot now imagine, for many centuries to come.

PART TWO

Sons and Heirs,

1880-1914

I am the Kallyope, Kallyope, Kallyope,
Tooting hope, tooting hope, tooting hope, tooting hope;
Shaking window-pane and door
With a crashing cosmic tune,
With the war-cry of the spheres,
Rhythm of the roar of noon,
Rhythm of Niagara's roar,
Voicing planet, star and moon,
SHRIEKING of the better years.
Prophet-singers will arise,
Prophets coming after me,
Sing my song in softer guise
With more delicate surprise;
I am but the pioneer
Voice of the Democracy;
I am the gutter dream,
I am the golden dream,
Singing science, singing steam.
I will blow the proud folk down,
(Listen to the lion roar!)
I am the Kallyope, Kallyope, Kallyope,
Tooting hope, tooting hope, tooting hope, tooting hope,
Willy willy willy wah HOO!

—*Vachel Lindsay* (*1913*)

3

La Belle Époque

THE DEATH OF DARWIN in 1882 and of Marx in 1883 signaled the passing of an age. Some members of their generation, notably Spencer, lived on into the post-Darwinian era. But *la belle époque* belonged to younger men, less systematic and certainly less ponderous, although no less brilliant. Many carried on the progressivist tradition in one familiar form or another; a few mounted the first major offensive against it since the conservative reaction to the French Revolution.

This last generation before the catastrophe of 1914 is the "nineteenth century" of living memory, a period brutally debunked in the 1920s and seen through a thick haze of nostalgia in the 1960s and 1970s. The great current vogue of such eccentric products of that generation as Nietzsche, Mahler, Strindberg, Gauguin, and Henry James gives it a dark luster, which it could not have had thirty or forty years ago. At the same time, the emphasis on its more convoluted and neurotic personalities, and especially on its prophets of doom, may by now have created a totally false impression of its mental climate. The prevailing mood was optimistic. *La belle époque* was an age of affluence, imperial power, and international peace. Not all shared in its wealth, and its peace was often violated by major industrial and international crises. But no growing civilization can thrive without some inner tension. That it blundered, by a fateful miscalculation, into a world war that modern military technology rendered far more destructive than any Great Power had wanted or expected it to be should not lead the historian to make too much of those inevitable moments of tension. From the point of view of events after 1914, Gerhard Masur is no doubt right in painting a picture of post-Darwinian Europe "basking in fruitful opulence under the autumn sun of its glory, ripe for the slaughter." [1] But from the point of view of the course to which the post-Darwinians not unreasonably thought themselves committed, the picture is rather overdrawn.

In a material sense, this was no autumnal society sinking into exhaustion and senility. The industrialized countries of Europe increased their combined populations by fifty percent between 1815 and 1870, and had nearly done so again by 1914. The United States recorded a tenfold increase over the same hundred years, much of it as a result of European immigration. Aggregate wealth more than kept pace with the demographic statistics. To choose just one index of industrial growth, the two leading European producers of coal and pig iron, Britain and Germany, tripled their combined output between 1880 and 1910. Real per capita incomes doubled during the same period throughout the West as a whole, although wages lagged appreciably behind capital gains. The systems of universal public education put in good working order during these years wiped out illiteracy. The trade unions, agents now of peaceful change rather than social revolution, rose from negligible proportions in 1870 to a world membership of fourteen million people by the outbreak of the First World War. Especially after 1895, employment was full, business and foreign trade flourished as never before, and the once popular safety valve of immigration to America was little used or needed by the populations of the most industrially advanced European countries. At the same time a new wave of empire building after 1870 had installed the European Powers as owners of ninety percent of Africa, more than half of Asia, and all of Polynesia. God had called upon Europe to civilize the world, said Wilhelm II: "We are the missionaries of human progress." Occidental civilization stood on the verge of fulfilling most of the dreams of the Enlightenment: the unification of the planet, the abolition of poverty, and the replacement of tyranny and superstition by freedom and literacy.

Some problems, of course, remained unsolved. But the faith in modern man's power to solve them rationally, which permeates almost every page of the last volume of the *Cambridge Modern History*, published in 1910, seemed well founded. Material and imperial expansion had been accompanied by a striking reduction in the incidence of violence, despite the breast-beating of the cruder sort of Darwinian pundit. Except for the Revolution of 1905 in Russia, no major social or political uprising of the type common between 1789 and 1848 disturbed the civil peace of Europe, and the foundations of the twentieth-century compromise between socialism and laissez-faire capitalism had been solidly laid. From 1871 to 1914, the Great Powers fought no major wars; ample diplomatic machinery existed to prevent such wars from breaking out, provided only that the will for peace held firm on each side. Crimes of violence had been much reduced, travel made safe, civil liberties secured, the cause of parliamentary democracy greatly advanced, and pain and disease brought under a measure of control far surpassing all previous standards.

It was in the life of the spirit, if anywhere, that the signs of *malaise* could be clearly read by the sensitive observer; and even here no one could be sure whether the sickness was a passing fever incidental to a too rapid growth, the birth pangs of a new civilization, or merely the inexorable decay of an old one. A "lengthy, vast and uninterrupted process of crumbling, destruction, ruin and overthrow . . . is now imminent," Nietzsche wrote in his *Joyful Wisdom*. With the final collapse of Christian faith after centuries of erosion, "some sun seems to have set, some old, profound confidence seems to have changed into doubt." But the philosopher found himself filled with wonder and expectation. "At last the horizon seems open once more, granting even that it is not bright; our ships can at last put out to sea in face of every danger; every hazard is again permitted to the discerner; the sea, *our* sea, again lies open before us; perhaps never before did such an 'open sea' exist." [2]

This sense of change, of apprehension mixed with a keen impatience to get on with the future, characterized much of the best prophetic writing of *la belle époque*. As in the early stages of the Renaissance and Reformation, in the transition to the Enlightenment in the last decades of the seventeenth century, and in the turbulence of 1789–1815, the West found itself passing through a great *crise de conscience,* perhaps more radical than any of its predecessors, in which familiar landmarks fell out of view and a confusing variety of ideas and values, old and new, struggled for acceptance in a world suffering from spiritual vertigo.

The central fact was unquestionably the almost sudden collapse of orthodox religious faith. This collapse had been on its way since the Renaissance, and for many thinking people some form or other of the new religion of humanity had largely supplied the need for religious orientation since the Enlightenment. Yet the breach between tradition and modernity had never seemed to widen to the breaking point. Most of the *philosophes* remained theists. The German philosophers of history temporalized the Absolute without robbing it of its transcendent spirituality. But by the middle of the nineteenth century, in the thought of such representative figures as Comte, Proudhon, Feuerbach, Marx, Büchner, and Spencer, the avant-garde rejected the very essence of Christian belief and, as it were, declared war on the Christian God. In J. M. Robertson's phrase, the "turning of the balance" from belief to scepticism came in the 1840s and 1850s on the Continent and the 1870s in Britain.[3] On one side of the divide, most educated people were either believing Christians or humanists who clung to significant elements of the traditional faith; on the other side, perhaps as many as half had broken decisively with that faith, if not publicly, then in their hearts.

By 1880 the effects of this desertion by the intellectual elite had begun to reach down deeply into every class. The struggle between the advocates of the evolutionary hypothesis and religious orthodoxy, which monopolized so much contemporary discussion, was only one symptom of the crisis, more an effect than a cause. In an earlier age it is easy to imagine a compromise between evolutionism and Christianity that would have papered over the cracks and left faith apparently intact. Several ingenious experiments along these lines were, in fact, made. But times had changed fundamentally since the seventeenth century. Many of the men of *la belle époque* surged restlessly forward to new creeds.

The post-Darwinians were torn between two broad opposing tendencies in this search for orientation. The first pulled them in the more familiar direction of a religion of science, reason, and humanity; and it commonly involved adherence to the idea of progress. Positivism, for example, enjoyed its greatest success as an organized movement between 1880 and 1900, although in this period it amounted to little more than a popularization of the ideas of Comte. Many natural scientists put forward speculative systems of philosophy that found a place for the idea of progress, often grounded in an interpretation of evolutionary theory. Most sociologists and anthropologists, following Comte and Spencer, continued to regard progress as a law of social development; evolutionist assumptions, again, typically guided their research. A positivistic faith in progress also informed such disparate movements as the new liberal theology, Marxist socialism, and social Darwinism.

The other broad tendency is less easy to capture in a single phrase. In literature and the arts generally it was "the Decadence" or "neo-romanticism," a hectic and often sapless revival of values associated with the romantic movement, a reaction against mid-century realism and, at times, against the belief in progress. But as H. Stuart Hughes argues in *Consciousness and Society,* these well-worn labels fail to do full justice to the serious thought of the era. He himself elects to see the newer tendencies as a "revolt against positivism," numbering among its leaders Sorel, Bergson, Pareto, Croce, Dilthey, Freud, and Max Weber. The unity of the movement stemmed from its objections to a crude reliance on the categories of natural science in the study of man, and from its penetrating interest in the role of non-rational factors in human thought and behavior. In this way, Hughes singles out two major critical themes in post-Darwinian thought, although there were others, and the intimate alliance he negotiates between anti-intellectualism, historicism, and the new sociology of Weber seems rather forced.[4]

But Hughes's thesis of a revolt against positivism is useful. If no one overarching persuasion unified the rebels, at least they all found a certain negative unity in their antipathy to some of the dominant modes of mid-nineteenth-century thought. They reacted against the vogue of materialism in metaphysics, the coldly rationalistic epistemology of positivism, the idea of "laws" in history, and the application of principles derived especially from the physical sciences to what Dilthey called the *Geisteswissenschaften,* the "human sciences." They discerned in the thought of such characteristic celebrities of the older generation as Comte, Büchner, and Spencer an absence of warmth and intuitive understanding, a tendency to reduce living truth to cut-and-dried fact, which offended them as much as the German idealists of the romantic era were offended by the rationalism of the Enlightenment.

The two opposing tendencies flourished side by side in these years, sometimes even in the same minds, although anti-positivism steadily grew in force and influence after the arrival of the new century. Roger Martin du Gard's novel *Jean Barois* mirrors the spirit of the age with extraordinary clarity: Christian orthodoxy, positivism, and anti-positivism meet and clash fatefully in the life history of a single representative European mind. The exhausted hero's death as a penitent Catholic also calls to mind the many intellectuals who returned to that ancient fold in *la belle époque,* perhaps sensing themselves, like Léon Bloy, "on the threshold of the Apocalypse."[5] But the center of the stage was still held by the positivists and their no less infidel rivals.

On the belief in progress the struggle between positivism and its opponents had little impact at first. Some anti-positivists, it is true, placed the faith in progress high on their list of *damnanda,* and of this more later, in Part Three. But others, notably the vitalists, while rejecting the positivist view of progress, proposed another of their own. It may also be argued that faith in progress remained strongest in the English-speaking world, which suffered least from social and political instability and the decline of Christian belief. Yet if the period 1880–1914 is taken as a whole, and the Anglo-Saxon contribution weighed fully, the *crise de conscience* down to World War I was more a phase of experiment and excitement than a time in which hope actually failed. The weight of opinion still inclined toward some form of belief in progress. The leading journals of the period, of the caliber of the *Mercure de France* and *The Westminster Review,* teemed with essays on progress and incidental references to it by the thousands. Most thinking people took progress for granted. They might find a *frisson* of masochistic pleasure in toying with the notion that perhaps, after all, it should not be taken for granted. But by and large, they believed.

4

The Cult of Science

THE GRAND ASSAULT on positivism launched by many thinkers and artists of *la belle époque* is impressive, but it becomes fully comprehensible only when measured against the surviving and, in some respects, the growing power wielded by science in the life of the mind in the late nineteenth and early twentieth centuries. The average educated man followed the progress of science with enthusiasm, gave thoughtful attention to the scientific sermons of Haeckel, Metchnikoff, and Huxley, relished the novels of Zola and Wells, and may even have agreed with Lord Morley that to science fell the duty of providing mankind with the religion of the future. Evolutionary theory at all levels, from the geological to the sociological, continued to excite public interest; even many self-styled antagonists of science found it impossible to resist importing evolutionary concepts into their thought.

The greatest triumph of positivism during these years was the emergence of the social and behavioral sciences as full-blown academic disciplines. Although Comte and Spencer had laid the groundwork, only in the closing years of the nineteenth century did large numbers of professional anthropologists, sociologists, and psychologists begin to make their appearance in the universities and bookstalls of the Western world. Nearly all of them, including those tagged by recent scholars as "rebels" against positivism, were devoted practitioners of the scientific method. They freely acknowledged Comte, Spencer, and others of the mid-century positivist generation as their intellectual forebears.

The belief in progress expounded by scientists and enthusiasts of science in the post-Darwinian generation followed closely the mid-century positivist approach to progress. Among the neo-Comteans—the public disciples of Comte and not a few other thinkers—progress was confined to man's history and had no cosmic dimension. Among the neo-Spencerians man's history continued biological evolution. For both camps progress was usually defined as a law, discoverable by scientific inquiry. Many scholars thought progress inevitable. Obstacles had to be overcome and battles fought against the enemies of progress, but its scientific true-believers assigned to humanity a splendid future.

The chief agency of progress, they contended, was the source of man's knowledge of the laws of progress: science itself. We may speak without exaggeration of a "cult of science," a cult at whose shrines many thinking men worshipped with no less zeal than in the 1850s or 1860s, and with the same fanatic intolerance of other gods. The British mathematician and biologist Karl Pearson, born in 1857, and a fair specimen of the positivist mentality of his generation, captured its religiosity with warm eloquence in his early lecture, "The Ethic of Freethought." Atheism, he warned, was not enough. By contrast "freethought" aimed at the destruction of supernaturalist dogma through the slow accumulation of scientific truth and the building up of a whole new world outlook. The true freethinker assimilated "the results of the highest scientific and philosophical knowledge of his day" and made his contribution to the solution of the deepest problems of life by working to bring to light new truth. In the labor of discovery freethought found "its noblest function, its holiest meaning. This pursuit of knowledge is the true worship of man." The new religion had to become a great living force, "strong in the conviction of its own absolute rightness, creative, sympathetic with the past, assured of the future, above all enthusiastic." Destined to be "the creed of the future," it would, in making man "master of his own reason," render him "lord of the world." [1]

The purest and most explicit devotees of science and the belief in progress as a scientific principle in the post-Darwinian era were the leaders of the Positivist movement founded by Comte. From its international headquarters in the rue monsieur-le-Prince in Paris, and from two rival centers in London, Chapel Street and Newton Hall, there issued a long stream of lectures, tracts, textbooks, and periodicals dedicated to the propagation of Comte's philosophy and his religion of humanity.[2] The last decades of the nineteenth century were the golden years of Positivism as an organized world movement. Money was raised by international subscription to erect a bust of Comte in the Place de la Sorbonne, where it still stands. Positivism flourished strongly for a time in Britain, and drew many supporters in Latin America, especially in Brazil. Comte's slogan "Order and Progress" was inscribed in Portuguese on the Brazilian national flag, an accomplishment perhaps unique in the history of philosophy.

In our period the most articulate spokesmen of Positivism were the leaders of the "orthodox" wing of the movement, Pierre Laffitte in France and Frederic Harrison in Britain: earnest, devout, and essentially conservative prelates of Positivism, whose acceptance, in practice if not in theory, of the near-infallibility of Comte prevented them from adding significantly to their master's thought. Their reverential attitude stood in the way of any real understanding or dynamic use of scientific method itself, so that under their leadership Positivism hardened into a creedal orthodoxy virtually impervious to change.

But they did much to disseminate Comte's idea of progress, as had the greatest of Comte's earlier disciples, Emile Littré, until his death in 1881. Laffitte's major work was a lucid series of lectures rearranging, clarifying, and, at points, expanding the Comtean synthesis, of which only the first course, his *First Philosophy,* ever appeared in print.[3] Here Laffitte presented a detailed discussion of the most general and abstract laws relating to both "human understanding" and "the world" found in Comte's philosophy. Laffitte fixed their number at fifteen. Comte's laws of the three states of knowledge, social activity, and social sentiment were treated as the seventh, eighth, and ninth laws, respectively. The laws of progress, like all the others promulgated by Comte, were immutable; even when man seemed to modify them, he was doing so in accordance with an inner orderliness, which might elude his fallible understanding but which ruled all the same. A perfect knowledge of the laws governing all things, not achieved even by Comte himself, would disclose that man, in actual fact, changed nothing. Everything was predestined.

Laffitte also never tired of pointing out that for Positivists progress was not revolutionary, but always gradual and continuous. As Comte had already maintained, the principle of order clearly took precedence over the principle of progress. "Progress is a succession of different states ordered to a certain end, and in this succession amelioration is realized gradually. . . . In short one must understand that this progress, although indefinite, cannot exceed all bounds. Our nature, like our situation, imposes a limit toward which the amelioration of the successive states ever tends." Charlatans liked to exploit the notion of progress to advance their own selfish schemes at the expense of humanity, but "for more than thirty years I have fully understood the importance of this danger, and I have pointed it out and struggled against it with all my power. More and more, I have appealed to the principle of stability in the face of proclamations of endless change." Social action, "which cannot be carried to a successful conclusion save by continuity of effort," had to be taken with "calm and dignity." [4]

The treatise on positive morality that Comte had planned but never lived to write was duly executed in 1881 by Laffitte, who used it to present not only his conception of ethics but also his philosophy of the future. The advent of a scientific morality, he proclaimed, would mark "the crowning-point of the evolution of our race." The laws of progress called man to his moral duty whether he willed it or not. We could no more withdraw from subjection to the power of humanity "than from subjection to the double motion of the planet, and still less can we control its working." Humanity forever tended, "by a vast co-operation, towards a better state in all forms of activity, without ever falling back, even in situations which seem to be the most desperate." [5] Comte's prophecy of a Western Republic would be fulfilled, and by degrees, as the various Eastern nations caught up with the West by passing through a similar process of organic growth, humanity would unite in a world republic. Much time could be saved and useless experiments forestalled with help from Positivist missionaries dispatched from the West. Eventually mankind would constitute a single immortal organic being, composed of the good works of all souls, dead and living, growing ever purer and better in all ways.

When Laffitte died in 1903 his place was taken by Emile Corra, who published dozens of Positivist tracts during his twenty-five years as international director of the movement. Corra evinced faith in the "ceaseless progress" and the "invariable youth" of Comte's philosophy and foresaw its destiny, "not only to expand throughout Western civilization, of which it is the most precious fruit" but also to convert, sooner or later, "all men everywhere and enjoy eternal life." Even the World War did nothing to shake his faith in the future.[6]

British Positivism stressed the religious aspects of Positivism at the expense of the scientific even more than did the movement in France. Its founder, Richard Congreve, had been an Anglican clergyman before his conversion to Positivism in the mid-1850s. He led the British movement so far in the direction of sacerdotalism that he was finally driven in 1878 to break with Laffitte and transform the Positivist headquarters in Chapel Street, London, into a full-fledged "Church of Humanity," with himself as priest. Unwilling to defy Paris, the majority of the British Positivists seceded to Newton Hall, just off Fleet Street, where they established a new center under the direction of Frederic Harrison. But even Harrison, although opposed to Congreve's idea of the immediate realization of Comte's plans for an institutionalized religion of humanity, looked on Positivism primarily as a religious movement. Whereas Laffitte was a professional teacher of mathematics and the history of science, Harrison was a barrister, who had received a classical education at Oxford and had been a devout Anglican with a profound interest in theology until his conversion to Positivism. "In passing from allegiance to Christ to Humanity," he wrote near the end of his life, "I have never known the sense of spiritual conflict within, or of doubt, of abandonment, of despair, of desertion. I can recall nothing but an impression of very gradual, quite peaceful, and almost unconscious evolution of mind, a happy widening of sympathy and deepening of interest and zest in life." Like Comte himself, Harrison had recapitulated in his personal life the spiritual pilgrimage of humanity described in Comte's law of progress. Since becoming a Positivist, "no shadow of doubt in general principles has ever crossed my mind."[7]

He was most effective as a propagandist for Positivism in the many articles he contributed to Victorian journals defending the Positivist creed and often carrying the attack to its enemies. His powers as a serious controversialist rivaled those of T. H. Huxley. Much like Huxley, although the two clashed bitterly, Harrison was especially concerned to distinguish his faith from the grandiose philosophies of the evolutionists. Positivism rejected "the inhuman nothingness presented by a blank infinity of Evolution. . . . The Universe is all very grand, but it is a mere background. . . . For purposes of human progress and happiness we must think and act as if the world revolved around our globe, and Man was its master and its ruler." The struggle for survival might lead to inhumane and degenerate tendencies in man, as in lower nature, but

> we are confident that humanity, which has overcome far more ominous antagonists, has ample resources within itself to counteract any tendencies which threaten its progress. . . . Endless progress towards a perfection never, perhaps, to be reached, but to be ideally cherished in hope, a hope which every stroke of science and every line of history confirms to us, and with which every generous instinct of our nature beats in unison—such is the practical heaven of our faith.[8]

For Harrison and his Positivist fellow-believers, Humanity—the distilled human goodness which "abundantly predominates" in the "inspiring record" of human progress—completely filled the space left in their hearts by the disappearance of the Christian God; they were fortified in their faith not only by the sacred texts of Comte but also by a panoply of hymns, prayers, and rites based on Christian models.[9]

As W. M. Simon suggests, "Hegel had immunized scientifically minded Germans against the appeal of historicizing metaphysics," [10] and Comte's system found only a small circle of admirers in central Europe. But the German-speaking world had its radical scientific humanists who believed in progress on the grounds of a deep faith in science itself, or in evolution, or in both. Earlier in the century the heroes of scientific thought had been thoroughgoing materialists, such as Ludwig Büchner and Jacob Moleschott. In the Wilhelmian era the torch was handed on to the biologist Ernst Haeckel and the chemist Wilhelm Ostwald, both distinguished scientists who took a lively interest in philosophy and public affairs.

Although they owed less to Positivism than to Spinoza and the materialist tradition in German thought, Haeckel and Ostwald reached a large international public with a message that radiated faith in science and human progress. Their most successful book, Haeckel's *The Riddle of the Universe,* was first published in 1899. Its theme, in a word, was *Kulturkampf*—the culture war between the "ignorant anthropism" of revealed religion and the monistic world-view of modern physical science. The latter had proved itself capable in the nineteenth century of solving most of the great riddles of the universe. The laws of physics revealed that all things from mud to man were composed of the same unitary substance, which everywhere obeyed the same iron laws and everywhere had two aspects, matter and spirit (or energy). The laws of biological evolution—Haeckel was Darwin's German "bulldog"—disclosed step by step how this universal substance had developed from the lowest to the highest forms of existence. Man, though "a tiny grain of protoplasm in the perishable framework of organic nature," was nevertheless the most "perfect," the most "highly organized," and the most "important" species on earth.[11]

Haeckel looked forward with enthusiasm to the ultimate replacement of the churches of the world by monistic "free societies" teaching a system of monistic faith and ethics rooted in the wisdom of science. The vogue of his thought in the years just before the First World War is nicely illustrated by Arnold Zweig's sarcastic portrait in one of his novels of a typical parvenu of the Wilhelmian era, General Schieffenzahn. Against the traditional Christian arguments of the aristocrat von Lychow, Schieffenzahn offered the new pieties of scientism. He was indignant that an "old stick" like von Lychow should "come to him with quotations from the Bible, as if Haeckel's *Riddle of the Universe* had never been written." [12]

Haeckel's younger compatriot Ostwald denied the reality of matter altogether, and argued that the whole universe consisted of energy in process of ceaseless evolution. Mankind had distinguished itself among all living beings by its superior ability to transform "crude" energy into "higher" energy and thereby conquer nature. In the Germany of Kant, Ostwald had the temerity to define the law of progress in terms of a new version of the categorical imperative, which he dubbed *der energetische Imperativ:* "So act that crude energy is transformed into higher with the least possible loss." [13] He could not doubt that mankind would eventually abandon its waste of precious raw energy in warfare and class conflict, and create a world of brotherhood and equality, in obedience to the second law of thermodynamics.[14]

Another popular prophet of positivism in central Europe was Max Nordau. Born of German-Jewish parents in Budapest, Nordau pursued a remarkable career during *la belle époque* as physician, novelist, Zionist, travel writer, and philosopher of culture and history. He earned much of his prophetic reputation with one book, *Degeneration,* a querulous psychological analysis of *fin-de-siècle* art and literature first published in 1892–93.

From Nordau's perspective, the famous "decadent" writers and painters and musicians of the late nineteenth century, from the disciples of Schopenhauer to Huysmans, Maeterlinck, and Nietzsche, were lunatics suffering from inherited degeneracy and hysteria brought on by maladjustment to the rapid tempo of modern living. But his prognosis was far from despondent. Humanity was still young, he wrote, and this period of racial fatigue would probably pass. The degenerate types, weak of body and mind, would disappear from the scene. "They can neither adapt themselves to the conditions of Nature and civilization, nor maintain themselves in the struggle for existence against the healthy." The despised bourgeois *paterfamilias* would triumph over the neurotic aesthete merely by outlasting and outbreeding him. Within a century, Nordau predicted the emergence of a race of men capable of living at the faster pace dictated by progress. Art would lose its pathological quality, and ultimately, in a still more rational and scientific age, "art and poetry will have become pure atavisms, and will no longer be cultivated except by the most emotional portion of humanity— by women, by the young, perhaps even by children." As demonstrated by psychology, the march of progress was

> from instinct to knowledge, from emotion to judgment, from rambling to regulated association of ideas. Attention replaces fugitive ideation; will, guided by reason, replaces caprice. Observation, then, triumphs ever more and more over imagination and artistic symbolism —i.e., the introduction of erroneous personal interpretations of the universe is more and more driven back by an understanding of the laws of Nature.[15]

Many years later, in *The Interpretation of History,* Nordau reviewed the history of the idea of progress, concluding that it had been somewhat overworked. Men had expected and seen too much. From the point of view of modern science, the universe had no purpose, nor could any purpose be rationally discerned in the history of life. He also detected no evidence of improvement in individual human intelligence, morality, or happiness. But there had been steady progress in extending the realm of will: in the gradual acquisition by the conscious, reasoning will of more self-control and discipline, made possible by advances in knowledge. This in turn had led to an increasingly more complete adaptation of the individual to his environment, not only for the gifted few, but also for the greater part of the human race. In the future Nordau foresaw the homogenization of humanity in a scientifically managed world community marked by the absence of the exploitation of man by man, the replacement of the supernatural religions by a scientific faith, and the progressive elimination of irrational and criminal behavior. He insisted throughout that he was concerned only with progress in science and technology, but most of the familiar hopes of the Enlightenment somehow found their way into the discussion.[16]

Some of the finest scientific minds of the age prophesied the conquest of decadence and evil by science in flights of speculation far more taxing than Nordau's. The great Russian bacteriologist, Elie Metchnikoff, who succeeded Pasteur as director of the Pasteur Institute in Paris in 1895, saw opportunities for unlimited progress through the prolongation of the average life span. Many diseases and disharmonies had already been overcome in man's rise to civilization. But by eating large amounts of yogurt and following a diet low in rich foods (farewell to *la cuisine française!*), men could live to well over a hundred years in excellent health. A society dominated by wise and healthy old men, freed of the pessimism bred by fear of illness and death, would confidently submit to the dictatorship of science, and through science, of progress.[17]

Even feminism inspired a scientific theory of progress. In England Alfred Russel Wallace, the codiscoverer with Darwin of the law of natural selection, admitted that little progress had occurred in human character or happiness since earliest times. Yet mankind possessed the divine spark, which assured its ascendancy over the lower animals and its eventual triumph. In one of the books of his old age, published in 1913, Wallace discovered the secret of future progress in the hearts and wills of women. More moral by nature than men, they would use their coming emancipation from financial dependence on the male of the species in the twentieth century to practice eugenics on a massive scale. Rejecting suitors of bad moral character, and choosing to have children, if at all, only by good men, they would institute "a system of *truly natural* selection . . . which will steadily tend to eliminate the lower, the less developed, or in any way defective types of men, and will thus continuously raise the physical, moral, and intellectual standard of the race."[18] Wallace also rejected the atheistic materialism of Haeckel and Ostwald, maintaining that man's soul was immortal and that human progress continued after death "in the spirit-world."[19]

37

The preoccupation of much twentieth-century sociology and anthropology with structural and functional analysis and with case studies of very recent or very primitive societies makes it easy even for specialists to forget that both sociology and anthropology began in the nineteenth century as sciences of human progress. Sociology originated in the 1830s in the work of Comte and Mill, and, later, of Spencer, whose *Principles of Sociology* (first volume, 1876) did much to fix the term "sociology" in scholarly and popular usage. Modern anthropology began, perhaps, with Darwin; its first great period was in the 1860s and 1870s, when many of its basic concepts and tasks were defined by Sir Edward Tylor, L. H. Morgan, and Adolf Bastian. An American sociologist, A. J. Todd, confidently reported in a textbook of sociological thought published in 1918 that "from Comte onward sociologists have pretty generally agreed that the only justification for a Science of Society is its contributions to a workable theory of progress." [20] Todd's statement would have suited the anthropologists of the post-Darwinian generation just as well.

With the concept of sociology and anthropology as studies of human progress went the pragmatic corollary that the social scientist had the sacred office of lighting the way to future progress. As the British mathematician W. K. Clifford pointed out, it was the duty of mankind to learn the laws that governed the social organism and use the knowledge acquired, like the engineer who exploited his understanding of water flows to construct irrigation works. "The use which the Republic must make of the laws of sociology is to rationally organise society for the training of the best citizens. . . . Those who can read the signs of the time read in them that the kingdom of Man is at hand." [21]

Not every sociologist or anthropologist in our period would have gone as far as Clifford, but the disposition to accept the doctrine of progress and to anticipate a better future for mankind, or for some specially favored segment of mankind, was built into the methodology of both disciplines. Scholars habitually assumed that human society and culture, studied scientifically, would exhibit a similar, or even identical, pattern of evolutionary development throughout the world. The evolutionist assumption, often coupled with organismic theories and analogies, almost inevitably brought the sociologist or anthropologist to a belief in the general progress of humanity.

Each of the major social theorists of the post-Darwinian generation approached the problem of progress rather differently, and for some progress was more an underlying assumption than something to be proved scientifically. The organicists, such as Paul von Lilienfeld, Albert Schäffle, and René Worms, accepted Spencer's definition of progress as the advance from simplicity to complexity, from the loosely formed, undifferentiated societies of prehistory to the highly integrated "super-organism" of modern civilization. But it was the modern societal organism itself, more than progress as such, that absorbed their interest; the state, in particular, was viewed as the highest manifestation of human progress, as the sum and synthesis of civilization. Emile Durkheim, the most brilliant French sociologist of his time, accepted in a general way Comte's view of history as a record of progress from superstition to reason, but could not subscribe to a theory of inevitable or automatic progress. Gabriel Tarde —echoing Bagehot—charted the progress of mankind from ages of warfare and commercial rivalry to a nascent age of "discussion," but his attention centered on the cyclical process of invention, imitation, and opposition by which social change in all ages occurred.

Yet for many theorists progress remained the central problem. The Belgian sociologist Guillaume De Greef, a syndicalist who sought to fuse the thought of Proudhon, Marx, Comte, and Spencer, devoted nearly three hundred pages of his *Social Transformism* (1895) to a careful history of the idea of progress.[22] He accepted Comte's laws of progress, but only if they were purged of their intellectualism and rooted in the economic realities of human life, as in Marxist theory. He anticipated a just, harmonious, syndicalist world society that would be able to maintain itself forever. Never before in history, he wrote, had civilization attained proportions "so vast and complex as in our time; never have its organs of coordination been so perfect; the progressive life of the social organism is thus better assured than in preceding civilizations and we can cry out, with burning conviction, to the generations present and future: confidence and courage!" By learning the laws of civilizational progress and retrogression, the mind of man could enable the species to seize firm control of its future. "The more humanity rises, the vaster become its horizons." No limits could be set to human possibilities.[23]

Similar visions of past and future may be found in the sociology of men such as Jacques Novicow, Lester Ward, and L. T. Hobhouse.* In anthropology, L. H. Morgan's thesis of the progress of mankind from savagery to barbarism to civilization exerted a wide

* See below, pp. 43–47, 50–54.

influence. Even a relatively cautious scholar such as the Austrian Julius Lippert, who substituted a multilinear for a unilinear scheme of cultural evolution, insisted on "the essential similarity of the course pursued by cultural development everywhere, a similarity due to the identity of the primary impulses and mental laws of all peoples." [24] The most popular of the younger anthropologists, Sir James Frazer, fully endorsed the evolutionist point of view. In *The Golden Bough,* published in several editions between 1890 and 1915, he advanced a theory of the mental progress of mankind from magic to religion to science, concluding that "the hope of progress—moral and intellectual as well as material—in the future is bound up with the fortunes of science." [25]

In the rapidly developing field of *Kulturgeschichte,* or cultural history, which spanned sociology, anthropology, and history, the idea of progress was often the organizing principle. The founder of *Kulturgeschichte* as a distinct discipline, Karl Lamprecht, limited himself for the most part to German history. His theory of cultural change was ultimately cyclical, rather than progressivist, anticipating Spengler. Nonetheless, Lamprecht saw the internal development of a culture, such as the German, as a history of progress toward freedom, with decadence setting in very late, and constituting for him the least interesting stage in the whole cycle.

Of greater interest to the student of the belief in progress was the thought of Lamprecht's contemporary, Franz Müller-Lyer, a scholar not perhaps of the highest rank, but a passionate evolutionist who brought together most of the elements of the belief in progress of his generation. His "phaseological" method followed with only minor variations the familiar strategies of earlier evolutionist sociology and anthropology. Throughout the world culture had evolved through a series of identical phases, along nine major parallel lines of advance: economic, familial, social, linguistic, scientific, religious, ethical, juridical, and artistic. For his phases of culture growth, Müller-Lyer relied on Morgan's triad of savagery, barbarism, and civilization, and divided each into two or three subphases. Progress itself had consisted of a steady advance in complexity and heterogeneity over simplicity and homogeneity—the Spencerian formula again—and the ultimate explanation for progress lay in Ludwig Gumplowicz's discovery that primitive social groups tended to change their ways of life only when jolted out of their accustomed routines by contact with other groups. For Gumplowicz, this contact had most often taken the form of conflict, but in Müller-Lyer's theory, the important point was not whether the contact was violent or peaceful, but its results. If it resulted in the amalgamation of the groups, "all the experiences which the separate groups had gathered together in the course of ages from different surroundings united into one whole and were raised to higher forms by mutual permeation and productiveness." In a word, progress occurred. Generally the more advanced and dynamic groups tended to impose their will on the more backward, and in early times chiefly by force, but progress was the fruit of the new life made possible by amalgamation. Nations rose and fell. Culture was immortal. "In fact," he added, "if one studies with enlightened eyes, the death of nations has often been the means of advance of culture. If culture is to progress it must subjugate all nations that oppose it, advancing over their dead bodies." [26]

Yet Müller-Lyer was not the ravening social Darwinist rejoicing in the elimination of the "unfit" that such a statement, read out of context, would seem to imply. Progress for him had two meanings: the progress of the community and the progress of individuals; cultural history down to the present had pitilessly but necessarily concentrated on the first kind, with scant attention to the second. Applying the benefits of social progress to the enhancement of individual life became the task, in fact, of the future. From the rigors of his "civilized" existence, man would rise to the higher level of "socialized" existence. The new order would guarantee the individual maximum freedom and organize work on the principle of rational cooperation, rather than private gain. The blind struggle for existence would end, and the world would become a single community of labor, ensuring immense and continuing progress along every line of cultural advance. Although "dark shadows still lie in the valleys," Müller-Lyer could report that "the mountain tops are beginning to grow rosy with the dawn; the social intellect has entered the zone of self-consciousness." An era of perfect culture was imminent, "in the light of which all the phases of our present half culture put together will seem like a kind of childhood of the human race." [27]

The problem of the relationship between progress and conflict explored by Müller-Lyer turns up in most of the sociological literature of the post-Darwinian generation. No other issue excited so much controversy. But cursory treatments of the theme of progress and conflict in late nineteenth-century sociology—above all, treatments in textbooks—tend to exaggerate the differences among thinkers. Believers in progress through interpersonal conflict are sharply contrasted with believers in progress through intergroup conflict. All conflict theorists (or "social Darwinists") allegedly stand in polar opposition to the so-called foes of Darwinism. The foes in turn divide into thinkers who held that conflict was essential in the state of nature but not in civilization, and those who rejected conflict altogether. These categories are logically satisfying, perhaps, but they prove not to be very helpful tools for classifying the thinkers themselves.

It would be fairer to say that nearly all sociological studies in our period found a positive correlation between both conflict and cooperation, on the one hand, and human progress, on the other, and that most looked forward to the reduction or elimination of conflict in the future, as a condition of further progress. Comte, not to mention Kant, Hegel, and Marx, had already clearly laid down the proposition that warfare and other modes of human conflict were historically essential to the making of modern civilization. Darwin's evidence from natural history supplied fresh fuel for the same proposition. It also encouraged some thinkers to maintain that violent conflict alone made progress possible, but few of them were professional social scientists.* The great problem for most professionals was how much and what kinds of conflict and cooperation were required for progress, and at what stages in man's history.

Some students of conflict, of course, did not accept the doctrine of progress at all. One of the toughest-minded, the Austrian sociologist Ludwig Gumplowicz, taught that societies rose and fell in accordance with invariable laws of nature, but that mankind as a whole neither advanced nor declined. "It could not be otherwise," he concluded, "for men are always the same, the elements of society are always animated by the same forces, the quality and quantity of these forces remain always the same." [28] Belief in the general progress of humanity was an ethnocentric illusion, the conceit of a society currently dominant, which measured all other societies past and present by its own parochial standards and imagined itself to be the culmination and indeed the purpose of history.

Most writers who sang the praises of conflict and competition in

* See below, pp. 108–10, and 122–25, for examples of conflict theory expounded by political thinkers.

human life rejected Gumplowicz's cynicism, and believed in progress. The flinty gospel of self-reliant individualism preached by William Graham Sumner in the United States, the glorification of interpersonal and interracial struggle in Benjamin Kidd's *Social Evolution* (1894), and Karl Pearson's celebrated lecture on the lessons of the Boer War (*National Life from the Standpoint of Science*, 1901), were all cast in the form of theories of general human progress. It should not be overlooked that in later life Sumner acknowledged the need for greater social cohesion in the more densely populated world of the new century, that Kidd was an ardent believer in democracy and the social value of religion, and that Pearson embraced national socialism. All preferred to view life primarily in terms of struggle, but none could dispense entirely with cooperation in some form or other, if only as a means to more efficient competition.

Still more typical of sociological thought in the late nineteenth and early twentieth centuries were doctrines of progress that followed Müller-Lyer's formula of finding conflict indispensable in its proper time and place but subordinate, either ethically or pragmatically, to cooperation and mutual aid. In this camp stood such notable thinkers as the Austrian sociologist Gustav Ratzenhofer and his American disciple Albion Small; Michelangelo Vaccaro in Italy; Jacques Novicow in France; T. H. Huxley, Peter Kropotkin, and L. T. Hobhouse in Great Britain; and the American sociologist Lester Ward.

Novicow, for example, made use of the concept of struggle to explain the whole course of cosmic evolution, from the fundamental processes of chemistry and astrophysics to the development of life and mind. But at the human level, the character of struggle progressively changed, from the rudimentary physiological struggle of savages to economic, then political, and finally intellectual conflict. Each higher stage in the series made possible a happier and fuller life, the goal of all evolutionary advance. Each was further divided into two substages, the first slow-moving and irrational, the second rapid and directed by reason. Not surprisingly, progress consisted of the movement of mankind up this scale to the final and highest stage, at which conflict would be primarily confined to competition among minds, carried on in the full light of sociological understanding. In all his work Novicow looked forward to a world federation of nations living together in peace, but also in an atmosphere of the keenest possible intellectual rivalry. Already, he warned, war had become a wasteful, obsolete form of struggle, except where force had to be used against primitive or retrogressive societies for the general good of mankind. "The barbarous anarchy in which we vegetate today will come to an end; nations will feel themselves as secure in the bosom of humanity as individuals in the bosom of the state." [29]

In William Graham Sumner's United States, Lester Ward—a sociologist no less brilliant and influential than Sumner himself—constructed a system of thought during *la belle époque* to which no two critics have been able to attach the same label. For Howard Becker, Ward is a conflict theorist; for Don Martindale, a positivistic organicist; for Richard Hofstadter, a collectivist, who abhorred social Darwinism and devoted the greater part of his work to the destruction of "the tradition of biological sociology." [30] The difficulty stems from the complexity of Ward's thought, which drew on such divergent sources as Comte, Spencer, Schopenhauer, and Ratzenhofer, and from the shifts in emphasis that occurred between his first work, *Dynamic Sociology,* published in 1883, and the system embodied in his last books, *Pure Sociology* (1903) and *Applied Sociology* (1906).

Born of poor parents in Illinois in 1841, Ward acquired the rudiments of a higher education in his spare time while working as a common laborer. After the Civil War he was able to enter the federal civil service and complete his education in night school. His earliest scholarly attainments were in the field of paleobotany, but he had taken a deep interest in sociology from the first, and his *Dynamic Sociology* was the result of work begun fourteen years before its publication in 1883. Like Comte, he conceived of sociology as the queen of the sciences. "Sociology, standing at the head of the entire series, is enriched by all the truths of nature and embraces all truth. It is the *scientia scientiarum.*" [31] This was "the age of science," and the sociologist was the master scientist, who alone could explain the evolution of society, and who put his indispensable vision and knowledge at the disposal of humanity for the social planning of the future.

Ward divided the human experience into two sharply opposed processes: the genetic process of nature, in which conflict predominated and all progress was slow, unconscious, wasteful, and unsure; and the telic process of mind, in which man took control of his own evolution and moved forward deliberately. Struggle in all forms

44

could not be denied its paramount place in the genetic process. By "synergy," the creative synthesis of antithetical forces in human society as in all the rest of nature, higher levels of being were drawn out of lower. Synergy operated in social evolution through race struggle and natural selection, and on these points, in his *Pure Sociology,* Ward made free use of the research of Gumplowicz and Ratzenhofer, which he called "without any question the most important contribution thus far made to the science of sociology." Of the various forms taken by struggle, war itself "has been the chief and leading condition of human progress." In particular, the dominant race in the world's history, the Indo-Germanic, "has never hesitated to employ force or resort to war" in its upward march. Much as pacifists might deplore it, Ward saw no reason to doubt that race struggle would continue in the future: "There seems no place for it to stop until, just as man has gained dominion over the animal world, so the highest type of man shall gain dominion over all the lower types of man. The greater part of the peace agitation is characterized by total blindness to all these broader cosmic facts and principles, and this explains its complete impotence." [32]

Although the genetic process had facilitated progress, in the Spencerian sense of more complex structures displaying greater differentiation and more complete integration of their parts, and in the Darwinian sense of the survival of the fittest, it was not the only agency at work in human betterment. Left to the play of natural forces, societies only too readily reached a point of equilibration, where they stagnated until fresh challenges from the environment or new creative forces came to disrupt the established order. Opposed to genesis was telesis, the direction of life by mind, and on the possibility of the ultimate replacement of the genetic by the telic process, Ward risked all his faith in the future of humanity. The telic factor in history had so far been mainly confined to the inventiveness of individuals, but he confidently predicted the progressive collectivization of telesis, by which control would pass from scattered individuals to society as a whole.

Ward drew the strongest possible distinction between natural and telic progress. "If nature progresses through the destruction of the weak," he pointed out, "man progresses through the protection of the weak." [33] "Nature" here, of course, included the genetic process as it operated in human society, and "man" meant "mind." In mind-directed evolution, struggle as an instrument of progress could be dispensed with, although it might well continue, as we have already noticed, for some time into the future, until the unification of mankind and the assumption of full telic control of social evolution. "The method of mind," Ward wrote in *Pure Sociology*, "is the precise opposite of the method of nature. The method of nature with unlimited resources is to produce an enormously redundant supply and trust the environment to select the best. . . . Only mind knows how to economize. . . . Mind sees the end and pursues it." As a result of the superiority of mental prevision over natural waste, mind-directed evolution took place at an incomparably higher velocity, and with incomparably lower social costs. "The law of telic phenomena seems to be a geometrical progression, every new structure breeding a brood of younger and better ones." One had only to contrast the millions of years required by nature to shape the hoof of the horse with the "mushroom growth" of modern industry. "How brief is the life of the factory, the steamship, the railway, the telegraph, the telephone, the bicycle (already in its dotage), the automobile! Yet most of these are giants, and if they do not stay it will be because a superior substitute will take their places." [34]

By and large, the evolution of society had been a record of progress, thanks in great part to the inventiveness of mind. Progress was self-evident in man's increasing capacity for happiness, in the improving status of women, in the sciences, and even in the fine arts. Ward also quoted with approval Charles Letourneau's judgment that history showed a steady advance in human benevolence.[35] Today men were more altruistic and sympathetic than at any time in the past, despite the continuation of the natural struggle for existence. Even more important, the recent rapid growth of governmental services and public education heralded a golden age of collective telesis, when social planning would replace conflict as the principal agency of progress. The laws of the sociocratic welfare state would be drafted by professionally trained sociologists. "Society, possessed for the first time of a completely integrated consciousness, could at last proceed to map out a field of independent operation for the systematic realization of its own interests, in the same manner that an intelligent and keen-sighted individual pursues his life-purposes." Human happiness would be organized without loss of liberty, and "no effort or expense would be spared to impart to every citizen an equal and adequate amount of useful knowledge." [36] But even this was only the beginning of man's illimitable future. Beyond sociocracy lay a more distant age of harmonious anarchy. "Just as reason, even in early man, rendered instinct unnecessary, so further intellectual development and wider knowledge and wisdom will ultimately dispense with both religion and ethics as restraints to unsafe conduct, and we may conceive of the final disappearance of all restrictive laws and of government as a controlling agency." [37]

Less optimistic but of much the same mind in his approach to the problem of conflict was the most eminent of Darwin's biologist-allies in Britain, Thomas Henry Huxley. An amateur sociologist and moral philosopher in later life, Huxley confronted squarely the moral issues that Darwin himself had fudged. No document of the controversy that swirled around the ethical implications of Darwinism is better remembered or more often quoted today than his Romanes Lecture of 1893, "Evolution and Ethics." A number of twentieth-century thinkers hostile to the faith in progress have represented him as an anguished prophet of despair crying in the wilderness of late Victorian optimism, but nothing could be more mistaken. His hopes soared to a lower altitude than those of some of his contemporaries, but he never attacked the idea of progress itself: only the thesis that the progress of civilization had in large measure resulted from the struggle for existence and natural selection.

Huxley's strictures were fundamentally those of the moralist, rather than of the biologist. At the center of his argument stood the premise of an absolute disjunction between what he chose to call the "cosmic process" and the "ethical process"—in Ward's vocabulary, the "genetic" and the "telic." Although the ethical process was clearly an evolutionary product of the cosmic, the two now confronted one another in unreconcilable opposition from the point of view of human values. The cosmic struggle consisted of a ceaseless struggle between organisms for the possession of the means of existence; the ethical process demanded the suspension of that struggle and its replacement by self-restraint and mutual assistance. Natural man was a wily, resourceful, insatiable creature, bent on maximizing his selfish pleasure at no matter whose expense. Following the higher call of sympathy and conscience, ethical man cared about the feelings of others and accepted limits on his own freedom for their sake. The one kind of man had evolved from the other by natural processes, but once ethical man came into existence, he immediately found himself at odds with the still powerful compulsions of the natural man deeply rooted in his own being. "The ethical progress of society depends," Huxley concluded, "not on imitating the cosmic process, still less in running away from it, but in combating it." [38]

Intimately bound up with ethical progress was the seizure of control from the blind forces of nature by human will and intelligence. Instead of bowing to the impartial inevitability of the cosmos, the mind of man by a vast effort of invention and cooperation had begun to transform the natural world into a garden. Unbridled competition had yielded little by little to intelligent planning. Fatalism and supernaturalism retreated before reason and science. In the modern world above all, it was science that had done most to enlarge the dominion of man over nature, to rationalize the social order, and to clear men's minds. It had revolutionized ideas of right and wrong by teaching "that the foundation of morality is to have done, once and for all, with lying; to give up pretending to believe that for which there is no evidence." Truthful men would be palpably better men, secure in the "real and living belief in that fixed order of nature which sends social disorganisation upon the track of immorality, as surely as it sends physical disease after physical trespasses." [39]

Unfortunately the conquest of nature had also meant the return in one sense of biological struggle. The improvement of the standard of life, accomplished by human ingenuity, made possible an immense increase in population, which exerted a corresponding pressure on available resources. In the most advanced societies, at least, the struggle for bare existence had largely come to an end, but its place had been taken by a struggle for the means of enjoyment, which all the aggressive instincts of natural egotism often rendered dangerously bitter. Huxley was not even sure that international conflict, as it occurred between societies still in a "state of nature" with respect to one another, could be dispensed with under modern conditions of life. He would say only that "the problem of the effect of military and industrial warfare upon those who wage it is very complicated." [40]

Huxley could in fact imagine no organization of society, even a world polity ruled by just laws, so perfect that it could abolish completely the tendency of men to struggle for wealth and power. "Every child born into the world will still bring with him the instinct of unlimited self-assertion." Even when he succeeded in repressing this instinct, to a greater or lesser degree, by learning "the lesson of self-restraint and renunciation," his success would necessarily diminish his happiness in life and abridge his freedom. Inevitably, "the progressive evolution of society means increasing restriction of individual freedom in certain directions." Opposed by nature, threatened by struggle, subject to error, haunted "by inexpugnable memories and hopeless aspirations," unable to explain the ultimate mysteries of the universe, man was ill-advised to dream of perfection. "The prospect of attaining untroubled happiness, or of a state which can, even remotely, deserve the title of perfection, appears to me to be as misleading an illusion as ever was dangled before the eyes of poor humanity." [41]

At the same time, Huxley was a passionate meliorist, even in his darkest moods. "The majority of us," he wrote, "profess neither pessimism nor optimism." He had no patience with the superstition that earth could be made into either a heaven or a hell. But it could be improved.

That man, as a "political animal," is susceptible of a vast amount of improvement, by education, by instruction, and by the application of his intelligence to the adaptation of the conditions of life to his higher needs, I entertain not the slightest doubt. . . . I see no limit to the extent to which intelligence and will, guided by sound principles of investigation, and organized in common effort, may modify the conditions of existence, for a period longer than now covered by history. And much may be done to change the nature of man himself. The intelligence which has converted the brother of the wolf into a faithful guardian of the flock ought to be able to do something towards curbing the instincts of savagery in civilized men.

Throwing aside youthful over-confidence and senile despair, we were bound to do our best, "cherishing the good that falls in our way, and bearing the evil, in and around us, with stout hearts set on diminishing it." [42] The "service of humanity" could be the only unassailably acceptable religion for rational men "so long as the human race endures." [43]

Huxley's faith appealed to thousands of post-Darwinian intellects, of which the best possible example was one of his own former students at the Normal School of Science in London, H. G. Wells. To his basically Huxleyan views of man, nature, science, and progress, Wells added cosmopolitan socialism and the quasi-Positivist conception of an emergent "racial mind" struggling to achieve self-con-

49

sciousness in the form of an integrated world civilization managed by scientists and engineers; but his thought remained fundamentally Huxleyan throughout his long career as a novelist and a publicist of science and socialism. Such hopes as men might reasonably entertain of improving their condition, he felt, depended on their success in replacing the irrational and savage disorder of nature by a social cosmos fashioned by human minds working in concert. As he explained late in life, in words that might just as well have come from his former teacher, "I am neither a pessimist nor an optimist at bottom. This is an entirely indifferent world in which willful wisdom seems to have a perfectly fair chance." [44]

Unhappily for his prophetic reputation, Wells became associated in the public mind with a naively optimist utopism that cost him many followers in the years after the First World War and, indirectly, diminished sympathetic understanding of Huxley's thought as well. His utopian novels, from *A Modern Utopia* in 1905 to *The Shape of Things to Come* in 1933, were all experiments in exhortation, intended to supply moral direction to human effort, and not confessions of faith in inevitable progress; but they were easily misunderstood. In any event, no British writer in the first quarter of the twentieth century did so much to awaken an exuberant faith in human possibilities as Wells, not only in his novels and journalism but also in his remarkable best-selling survey of human progress, *The Outline of History,* first published in 1919–20.[45]

One final instance of a theory of progress grounded in opposition to the cruder forms of Darwinism is provided by the thought of another Englishman, L. T. Hobhouse. He is still honored in Britain as the founding father of British academic sociology and the first holder of a chair in sociology at the University of London. In three books published between 1901 and 1913—*Mind in Evolution, Morals in Evolution,* and *Development and Purpose*—Hobhouse undertook to supply a synthesis of biology, sociology, and ethics in the great systematizing tradition of nineteenth-century thought. His work sums up the contribution of the social science of his generation to the development of the idea of progress. Although its roots are positivistic, it also vibrates sympathetically in some respects with the anti-positivism of *la belle époque,* which will be investigated in the next chapter.

As he explained in the Introduction to *Development and Purpose,* Hobhouse had been attracted in his student years to the evolutionary approach of Spencer, but dismayed by Spencer's failure to take account of the qualitative difference between lower nature and the mental and moral nature of man. Spencer degraded the human mind to the rank of "an organ like the lungs or the liver evolved in the struggle for existence." A little reflection sufficed to show that "if progress means anything which human beings can value or desire, it depends on the suppression of the struggle for existence, and the substitution in one form or another of social co-operation." [46] Hobhouse missed this respect for both mind and the sociality of mankind in Spencer and in the Darwinists, but he found it readily enough in the new idealism just then beginning to dominate academic philosophy in Britain, above all in the work of T. H. Green. His task as he saw it was to fuse the humanistic insights of idealism with the empiricism and evolutionism of Spencer. This program he faithfully carried out between 1901 and 1913. He felt bound, as a social scientist, to build on empiricist foundations; but his own deeply-held moral convictions compelled him to view the evolutionary process, as it were, from the top down, instead of from the bottom up.

The central concept in his sociology was "Mind," defined functionally as the power in living things that organizes both individuals and species by correlating their parts. He traced Mind far back into the nonhuman beginnings of natural history, dissenting from Huxley's thesis of an absolute gulf between the rational and moral behavior of man and its rudiments in animal life. Nevertheless, the only evolution that could be called progress was the "orthogenic" evolution of Mind. This single line of development, from struggle, accident, and purposeless proliferation to the establishment of control over the conditions of existence by the harmonizing and unifying work of Mind, alone constituted progress.

On the other hand, obviously, Mind became fully conscious and rational only with the emergence of man. Hobhouse described Mind at the human level as passing through four successive stages of evolution, along a variety of separate but closely related lines. As an instrument for acquiring knowledge, Mind advanced from the unformed concept to the construction of a "common sense" order of empirical reality, to the abstract construction of a higher reality (in classical thought), and finally to the critical reconstruction of experience of modern philosophy. As an instrument for gaining control over the natural environment, it progressed from the use of magic to the rise of handicrafts to the metaphysically based science of antiquity to modern experimental science. At the level of religious conceptions, the sequence ran from animism through anthropomorphic theism and metaphysical idealism to the modern idea of the "harmony of life." The development of Mind as Will, or ethical progress, similarly followed a four-stage pattern, studied at great length in Hobhouse's most influential work, *Morals in Evolution*. Ethical conceptions were first grasped intuitively and preserved by custom; in the second stage, categorical rules were laid down in the form of law; in the third, a higher order of religious belief established a spiritualized ethic out of touch with actual conditions of life; and in the fourth, a rational and realistic ethic of evolutionary development had appeared, pointing toward a harmonious world order. The same scheme of development explained social, legal, economic, and artistic evolution.

Much like Comte before him, Hobhouse repeatedly warned his readers that these were only hypothetical stages, which in the actual experience of the race often overlapped in time and were subject to frequent interruption. Reversion to earlier stages might occur. Evolution along one line sometimes entailed a failure of progress along some other, as when the establishment of authoritarian empires in antiquity resulted in a certain loss of personal freedom. Individual societies rose and fell. All the philosopher could certify was that "when we consider the life of humanity as a whole and compare our own civilisation with the whole series of earlier forms, together with their survivals at the present day, there appears . . . a certain net movement": in short, progress.[47] Mind advanced, growing steadily in unity of purpose and scope of understanding and control. Struggle, which had its humble and inevitable role to play in the drama of animal existence, little by little yielded to the superior methods of Mind.

Moreover, it could clearly be seen that the rate of mental progress accelerated geometrically. "Of the total growth of mind in scope and power during the existence of the human race, at least one half must be assigned to the comparatively short period from the beginnings of Greek history to the present day." Even in this "short period," the period of the third and fourth stages, the human race as a whole had acquired no true consciousness of its history, its unity, and its possibilities for deliberate and collective self-development until the modern era. Only in recent times, with the appearance of modern natural science, psychology, and sociology, and the ethic of free organic association in the cause of world order, did "the threads begin to be drawn together to weave the larger purpose." [48] It became the supreme task of the behavioral and social sciences to make mankind conscious of itself, and therefore to put the future of the race under the control of reason. Like Ward, whose influence he gratefully acknowledged, and like Wells, whose thought ran parallel to Hobhouse's without quite touching it at any point, he looked forward to the rule of the mind of the race as the goal of all earlier progress.

> Mind grasps the conditions of its development that it may master and make use of them in its further growth. Of the nature of that growth, whither it tends and what new shapes it will evolve, we as yet know little. It is enough for the moment to reach the idea of a self-conscious evolution of humanity, and to find therein a meaning and an element of purpose for the historical process which has led up to it.[49]

Although he held that no law made progress inevitable, Hobhouse could not bring himself in the final analysis to believe in the possibility of human failure. By circular arguments, he equated rationality, goodness, organicity, and harmony. "Evil," he affirmed, "is merely the automatic result of the inorganic. Physical evil results from the impact on the spiritual order of natural causes which intelligence has not been able to subordinate to its ends, moral evil from the clashing of purpose in minds which have not been brought into an organic unity." Evil was self-destructive, whereas "the elements of goodness, of rational harmony, in the long run support and further one another, and this upon the whole at an accelerating rate in proportion as they have already acquired organic union." [50] An "impulse to harmony" dominated the entire evolution of Mind, "and the rationality of the process is the guarantee of its ultimate success." [51] As Mind constantly enlarged the sphere of its self-knowledge, it constantly increased its power to control its further growth.

Nor was this all. In the onward orthogenic evolution of Mind, Hobhouse discovered the meaning and final cause of the cosmos itself. He felt able to answer, in a tentative way, the "fundamental questions" of life. Philosophical analysis made clear, and science offered much confirming evidence, that evolution was a teleological process, directed by Mind in its pursuit of self-realization. Seen at this level, Mind was nothing less than God, finite and emergent, of which "the highest known embodiment is the distinctive spirit of Humanity." Giving oneself to the service of so vast a world purpose made life meaningful and worth living again, in the teeth of the failure of traditional metaphysics and revealed religion. At its outer edges, Hobhouse's thought clearly resembled the emergent evolutionism of his friend Samuel Alexander.[52] *

The First World War shook Hobhouse's confidence, but not decisively. In 1904 he had called alarmed attention to the spread of anti-humanistic ideologies hostile to freedom and reason in his *Democracy and Reaction,* and he took the war as evidence of the wisdom of his warnings. Class conflict and national rivalry had created problems that Europe, at least, seemed temporarily unable to solve. "We must face the possibility," he wrote in 1924, "of a reversion of Europe to less civilised conditions." But it did not follow "that if our present civilisation fails the cause of progress is lost." After each experience of the race, no matter how costly, something was passed on, "and the succeeding effort begins at a higher remove, and in its culmination reaches a higher point. In our own time it is probable that the scientific tradition is strong enough to maintain itself, and if that is the case its application to society will ultimately yield that clarity and unity of purpose which for the moment mankind seems to have just missed." Provided only that men came to grasp the essential rightness and power of Mind, there was every reason to believe that "in ethics good, as in science truth, will prevail." [53]

* See below, pp. 72–73.

5

The Will to Power

I N THE STRICTEST SENSE, the "revolt against positivism" of *la belle
époque* was a philosopher's revolt. Social and behavioral scientists may have quarreled with some of the methodology of positivism, or with the rationalistic anthropology and psychology inherited by nineteenth-century thought from the Enlightenment. But
they remained scientists, deep in the debt of positivism. It was quite
otherwise with the philosophers. Academic and amateur alike, they
sought to transcend the world outlook of science, to enrich or even
to uproot its mechanistic and materialist assumptions, on behalf of
a "higher" wisdom inaccessible to the positive sciences. Dominated
by the great archetypal figures of Friedrich Nietzsche and Henri
Bergson, philosophy became the most formidable stronghold of
anti-positivism in *la belle époque.*

Yet the philosopher's revolt against positivism rarely took the
form of a revolt against the belief in progress. On the contrary, in no
field of thought did progress find more joyful prophets than philosophy. Most of the leading schools—vitalist, spiritualist, neo-
idealist, neo-Kantian, pantheist, pragmatist—proclaimed the good
tidings of the faith in progress with enthusiasm. Progress, they
urged, was not the blind, involuntary, and mechanical process described by positivist thought. It occurred in history, and perhaps in
all the cosmos, through the striving action of spirit, mind, will, vital
force, or emergent deity, a restless energy at the heart of things, as
expansive as Western civilization itself in this highest age of its
world power. Late nineteenth- and early twentieth-century philoso-

phers often deliberately turned back for inspiration to the romantic age—to Goethe, Hegel, and Schelling. If they betrayed the influence of positivism at all, it was in their fascination with the idea of phylogenetic evolution: but even here, they tended not to accept Darwinian explanations, or thought them incomplete and question-begging.

For the most representative philosophers of the age, the great villain and *bête noire* was not Darwin but Herbert Spencer. While sociologists continued to honor Spencer, philosophers exposed him to ridicule and contempt. Thinkers like Spencer had sought to explain away the whole universe as mere machinery, devoid of mind or purpose or will. How could one speak of real progress in a world where aspiration counted for nothing, where everything was predictable, and therefore closed? The philosophers preferred to see the world as self-creative, advancing through the volition of spiritual forces, and therefore open. Just as the individual man shaped his future life by acting according to his ideas of what it could and should be, so mankind—or the cosmos—evolved from lower to higher stages through the operation of immaterial forces, through what Alfred Fouillée, an eminent French philosopher of the period, termed *idées-forces,* centers of psychic energy uniting will, thought, and action. "Nature always repeats herself by mechanical means," he added, "but she changes by means of mind." [1]

Yet the philosophers encountered a paradox, in their discussions of progress. How could one know that the mindful change and struggle in life constituted progress, and how could one speak of purpose, in an open universe, where anything might happen, and where the goals of progress themselves were subject to unlimited revision? In the end, most philosophers of *la belle époque* simply affirmed on faith the equation of change and progress, but they preferred to impose no final ends on human and cosmic evolution. In visions of the ultimate they showed little interest. They could imagine fantastic achievements, a transfigured humanity, even evolution beyond humanity, but their realities were all this-worldly, on the earthly and temporal side of transcendence.

A plausible defense might also be constructed for the argument that philosophy, and above all the philosophy of the academy, was in search of a *raison d'être* near the end of the nineteenth century. Psychology and sociology had usurped some of the traditional functions of philosophy; the all-encompassing positivist systems of Comte,

Spencer, and Haeckel represented frontal attacks by science that removed nearly every reason for the continued existence of philosophy as a distinct discipline. In an age dazzled by the accomplishments of science, it behooved the professional philosopher to cry as loudly as possible that science was not "enough."

In some cases, no doubt, philosophers of *la belle époque* adopted anti-positivist views in an unconscious effort to justify the existence of their profession. But in other cases, such an argument explains nothing. Before considering the leading academic philosophers of the period, all of whom had in some sense a personal stake in the revolt against positivism, it may be instructive to explore the mind of that most unprofessorial of late nineteenth-century philosophers, Friedrich Nietzsche. Although he craved the attention of the academic world, spent part of his early life in its service, and predicted (in *Ecce Homo*) that university chairs would one day be established for the study of *Thus Spake Zarathustra,* his work bears none of the odors of academic philosophy. It is the antithesis of orthodoxy and professionalism. Yet no books written between 1880 and 1914 so completely incarnate the nervous, striving spirit of the age as the books of Nietzsche.

Since his death in 1900, his reputation has grown steadily. He has been the subject of thousands of critical works, which have offered a rich profusion of conflicting interpretations. Just before the First World War, the earliest Nietzsche critics saw him as an apostle of atheism, the martial virtues, and the conquest of man by superman. The fierce aphorisms of what was supposed to be his last testament, *The Will to Power,* edited from unpublished notes under the supervision of his sister, Frau Förster-Nietzsche, did service as the reputed key to his earlier works, above all to *Thus Spake Zarathustra.* But as Walter Kaufmann has remarked, even without *The Will to Power* Nietzsche might well have "supplied a Darwin-conscious age with a convenient tag for its own faith in progress." [2] For some time after the First World War, he continued to be closely identified in the public mind with the cult of progress through violence and irrationalism. The Nazis, in particular, made much of him, representing him as the prophet and forerunner of the Nazi movement itself. Even those who saw more deeply into the meaning of the Nietzschean "will to power" hailed him as the great teacher of a joyful, positive faith, which destroyed only to create.

His impact on the English composer Frederick Delius is a good case in point. While still a young man, in 1890 or thereabouts, Delius happened upon *Thus Spake Zarathustra* in a friend's library. As he told his amanuensis, Eric Fenby, in his old age, Nietzsche's volume "never left his hands until he had devoured it from cover to cover. It was the very book he had been seeking all along, and finding the book he declared to be one of the most important developments of his life. Nor did he rest content until he had read every work of Nietzsche he could lay his hands on." [3] The musical result of Delius' exposure to Nietzsche was *A Mass of Life,* completed in 1905, a setting for soloists, chorus, and orchestra of parts of *Thus Spake Zarathustra,* suffused with all the radiant spiritual energy of a Bach Passion. As the title indicates, it was a pagan celebration of life: the life of suffering and creation, the life of the higher man intoxicated by the vital force that he feels surging up within himself from the mysterious depths of *natura creatrix.*[*]

More recently, Nietzsche has come to figure as the founder, with Kierkegaard, of modern existentialism. Emphasis is thrown on the less synthetic aspects of his thought: his anguish and despair, his reflections on the "death of God," his contempt for the conventional idea of progress, his insistence on the supreme value of the individual man and the individual moment. The studies of Karl Jaspers and Walter Kaufmann provide an excellent introduction to this revaluation of Nietzsche's place in thought. It would be difficult indeed to find more than a few general accounts of existentialism that do not have their chapters on Nietzsche.[4]

In any event, the current fashion is to look on Nietzsche as an archenemy of the nineteenth-century belief in progress. He was no more deceived by the myth of progress, it is said, than by any of the other comfortable, self-congratulatory beliefs of the bourgeoisie of his time. Kaufmann traces his antipathy to any sort of conception of improvement in historical time as far back as the second and third *Untimely Meditations* of 1874, and crowns his argument with a choice quotation from *The Antichrist* (1888). "Mankind does *not* represent," Nietzsche had written, "a development towards a better, stronger or higher type, in the sense in which this is supposed to occur to-day. 'Progress' is merely a modern idea—that is to say, a

---

[*] *Thus Spake Zarathustra* also served as the inspiration for a remarkable tone poem (1896) by Richard Strauss, which contains some of the most brassily heroic music ever written; Stanley Kubrick used it to good effect in his science-fiction film *2001: A Space Odyssey.*

false idea." [5] Nietzsche's doctrine of the "Eternal Recurrence," an adaptation of the cyclical philosophy of history of classical thought, expresses in the clearest possible language his rejection of evolutionary optimism; as for the *Übermensch,* the Superman preached by Zarathustra, everything in Nietzsche's work suggests that he was referring only to the possibility of a superhuman self-transcendence on the part of certain rare, exceptional individuals. There had been supermen in the past, and there would be others in the future: nothing more. Nietzsche never imagined for a minute that mankind as a whole could achieve superhumanity.

All this is true as far as it goes. By stressing certain features of Nietzsche's thinking and not others, the scholar can make it seem absurd that anyone ever mistook him for a prophet of progress. But G. A. Morgan's *What Nietzsche Means* comes closer to the mark, perhaps, in pointing out that while Nietzsche ridiculed the idea of progress that was fashionable among his own complacent contemporaries, he had "a philosophy of the future" all the same, with a message for all humanity and the vision of a higher world civilization in the making.[6] Much depends, in this reading of Nietzsche, on the significance one is willing to attach to *The Will to Power.* Without using it, such an interpretation becomes considerably more difficult. Even if it is conceded that Nietzsche could not have wished his notes to be arranged as his sister and his editors finally arranged them, that he did not plan to include all of them in any single book, that at least some express judgments he might later have abandoned, and that by 1888 he had decided not to write *The Will to Power* at all, still they are Nietzsche's work. They provide invaluable evidence about a book that Nietzsche had long intended to write, and in which he had hoped to tear the veils from his philosophy and speak his mind clearly. They are of even greater importance, for this reason, than some of the books published during his lifetime.[7]

Nietzsche's philosophy of the future had its roots in his concept of "the will to power." Following Lamarck rather than Darwin, he insisted that creative growth occurred in living creatures only when they strove actively to achieve power through self-overcoming acts of will. This striving was the very essence of life, as he proclaimed in *Beyond Good and Evil,* and it could not help but involve exploitation and conquest, although mastery over others was not the goal of the struggle. Nothing was won easily. "A *species* originates, and a type becomes established and strong in the long struggle with essentially constant *unfavourable* conditions." [8] In man, the possibility emerged of a higher life than any animal could reach, because man alone was capable of carrying the struggle for power into the labyrinthine depths of his own soul. Man in society imposed limits on the outward flow of will, directing the surplus into his own inner world, where it generated tensions unknown in the simpler sphere of animal life. It was from this inner conflict, this soul-sickness peculiar to man, that he suffered, created, and in exceptional conditions won victories of mind and spirit wholly unimaginable to the lower orders of nature.

Man alone, in all the universe, still possessed unlimited potentialities. He stood unfinished, free to evade the mindless mechanisms of animal life and to create himself. But Nietzsche always insisted that the higher path lay open only to the few, rare, higher men, and never to the species as a whole. The majority of men might presumably hope to surpass the lower animals, but most men remained all too human, clinging to a way of life that they lacked the power of will to transcend. To become superhuman, to follow the path pointed out by Zarathustra, the path to infinite freedom, was possible for only a handful of men scattered through history.

Nevertheless, a certain pattern could be discerned in these seemingly random appearances: the shooting stars tended to come in showers. In the same passage in *The Antichrist* in which he denounced progress as a merely modern and false notion, Nietzsche defined the *Übermensch* as a "lucky stroke," and added that "even whole races, tribes and nations may in certain circumstances represent such *lucky strokes.*" [9] A whole age might succeed, in that it produced a relative abundance of great men. There were, in fact, two such ages in Western history: classical Greece, especially in the time of Aeschylus and the pre-Socratic philosophers; and Europe during the Renaissance. From his descriptions of both ages, it is clear that Nietzsche fully understood the importance of *milieu* and *Zeitgeist* to the fulfillment of genius: it was the whole age that he extolled, and not just its highest men. From the hints strewn all through Nietzsche's writings Morgan derives the firm outlines of a cyclical theory of culture. [10]

But for Nietzsche, unlike Spengler, cultures could be measured against a single universal standard. To what extent did they liberate the *Übermensch* and propel him upward? The Greeks soared higher than the Italians and Frenchmen of the Renaissance, proving that "mankind does not advance in a straight line," but for Zarathustra, as for Nietzsche, there appeared the vision of a mightier effort in the future, the Great Noon of a unified world order ruled by the highest men yet evolved, which might be man's last burst of creative energy before, in the inevitable course of cosmic events, he sank into exhaustion, and all things returned to their beginning. In the past, the great ages were splendid accidents, but now that Zarathustra had spoken, it might be possible to achieve superhumanity with eyes, as it were, open. "This kind of accident," Nietzsche wrote, "must now be *consciously* striven for." [11] The new philosopher who would come forward in response to Zarathustra's challenge would strive "to teach man the future of humanity as his *will,* as depending on human will." He would see "at a glance all that could still *be made out of man* through a favourable accumulation and augmentation of human powers and arrangements; he knows with all the knowledge of his conviction how unexhausted man still is for the greatest possibilities." [12]

The Will to Power contains a detailed forecast of things to come: the appearance of "new barbarians" from "the heights" in the twentieth century, who would overturn the decadent bourgeois society and impose a world empire; the establishment of an aristocratic order managed by philosophers in the interests of the higher men; hitherto unequaled spiritual triumph for the new race of supermen; and only after this, the unavoidable final collapse. Nietzsche did not suppose, of course, that this greater Renaissance was inevitable. It depended on the will to power of individual men, marshalled against the herd instincts of the masses: if these individual men failed the world would still be unified, but it would come to resemble the Chinese empire, a world of ants swarming mindlessly to their inglorious doom. Nevertheless, the possibility of at least one more pyrotechnic outburst of overwhelming will to power existed. The very sickness of modern civilization, the death of God and the collapse of all established truths and values, had created an unprecedented opportunity for human freedom. In *Thus Spake Zarathustra* and his subsequent attempts to elucidate its message, Nietzsche affirmed that he had put to mankind the terms of its last great challenge. He was the seer who saw all; the path ahead lay steeply up or steeply down.

This was clearly no orthodox doctrine of progress, but it contained many of the most vital ingredients of that doctrine. It was progress for selected individuals and not for the species. It was progress without historical continuity. It had no laws and no limitless future, although within its own time it knew no fixed bounds. But even Herbert Spencer had decreed an end to all things, an eternal return to fresh beginnings no less thorough and final than Nietzsche's. More important for the historian of ideas, Nietzsche's faith was so interpreted by thinking people in the post-Darwinian generation that it did fall somehow within the progressivist tradition, and this alone would qualify it for our serious consideration. As they saw him, Nietzsche was a yea-sayer and a futurist: he had the messianic instinct, the sense of great things to come, of the authentic prophet of progress. If he transcended the limits of his own intellectual *milieu,* he was also its unmistakable son and heir.*

Nietzsche's ten years as professor of classical philology at the University of Basel (1869–79) did not make an academic philosopher of him,** although he taught Greek philosophy as well as literature. In poor health, he retired from university life, and wrote his major philosophical works in the 1880s as a private citizen. The true academicians of German philosophy in the post-Darwinian generation were men like Wilhelm Wundt, for forty-two years professor of philosophy at Leipzig, and Rudolf Eucken, who spent forty-six years at Jena. Both expounded philosophies of progress that rejected materialism in favor of a concept of the world as will or spirit in process of self-realization through time. The end of life, as Wundt argued in his treatise on ethics, was not individual happiness or survival, but the universal progress of spirit, "the development of all the physical forces of mankind in their individual, social and humanitarian functions, a development that progresses beyond every stage once attained and proceeds to infinity."[13] At still another great German university, Marburg, the cause of progress through the full cultivation of the powers indwelling in the human spirit was championed by the neo-

---

* George L. Kline makes the point that Nietzsche agreed with Marx, even the "young Marx," and by the same token disagreed with Kierkegaard, in holding that present-day man was instrumental to the creation of a higher future order. Kline, "Was Marx an Ethical Humanist?" *Akten des XVI. Internationalen Kongresses für Philosophie* (Vienna, 1968), II: 71.

** That vitalism could be made academically respectable in Germany, however, is borne out by the work of Hans Driesch, who turned from biology to philosophy in later years and taught a strongly voluntaristic meliorism at Heidelberg and Leipzig. See his *Man and the Universe,* tr. W. H. Johnston (New York, 1930).

Kantian school of Hermann Cohen and his younger colleague Paul Natorp. Cohen exhorted his fellow men to abandon the selfish search for personal salvation or immortality and work in concert for the realization of the kingdom of heaven on earth. No bounds could be set to the creative power of the unified spiritual will of humanity.[14]

French philosophy, in the final years of *la belle époque,* produced a worthy rival of Nietzsche in the person of Henri Bergson, whose academic credentials were impeccable. But Bergson was no splendidly isolated thinker in the Nietzschean style; rather, he came at the end of a considerable line of vitalist philosophers in France, who are now as little read as Wundt, Eucken, and Cohen. One of them, Alfred Fouillée, has already received mention above. His most important work, *The Evolution of Idea-Forces* (1890), rebelled vigorously against the cosmology of Spencer. Mechanistic explanations of reality, such as Spencer's, could not unravel the dynamics of evolution, Fouillée pointed out. Even Spencer's definition of progress as movement from homogeneity to heterogeneity was at best a half-truth, obscuring the higher tendency in the cosmos toward ever-increasing interdependence. In a later work Fouillée traced three stages in the moral progress of humanity: an era of primitive synthesis, in which the individual and the social good were not distinguished; an era of analysis, in which conscience, detached from the collective tribal will, struggled to reconcile the two; and a coming era of total synthesis. "It is the complete union of individuality and sociality that is the final end of moral history," he foresaw. "The future will find it equally false to slight the worth of the individual *and* that of society, composed of individuals sharing sympathies and ideas." Even in modern states, the mutual reinforcement of sociality and individuality was quite apparent. "The more individuals are related one to another and joined together, the more we see developing in them true indivduality." [15]

Emile Boutroux, professor at the Sorbonne from 1888 to 1921, and teacher of Bergson, furnishes another example of the evolutionary optimism that dominated much of French philosophy after Darwin. His cosmology represented evolution as a pilgrimage, unfolding in limitless time, from the absolute unfreedom of matter in a state of primordial inertia to the absolute freedom of mankind's unimaginable future. Boutroux challenged his contemporaries to transcend the life of the brute and the life of the intellect. Without abandoning either body or brain, mankind could give supreme meaning to both by embracing the religion of the spirit, which called upon man to strive freely for perfection in faith, hope, and love. The truly religious man did not hesitate to risk present security for the sake of progress. "What is progress, that lever of the modern mind," he asked, "save the right of the future over the present?" From science and religion, each performing its appointed tasks, "will spring a form of life ever ampler, richer, deeper, freer, as well as more beautiful and more intelligible. But these two autonomous powers can only advance towards peace, harmony, and concord, without ever claiming to reach the goal; for such is the human condition." [16]

Of the several French philosophers whose thought bridges Nietzsche and Bergson, the most interesting is Fouillée's stepson, Jean-Marie Guyau. He was born in 1854, only five years before Bergson, and started his professional life as a teacher of philosophy in a provincial *lycée*. Poor health forced him to resign not long after his appointment, and in a few years, at the age of thirty-three, he was dead, leaving behind him nine books and several essays in *La Revue philosophique*. He had an insight of sometimes Nietzschean intensity into the spiritual crisis of the nineteenth century, and at least a little of Nietzsche's flair for revealing metaphor. There, perhaps, the resemblance ends.

For Guyau, as for Nietzsche, the positive religions had run their course, and this was the great fact of the century. Science had made nonsense out of all forms of supernaturalism. Dogma, miracle, and myth had been exploded for all time, and with their collapse, Guyau also predicted the collapse of the systems of morality founded upon them. "Symbolic" interpretations of traditional beliefs could not save them from the final catastrophe, since to render dogma as symbol was to rob it of its exclusiveness and authority. At first this irrevocable failure of religious faith might cause severe unrest, a sense of exile and loneliness, but in time even the masses of mankind would be able to make the necessary mental adjustment; they would "get over" religion as surely as the Bretons would one day "get over" buckwheat *crêpes*. The history of man's religious life, reviewed from a modern perspective, revealed an inexorable evolutionary advance toward "greater independence of spirit" and the final result would be non-religion, "a higher degree simply of religion and of civilization," the complete transcendence of the positive religions by absolute liberty of belief. Buddha, Jesus, and Luther had all been forerunners of this coming age of conscience. Guyau anticipated a humanity composed entirely of self-determined Luthers. "The very progress of intelligence and conscience must, like all progress, proceed from the homogeneous to the heterogeneous, nor seek for an ideal unity except in an increasing variety." [17]

At the same time, Guyau did not suppose that men would move further and further apart. Although each mind would have its own finely discriminated set of personal beliefs, the objective distance between each mind would actually shrink, as the number of possible world-hypotheses was reduced by the progress of science. Errors, once exposed, were banished permanently from history, since a fully conscious step toward truth "renders impossible a step backward." He could predict without inconsistency "free and continuous progress toward ultimate unity of belief on the most general subjects of human inquiry." [18] One of his most Nietzschean metaphors represented mankind in its present spiritual distress as a ship whose rudder and mast had been torn away by the fury of the elements, a ship drifting purposelessly on the high seas. And yet the ship might still "arrive at an unknown goal, which it will have created for itself." Man had to assume control over his own destiny, both individually and collectively.[19]

But Guyau could no more let matters rest there than Nietzsche could. To satisfy his own personal thirst for meaning, he also elaborated a world-view, a type of vitalistic monism, which by implication could actually provide the metaphysical framework for that "ultimate unity of belief on the most general subjects of human inquiry" that was the goal of intellectual progress. Like Bergson, Guyau insisted on the creative power of life itself, conceived as an onrushing stream of vital force that developed and improved the organism by an inner compulsion, wholly inexplicable in mechanical terms. "There is in the living being," he wrote, "an accumulation of force, a reserve of activity, which spends itself not *for the pleasure of spending* itself, but *because spending is a necessity of its very existence.*" [20]

This view of life led Guyau to what he called a philosophy of morals independent of obligation or sanction, a purely autonomous, naturalistic ethics, in which he embedded a theory of progress. Life not only impelled the organism to satisfy its need for food, he declared; it also gave organisms extra energy, as a margin of safety, which found expression in play and love, and in those individual variations on which nature seized to improve the racial stock. The higher organisms, culminating in man, were more and more individualized, creative, and capable of emotion. In man himself, of all creatures the most richly endowed with superabundant natural energy, life in its upward thrust had created a being for whom religion, science, philosophy, art, and work were all as necessary as food and drink. One did not have to search for the categorical imperatives of some supernatural authority in order to know the good. "Life makes its own law by its aspiration towards incessant development; it makes its own obligation to act by its very power of action." The formula, then, was simple: *je puis, donc je dois* (I can, therefore I must). Man was also by nature a social animal, and the higher he advanced up the scale of evolution, the more he needed the company of his fellows to find happiness. "By virtue of evolution our pleasures become wider and more and more impersonal. . . . Our environment, to which we better adapt ourselves every day, is human society, and we can no more be happy outside this environment than we can breathe beyond the atmosphere of the earth." We all felt within ourselves "a kind of pushing of moral life, like that of the physical sap." [21]

Guyau had few doubts about the future. "Occidental nations," he pointed out, "or rather the active people in the world, to whom the future belongs, will never become converts to pessimism. Whoever acts, feels, has power, and to be strong is to be happy." The fantasy of an external providence would be replaced by the idea of man, and every individual member of the race, as his own providence. "We shall love God all the more that He will be, so to speak, the work of our own hands." History stood under no obligation to repeat itself, and man was not doomed to failure, as other races had failed. It was easy to imagine the evolution of completely self-conscious beings wise enough to avoid evolutionary blind alleys and pitfalls, able to prevent evolutionary retrogression, and preserve their kind for eternity. Guyau even foresaw the possible conquest of death through the development of telepathic racial intercommunication. An unbreakable psychic ring of love and sympathy could keep all minds alive forever. "Every individual consciousness may come to survive as a constituent part in a more comprehensive consciousness. . . . Every soul will be reflected and mirrored in every other." [22]

From Guyau it is but a short step to the doctrine of progress taught by Henri Bergson. Without question his *Creative Evolution,* published in 1907, was the most widely read and broadly influential book by a philosopher in the decade preceding the First World War. It united many of the leading tendencies in French thought—the vitalism of Guyau, the actionism of Blondel, the anti-positivism of Boutroux, the theological liberalism of the Catholic modernists—in a synthesis that seemed firmly grounded in the great truths of evolutionary biology, explained them in a more soul-satisfying way than could either the neo-Darwinians or the neo-Lamarckians, found a high place for the life of the spirit, evoked a sense of cosmic purposefulness without resorting to finalism, and at the same time preserved at least some of the ancient dignity of reason and intellect. Bergsonism offered solace and inspiration for men of every persuasion, and it pumped fresh life into the progressivist faith at a time when it was growing increasingly vulnerable to sceptical attack.

Bergson was not a prophet without honor in his own country or in his own time. Rising fairly rapidly through the academic ranks, he achieved election to the French Academy, won a Nobel Prize, and for the last forty-one years of his life held the coveted chair of philosophy at the Collège de France. By the time of his death in 1941, he had long ceased to speak for the avant-garde, but in the last years of *la belle époque,* he was indubitably *le cher maître* of the forward-thinking young, not only in France but throughout the Western world.

Creative Evolution was the outgrowth of many years of earnest thought. One important early book, *Matter and Memory,* with its characteristically post-Darwinian philosophy of time, decisively shaped Proust's world-view. The reputation that earned Bergson his chair at the Collège de France in 1900 was founded on this work, and on still another, earlier effort, *Time and Free-Will,* neither of which dealt directly with evolution or progress. Here Bergson was concerned only to show the inadequacy of intellectual analysis as an instrument for understanding consciousness and, indeed, all life processes, which unfolded in time and therefore partook of duration as well as the three spatial dimensions. Life had to be grasped in its wholeness and temporality by means of intuitive insight; analysis, by dismembering and freezing reality, missed all that really mattered.

At first Bergson confined the application of his thesis primarily to the problems of human psychology and epistemology. But in 1903, with his *Introduction to Metaphysics,* and then in 1907, with *Crea-*

tive Evolution, he converted it into a philosophy of evolution that was simultaneously a philosophy of progress. He agreed with the Darwinists that the natural selection of variations arising in the germ plasm accounted for evolution much more convincingly than the Lamarckian theory of the inheritance of characteristics acquired in life as the result of the striving or the will of the individual organism to change. But Darwinism could not explain the origin of the variations themselves. Entering insightfully into the internal reality of the life process, Bergson discovered a principle at work in evolution that could not be broken down by analysis or represented by a mechanical model. Life, at bottom, was creative; since chance alone could not conceivably have produced any of the great leaps forward in natural history, one had to picture an *élan vital* at the heart of things, a vital impetus or life force, struggling against the inertia of matter, transcending individual wills, and driving life onward.

This creative force did not labor to carry out a divinely preconceived plan for the cosmos. It threw out many different lines of advance, most of which failed after a time and led either to the extinction or to the stagnation of the species involved. But its experiments were not pointless or frivolous. While matter, as contemporary physics abundantly demonstrated, was actually running out of energy and approaching exhaustion, life constantly renewed itself. It was never daunted by failure. No matter how often individual species collapsed, or fell into a changeless mechanical routine hardly different from death, life continued to push forward, acting now through one species, now through another. It was, in fact, the diametrical opposite of matter; and progress could best be described as life's success in achieving steadily higher degrees of freedom from the mechanicalness of matter without, at the same time, shutting off the possibility of still further creative development in the same direction. Man, in whom the *élan vital* had for the first time won a measure of self-consciousness and thereby vastly multiplied its freedom, was of necessity the most advanced creature in the known universe.

Before the emergence of man on earth, evolution had taken several quite different paths, always striving by diversification to improve on itself and to open up new possibilities. The first main thrust of life had brought into being the vegetable kingdom, whose species were relatively immobile, and this very immobility had prevented them from achieving much true independence of their environment. The second thrust produced the invertebrate animals, who had developed the powers of instinct to the highest degree, and the third, the vertebrates, who owned their triumphs primarily to a phenomenal expansion of the faculty of intellect. Progress among the invertebrates had carried as far as the social instincts and there had stopped dead. Among the vertebrate animals, man alone was still capable of progress, and Bergson could conclude that of all the paths taken by life in its upward struggle, only that which led to man "has been wide enough to allow free passage to the full breath of life." [23] Instinct was safe, but limited; intelligence opened up endless possibilities of advance.

And yet instinct had one great advantage over intellect. It was life-directed. Albeit unconsciously, it perceived relations and forms in their wholeness. The insect did not pause to analyze. His actions, though involuntary, sprang from an inward sympathy unattainable by the conscious, reasoning mind. By its very nature, intellect could not help chopping up the data of experience into discrete particles and mentally reassembling them as lifeless machinery. All this Bergson had argued before, in *Matter and Memory*. Intellect allowed to run absolutely free would in the end no doubt transform the world into a gigantic factory. Mechanism would have the last word. But fortunately for man, he was not entirely dependent on his intelligence for his salvation. The development of consciousness had made possible further evolutionary progress in the instinctual faculties, dwarfed and suppressed, but still far from extinct in man's nature. The interpenetration of consciousness and instinct could generate the power of spontaneous insight, or intuition, and here lay man's best hope for open-ended progress.

Intuition was instinct brought up to the level of self-consciousness, the very gift that had allowed Bergson himself to discover the *élan vital*. In the intuitive faculty, instinct became not only conscious, but also disinterested and emancipated from the pressure of practical need. In the free play of intuition lay the source of true creativity, of invention, discovery, aesthetic insight, and, therefore, in the highest sense, progress. Given the fullest possible development of the intuitive faculty, no limits could be set to human progress. Man would see much more deeply than ever before into the mysteries of the universe, surmount the barriers of mortality, win undreamed of freedom. "The animal," Bergson concluded, "takes its stand on the plant, man bestrides animality, and the whole of humanity, in space

69

and in time, is one immense army galloping beside and before and behind each of us in an overwhelming charge able to beat down every resistance and clear the most formidable obstacles, perhaps even death." [24]

As Gerhard Masur has noted, much of Bergson's popularity—and there is the clearest parallel here with the public reception of Nietzsche—stemmed from an easy misunderstanding. Many of his readers wrongly equated the *élan vital* with "the titanic drive which impelled Western civilization toward shores yet unexplored." [25] By an almost complete inversion of his meaning, it could be made to stand as a philosophical explanation and endorsement of imperialism, capitalism, industrialism, and Western material superiority generally. But his failure in *Creative Evolution* to draw out the specific moral and social implications of his thought made such a misinterpretation almost inevitable.

Only a quarter-century later, in *The Two Sources of Morality and Religion* (1932), when the climate of opinion in Europe had turned decisively against evolutionism, was Bergson able to repair these omissions, and by then he no longer had the ear of the younger generation. As some books are written ahead of their time, *The Two Sources of Morality and Religion* was written after its time; in most respects, it is patently a product of *la belle époque*.

Bergson prepared himself for his task in a way that few early admirers of *Creative Evolution* would have expected. He studied the literature of the sociology of religion, but he also spent much time absorbing the works of the great Catholic mystics, from St. Bernard to St. Teresa and St. John of the Cross. The similarity between intuition, as he defined it, and the mystical experience presented itself forcefully to him. Beyond this, he came to see a growing resemblance between the *élan vital* and the creative God of the Judeo-Christian tradition. It would be difficult to maintain that Bergson met the demands of the Christian faith less than half way. He died, at least in his own eyes, a believing Catholic.

The Two Sources of Morality and Religion, applying the dualistic epistemology of *Creative Evolution*, contrasted the protective "morality of pressure" with the progressive "morality of aspiration," and the static religion of "closed societies" with the dynamic and unifying religion of "open societies." In static religion, man searched for safety in consoling myths and the arts of magic; the method of dynamic religion was nothing less than the ancient path to understanding of the great religious mystics. The mystic, in effect, drew upon the powers of the *élan vital* to become one with life. The result was inward serenity and joyful, liberating love. The man capable of such experience was the true *Übermensch*. "If all men, if any large number of men, could have soared as high as this privileged man, nature would not have stopped at the human species, for such a one is in fact more than a man." The testimony of the mystics convinced Bergson that what he had been calling the *élan vital* was actually the God of the mystical faith. In *Creative Evolution* he had been content to point to obvious biological facts; now he went further to suggest that the divine life-force had created the universe for love's sake. Creation appeared to him "as God undertaking to create creators, that He may have, besides Himself, beings worthy of His love." The universe was "the mere visible and tangible aspect of love and of the need of loving." [26] Matter had been created to support life, and life had evolved in order to produce creative beings capable of love— men, and others like them scattered throughout the cosmos.

Although modern Western civilization seemed demonically possessed by greed and the quest for animal pleasure, Bergson trusted that things would work out for the best in the end. Western materialism might only serve the higher cunning of the life-force. Progress, he remarked, often occurred by oscillation. Societies pushed one tendency to extremes and then veered around to its opposite, carrying it to extremes in turn. Each swing of the pendulum often resulted in certain permanent gains for humanity, by "the law of twofold frenzy." There was no inevitability about such oscillation, and no guarantee of progress. Still, the West might even now be on the verge of a return to the ascetic ideals of earlier ages. The lust for luxuries and pleasures, once they became too common, might give way to an overpowering sense of *ennui*. At the very peak of materialistic frenzy, one could imagine a sudden shift back toward the simple life. Although men would preserve the lasting accomplishments of mechanism and use them to facilitate spiritual emancipation, no longer would they crave the endless heaping up of pleasure on pleasure so characteristic of modern times. Science might even add to our store of spiritual knowledge by engaging in psychic research as intensively as, in the past, it had concentrated on physical research. But however and whenever progress occurred, it would always come as the result of human choice, freely made out of the interior resources of the life-force. Men could not shift the responsibility to some external fate. Upon them fell the decision of whether to be satisfied with mere existence or to put forth "the extra effort required for fulfilling, even on their refractory planet, the essential function of the universe, which is a machine for the making of gods."[27]

After Bergsonism the most academically respectable movement in early twentieth-century French philosophy, the immanentist idealism of Léon Brunschvicg, also helped keep the doctrine of progress alive. Brunschvicg defined reality as the persistent progress of consciousness toward the rational fulfillment of its freedom and unity. His teaching envisioned, in philosophy, the triumph of pure reason, foreshadowed in the science of mathematics; and, in religion, a Third Testament, the emergence of a faith founded on the awareness of God as the immanent Word who makes possible all spiritual life. Even Brunschvicg acknowledged Bergson's special eminence. He devoted the last historical chapter in his treatise on the progress of the spirit in Western thought to a study of Bergson's philosophy, and dedicated the book itself to him, "as a token of affectionate admiration for the man, of deep gratitude for the works."[28] *

British philosophy after Darwin was dominated first by the largely ahistorical idealism of T. H. Green and F. H. Bradley, and then, early in the twentieth century, by the analytical philosophy of G. E. Moore and Bertrand Russell. Neither school gave much official attention to the problem of progress, although Green found a place for it, defined as the ever-fuller realization of personality within historical time, in his *Prolegomena to Ethics* (1883).

Samuel Alexander, professor at Manchester from 1893 to 1924, was clearly the most articulate of the post-Darwinian British philosophers of progress. His first book, *Moral Order and Progress* (1889), posited a system of evolutionary ethics with strong overtones

---

* Attention is also called to the work of Louis Weber, especially his *Vers le positivisme absolu par l'idéalisme* (Paris, 1903); and *Le Rythme du progrès* (Paris, 1913). In the latter, Weber proposed a "law of two states" to supersede Comte's law of the three states: progress occurred as a result of alternating epochs of technique and speculation (*Le Rythme du progrès*, pp. 301–305). Yet another approach was that of André Lalande, who contended that progress issued not from the differentiating effect of evolution but from the synthesizing effect of spiritual life, which he termed "dissolution" or "involution." See his *L'Idée directrice de la dissolution* (Paris, 1898), and *Les Illusions évolutionnistes* (Paris, 1931).

of Hegel and Spencer. In *Space, Time, and Deity* (1920), a far more ambitious and original work, he expounded a theory of cosmic progress, made possible by the struggles of "a nisus in Space-Time which, as it has borne its creatures forward through matter and life to mind, will bear them forward to some higher level of existence." This next highest empirical quality scheduled for appearance in the cosmos Alexander defined as "deity," but nothing could be known of it except by creatures who possessed it already. Man could no more anticipate the nature of what was deity for him than animals could have anticipated mind, which, as subsequent evolutionary events proved, constituted deity for them. At the same time, deity was not to be confused with God. The universe was not the workshop in which God was being built, but God himself in process of self-development. "God is the whole universe engaged in process towards the emergence of [deity], and religion is the sentiment in us that we are drawn towards him, and caught in the movement of the world to a higher level of existence." [29]

The crucial concept here, perhaps, is the "nisus" in space-time, the striving of the primordial stuff of the cosmos to create ever-higher orders of being out of itself. Like Bergson's *élan vital,* it was nowhere clearly accounted for in Alexander's work; but he could not doubt that it would go on ceaselessly creating. The creative process was not, fundamentally, mechanical, but spiritual, and therefore infinitely self-renewing.

The South African statesman and philosopher J. C. Smuts took much the same view of progress in his *Holism and Evolution,* another book of the 1920s whose world conception and intellectual origins belong to the prewar period.* Smuts's special contribution was to point out that the higher empirical qualities that emerged in evolution were not merely "higher" but were also characterized by increasing wholeness, or organicity, from physical mixtures to chemical compounds to living organisms to minds to human personalities, unifying mind and body in a higher organic synthesis. At the center of the process stood the principle of action, which was not confined to the structures of space-time, "but continually overflows into their 'fields' and becomes the basis for the active dynamic Evolution which

---

* *Holism and Evolution* grew out of an unpublished essay written in 1910. Smuts did not get the chance to make a book of it until his party's defeat in the South African elections of 1924.

creatively shapes the universe." Not even the recent Great War could rob us of our confidence in the continuing creativity of the holistic cosmos.

> It is the nature of the universe to strive for and slowly, but in ever-increasing measure, to attain wholeness, fullness, blessedness. . . . The holistic nisus which rises like a living fountain from the very depths of the universe is the guarantee that failure does not await us, that the ideals of Well-being, of Truth, Beauty and Goodness are firmly grounded in the nature of things, and will not eventually be endangered or lost.[30]

Alexander and Smuts had their followers, but the only British writer of the age who came even close to equaling the popularity of Nietzsche and Bergson as a prophet of vitalism and its doctrine of progress, was Bernard Shaw. Not a philosopher and not even a thinker of any real originality, Shaw nonetheless popularized vitalism in Britain with intelligence and *panache*. Drawing the substance of his thought from Nietzsche, the neo-Lamarckian tracts of Samuel Butler, and, much later, Bergson's *Creative Evolution,* he gave it literary form in two great, if unplayable, plays: *Man and Superman* (1901) and *Back to Methuselah* (1921).

Shaw had no more patience than Nietzsche with the self-congratulatory faith in progress of the ordinary bourgeois of his time. "The more ignorant men are," he observed in the notes to *Caesar and Cleopatra,* "the more convinced are they that their little parish and their little chapel is an apex to which civilization and philosophy has painfully struggled up the pyramid of time from a desert of savagery." [31] John Tanner, his mouthpiece in *Man and Superman,* took three chapters in "The Revolutionist's Handbook and Pocket Companion" to demolish the dogma of progress. Men were not one jot or tittle better, he protested, than they had been at the dawn of history. Progress had not occurred and never could, so long as man remained man.

Up to this point Shaw seemed to concede even less than Nietzsche had to the popular faith in the superiority of modern times. As a metaphysician, however, or rather as a metabiologist, he had hopes as high as those of any thinker of his generation. His optimism issued not from faith in man, but from confidence in the power of the "Life Force" proclaimed by Don Juan in Act III of *Man and Super-Man,* and later identified with Bergson's *élan vital* in the Preface to *Back to Methuselah*.[32] All the progress that had so far taken place in the cosmos had occurred in the evolutionary process, and evolution had not gone about its work blindly, whatever the Darwinists might argue. The microscope and the dissecting knife were powerless to discover the ultimate source of nature's inventiveness. If life in its upward march displayed "incessant aspiration to higher organization, wider, deeper, intenser self-consciousness, and clearer self-understanding," this could be explained only by the presence of a great will to power in the evolutionary process itself. Life was "the force that ever strives to attain greater power of contemplating itself," and in man it had at last succeeded in creating a being capable of self-consciousness. "My brain," said Don Juan, "is the organ by which Nature strives to understand itself." [33] But the progress achieved so far in geological time fell desperately short of nature's goal. Man's capacity, such as it was, for the contemplative life had not yet negated his deeply rooted animal nature. He remained, as always, pleasure-loving and lazy, only in the most exceptional instances breaking through to the fulfillment of his higher potentialities.

The only solution was to do away with man, everywhere and forever, and put in his place a race of supermen. Shaw rejected the elitism of Nietzsche, but not his definition of the superman. He was the man who had command of himself, the man who could master the impulses of his lower nature in order to develop to their fullest extent the higher powers implanted in his inner being by the Life Force. Hence the symbolism of hell and heaven in *Man and Superman:* "To be in hell is to drift: to be in heaven is to steer." Don Juan had tired of the endless round of self-indulgence in hell. He was bent on going to heaven "because there I hope to escape at last from lies and from the tedious, vulgar pursuit of happiness, to spend my eons in contemplation." [34] So, ultimately, all men would have to choose. Darwinian factors of a sort might help: there would be a tendency, under conditions of general prosperity, for the weak and the self-indulgent to die off without reproducing themselves, through an excess of luxurious living and an unwillingness to accept the burden of childbearing. The chief instrument of man's salvation would be his own will to power, in the Nietzschean sense. As the Serpent explained creative evolution to Eve in *Back to Methuselah,* "You imagine what you desire; you will what you imagine; and at last you create what you will." [35]

Back to Methuselah, Shaw's second attempt to present his vitalist philosophy of progress as a dramatic legend, must be counted as a Shavian *Divine Comedy*—he himself called it a "Metabiological Pentateuch." To play it through would require something like nine hours. The curtain rises in Eden and falls for the last time in A.D. 31,920, by which time men had learned by sheer force of will to live for thousands of years. They were born from eggs fully grown, and after four years of love, play, and art (the hell of *Man and Superman*) they changed into sexless Ancients and, like Don Juan, spent their eons in contemplation. Man had become Superman, no longer preoccupied with the gratification of his animal needs and the frivolities of adolescence.

But the Ancients were not satisfied even now. Their next task was to dispense with bodies altogether and become pure spirits. As Lilith, the Life Force personified, announced in the final scene of the play, "after passing a million goals they press on to the goal of redemption from the flesh, to the vortex freed from matter, to the whirlpool in pure intelligence that, when the world began, was a whirlpool in pure force." She vowed not to "supersede them until they have forded this last stream that lies between flesh and spirit, and disentangled their life from the matter that has always mocked it." The cosmos would be filled with self-conscious spiritual life; but beyond that, Lilith herself could not see, and did not choose to see. "It is enough that there *is* a beyond." [36]

Like all utopias, *Back to Methuselah* can be read on one level as the utopographer's puerile dream of a world peopled by idealized images of himself, even if Shaw was far too sophisticated to expect his readers to take it seriously in all its details. On another level, however, it is quite clearly a sober piece of sermonizing and a significant elaboration of the idea of progress embodied in *Man and Superman.* All the familiar, and by 1921 one might almost say hackneyed, themes of post-Darwinian metaphysics come into play.

Transplanted to American soil, most of the European schools of anti-positivist philosophy fared rather poorly. In their place, and fulfilling many of the same purposes, emerged pragmatism, a native American movement still influential in many areas of American life and thought. Founded at Harvard in the 1870s and 1880s by C. S. Peirce and William James, pragmatism approached the problems of logic and knowledge in a new way. It declared war on many of the philosophical tendencies of the main part of the nineteenth century, notably monism and determinism. With what seemed like impertinent contempt for abstract truth, it insisted on relating the meaning of philosophical propositions to their logical and real consequences, and soon earned the reputation of having mounted an American offensive against the wisdom of the Old World. The offensive cut both left and right, attacking the dogmatic claims of positivism on the one side and of idealism on the other. In the America that William Jennings Bryan described in 1900 as "the supreme moral factor in the world's progress and the accepted arbiter of the world's disputes," [37] pragmatism could be regarded almost as a precious national resource, establishing for the Republic the same eminence in the world of thought that she had already won in business and government.

The parallels with European thought are perhaps no less remarkable than the differences. Now that the philosophical dust of the late nineteenth and early twentieth centuries has settled, the close affinity between American pragmatism and some of the newer movements in European philosophy seems obvious—especially between pragmatism and vitalism.

Like Nietzsche and Bergson, the pragmatists were essentially yea-sayers and meliorists. This is clear above all in the case of Peirce. An outstanding mathematician and logician, Peirce was also a careful student of the nature-philosophy of Schelling. From Schelling and from his own studies of nature, he became convinced that the universe was in process of continuous free growth toward certain evident goals. Gradually, as spontaneity yielded to habit, the universe became more rational and more orderly, and the final result would be an "absolutely perfect, rational, and symmetrical system," in which God would stand entirely revealed to his creation. In frank opposition to the Darwinian "Gospel of Greed," he also argued that the agency of progress in cosmic evolution was *Agape,* the sentiment of loving sympathy, which alone gave moral direction to the good work of freedom and habit. He concurred with the Sermon on the Mount: "Progress comes from every individual merging his individuality in sympathy with his neighbors." [38]

Of much greater moment to thinkers on both sides of the Atlantic during *la belle époque,* however, was the philosophy of William James. As a moral philosopher, James made quite tangible contributions to his generation's sense of the openness of the future. Contradicting Spencer and agreeing with Bergson, he contended that life could not be reduced to laws. Man was a willing animal, who could bend fact to his heart's desire. The evolutionary philosopher who affirmed that the good was that which is predestined to prevail evaded the whole issue. Only a "herd of nullities" could be impressed by such a test of goodness. If we so chose, we might come to the aid of the men or forces "predestined" to win, or we might take advantage of our alleged prescience to strangle them in their cradles. Either way, as willing, acting beings, we might well determine ourselves who or what eventually prevailed. With the supreme confidence of a young, growing society, James wrote: "For again and again success depends on energy of act; energy again depends on faith that we shall not fail; and that faith in turn on the faith that we are right,—which faith thus verifies itself. . . . The thought becomes literally father to the fact, as the wish was father to the thought." [39]

James's universe also contained a God, but a God who fought and struggled for the good alongside man, a being less than omnipotent, "who calls us to co-operate in his purposes, and who furthers ours if they are worthy. He works in an external environment, has limits, and has enemies." [40] In sum, James's cosmology stood at opposite poles from what he felt to be the arid perfections of determinism and monism. It was a universe of persons, divine and human, engaged in the ceaseless struggle to make life better, although James once admitted his inability to declare with confidence that "the total sum of significances is positively and absolutely greater at any one epoch than at any other of the world." [41] This last qualifying remark, if interpreted strictly, would of course compel us to strike his name from the roll of believers in progress altogether, despite the obviously meliorist thrust of most of his thought. Unlike many of his contemporaries, James had caught at least a glimpse of the insuperable bookkeeping problems involved in the classical theory of net progress.

In the third major figure in the history of pragmatism, John Dewey, the voluntarism of Peirce and James was linked to a positivist reverence for science, thus almost bringing us back, full circle, to the thinkers of the preceding chapter. Born in 1859, Dewey began his academic career in the Middle West, at Michigan and Chicago. In 1904 he moved to Columbia. Like his contemporaries Shaw and Wells, he lived on long beyond his "time," and like Shaw and Wells, he always remained a thinker essentially of the post-Darwinian generation, although he wielded considerable influence both in academic philosophy and in educational circles in the United States until the 1930s and 1940s. Many of his most important books, including *Human Nature and Conduct* and *Experience and Nature,* were published after the First World War.

The positivist element in Dewey's thought, his belief that science was the instrument by which human will could both regulate and give effect to its aspirations, led him to dissent from the scruples of his master, William James, on the question of progress. As James was concerned preeminently with the individual, Dewey focused on society. He had as much to say about education, politics, and social reform as he did about logic and psychology, and perhaps more. Through it all, he insisted on the paramount importance of science and the scientific method. "Until men got control of natural forces," he wrote in 1916, "civilization was a local accident. . . . Any civilization based mainly upon ability to exploit the energies of men is precarious; it is at the mercy of internal revolt and external overflow." But the scientific conquest of nature had now given us a "sure method" for achieving continuous progress. From now on, "wholesale permanent decays of civilization are impossible." [42]

Of course science in and of itself could do nothing. As a good pragmatist, Dewey shared James's distaste for Spencerian determinism. The possibility of progress was open to man because he was educable, and because he had learned to control his environment through the sciences. But no external process, evolutionary or otherwise, guaranteed success. "Progress is not automatic; it depends upon human intent and aim and upon acceptance of responsibility for its production. It is not a wholesale matter, but a retail job, to be contracted for and executed in sections." Those who hoped that man himself could be changed through the beneficent operations of natural selection labored under childish illusions. Human beings today were born with the same emotions and powers as savages. "Since the variable factor, the factor which may be altered indefinitely, is the social conditions which call out and direct the impulses and sentiments, the positive means of progress lie in the application of intelligence to the construction of proper social devices." War, for example, could be ended not by reducing man's bellicose instincts but by creating institutions to restrain them and, at the same time, to give maximum effect to his altruistic instincts. What we needed now was simply the willingness to use the means at our disposal; nothing else would avail. If we wanted progress, he concluded, "we can have it—if we are willing to pay the price in effort, especially in effort of intelligence. The conditions are at hand." [43]

As Dewey rejected determinism, so he also rejected—in the authentic spirit of the pragmatist movement—speculative thinking about "ultimate goals." "There is something pitifully juvenile," he noted in *Human Nature and Conduct,* "in the idea that 'evolution,' progress, means a definite sum of accomplishment which will forever stay done, and which by an exact amount lessens the amount still to be done, disposing once and for all of just so many perplexities and advancing us just so far on our road to a final stable and unperplexed goal." If all humanity wanted was stability and peace of mind, it would do better to embrace the primitivist utopism of Rousseau and Tolstoy. If it meant anything, progress meant struggle, conflict, trouble, and frequent disappointment. To care only about some unimaginable future bliss of perfection was to surrender oneself to illusion. Rather, man should concentrate on the needs and possibilities of the moment and find satisfaction in the tangible efforts that lay within his present power to put forth. "Instruction in what to do next can never come from an infinite goal, which for us is bound to be empty. It can be derived only from study of the deficiencies, irregularities and possibilities of the actual situation." [44] *

* Pragmatism also won a foothold in British intellectual life, especially in the work of F. C. S. Schiller at Oxford, who espoused a pragmatist conception of progress in his *Humanism* (1903) and *Studies in Humanism* (1907). But Schiller was never well accepted at Oxford, and in later life he moved to the University of Southern California.

6

The Theology of Progress

THEOLOGY is the science of God. But in the generation after Darwin, no great gulf separated God from his creatures. He was a God of vitality, action, adventure, and, above all, progress: a go-ahead sort of God, a scoutmaster of colossal proportions, a gallant commander in the battle for modern civilization. We have already met him in several other contexts, in Bergson's *élan vital,* in the emergent evolutionism of Alexander, in the thought of William James. The God of *la belle époque* was also the "invisible king" described by H. G. Wells during a brief Jamesian flirtation with theistic imagery (which he later repudiated), a God who

> looks not to our past but our future, and if a figure may represent him it must be the figure of a beautiful youth, already brave and wise, but hardly come to his strength. He should stand lightly on his feet in the morning time, eager to go forward, as though he had but newly arisen to a day that was still but a promise; he should bear a sword, that clean, discriminating weapon, his eyes should be as bright as swords; his lips should fall apart with eagerness for the great adventure before him, and he should be in very fresh and golden harness, reflecting the rising sun.[1]

The line between theology, science, and philosophy in the period from 1880 to 1914 is often difficult to draw. The theological systems of the late nineteenth and early twentieth centuries were powerfully influenced by secular thought. Positivism and anti-positivism, evolutionism and vitalism, and the climate of confidence in man and his civilization all had their effect. Philosophers who proclaimed themselves spiritualists or scientific theists or emergent evolutionists flourished alongside eclectic theologians who proclaimed themselves modernists or liberals or social gospellers, all believing more or less the same things. There was much talk of the evolution of religion, the religion of evolution, practical or ethical Christianity, modern faith, religious experience, the progress of dogma. Not all theologians looked respectfully on the efforts of philosophers, just as many philosophers had no interest in theology, but the substance of their hopes for the future was often the same. However much or little supernatural powers might be implicated, their world was this world, and their faith was in temporal progress. From the perspective of the neo-orthodoxy of the post-1914 years, the collapse of traditional Christian faith could be seen just as clearly in the compromises with secular thought negotiated by the theologians of the post-Darwinian generation as in the anti-Christian or post-Christian systems of its scientists and philosophers.

As might be expected, some of the most impressive work was done in Germany, which had produced most of the leading figures in at least Protestant theology and biblical scholarship since the early nineteenth century. From Friedrich Schleiermacher on down to Albrecht Ritschl, German theology had taken what might almost be called a pragmatic approach to faith, minimizing dogma and speculative thought, and laying greatest emphasis on the meaning of religious experience to the living believer. Faith was a matter of personal experience. Schleiermacher's own faith suggests the sensibility of the romantic movement, whereas Ritschl's has the dry and prosaic quality of positivism, but they were both primarily concerned with the impact of the divine Word on the inner man. This same emphasis continued, with modifications, into the post-Darwinian generation, where it harmonized with some of the leading tendencies in secular thought, so that of all the national schools of theology, perhaps the German had the least difficulty in accommodating itself to extra-theological influences.

During our period the thinker whose influence carried the most weight was Ritschl. Although his principal works, including *The Christian Doctrine of Justification and Reconciliation,* had appeared before 1880, and he himself died in 1889, the Protestant theologians of the next generation were nearly all Ritschlians. Something should be said about the master before turning to his disciples.

In the very briefest terms, Ritschl's theological program was to free theology from the tyranny, so obvious to a German thinker, of abstruse speculation, to bring it down from the Alpine wilderness of dogma and metaphysics to the everyday life of man. Religion had to be made, in a word, existential; or, in another word, pragmatic. Although the comparison is seldom made, Ritschl was reacting against Hegelianism as Kierkegaard had done (with certain obvious differences), or as James and Schiller would react a few years later. He wanted at all costs to save his faith in a personal God and Christ, but this could happen only through discovering God and Christ acting in history and in his own life. Faith, for him, had to be concrete rather than abstract. In Ritschl's case, he was led to a theology rooted in the positive historical record of the New Testament, which proclaimed the living Christ, defined God as love, and prophesied the coming of God's kingdom, in which humanity would be enabled by divine grace to live a life of brotherhood and love. The traditional dogma of original sin, the mythology of interceding saints and ministering angels, the concept of eternity, the idea of an omnipotent and transcendent absolute deity, all belonged either to superstition or to metaphysics. Man was the good child of a good and loving Father, who had made himself flesh in Christ in order that his will might be done on earth as it was in heaven. The task of the Christian religion could not be more clear: to perfect the moral order here below. Of transcendence we could, properly speaking, know nothing.

As James P. Martin has pointed out in his recent study of the doctrine of the Last Judgment in Protestant theology, Ritschl's identification of the kingdom of God with a universal moral order that could be achieved in time on earth strongly suggests the secular faith in progress. It stands especially close to the vision of Immanuel Kant, but it also reflects to some degree the radical this-worldliness, the pragmatic temper, and the pervasive optimism of the nineteenth century.[2] By the 1880s Ritschlianism was the universally acknowledged new wave in German theology and soon captured many adherents in the other Protestant countries.

In Germany the leading Ritschlians included Wilhelm Herrmann, a colleague of Hermann Cohen at Marburg, and Adolf von Harnack at the University of Berlin. Although some Ritschlians, notably Julius Kaftan, insisted that earthly righteousness could only be prelusive to a greater divine kingdom literally grounded in eternity,

Herrmann retained Ritschl's emphasis on the life of this world. The focus of his theology was what he termed the "impact of Christ." Christ changed men by laying hold of them in this life, in historical time, giving them the inspiriting power to work for the coming of his kingdom. "Not to despair of the world, and not to despair of ourselves, because Jesus Christ is a real part of our world: that is the beginning of Christian faith." [3]

The best known of the Ritschlians outside the circle of professional theologians was undoubtedly Adolf von Harnack. The son of a distinguished Luther scholar, Harnack advanced rapidly through the German academic hierarchy. After occupying chairs at the universities of Giessen and Marburg, he moved to Berlin, where he taught from 1889 to 1924. He was actually not a theologian, by his own admission, but a church historian. His most ambitious project was *The History of Dogma,* a study of the history of church doctrine from earliest times to the sixteenth century, first published between 1886 and 1890. In keeping with the Ritschlian view of dogma, Harnack demonstrated that the simple message of Jesus, in both its spiritual and ethical dimensions, had been mutilated in the course of time as a result of the evolution of a complex, over-sophisticated, Hellenized body of doctrine largely alien to the original gospel. In particular, the Christology of the fathers of the early Church, with its mixture of gross superstition and abstruse philosophy, could no longer be accepted. Jesus was divine only in the sense that he had experienced within himself an unprecedentedly full awareness of the fatherhood of God.

Even more influential in Harnack's own lifetime than his *History of Dogma* were his lectures on the essence of Christianity, delivered at the University of Berlin in 1899–1900, and published as a book in 1900. If he did not grasp the essence of Christianity in these lectures, he did at least present the essence of Ritschlianism. Once again the emphasis fell on the historical Jesus of the historical Gospel. The Gospel, to be sure, was not a philosophy of progress. To make what Christianity "has done for civilisation and human progress the main question, and to determine its value by the answer, is to do it violence at the start." [4] Religion spoke not to mankind, but to man, the individual. Jesus ministered to the inner needs of the individual, and the kingdom of God could be understood best as a spiritual reality, establishing itself within the soul. With the purely mundane life of civilization, with (for example) the arts and sciences, Jesus was not directly concerned.

At the same time, he who was moved by the Gospel could not remain indifferent to the world around him. Those who interpreted Jesus as a prophet of doom were quite mistaken: the man Jesus was an optimist, who declared God's own war against misery, poverty, and despair. Although the Gospel contained no concrete social or political program, it fired all who embraced it with brotherly love and the will to struggle for righteousness. The Gospel, then, was a social Gospel. "The Gospel aims at founding a community among men as wide as human life itself and as deep as human need. . . . Its object is to transform the socialism which rests on the basis of conflicting interests into the socialism which rests on the consciousness of a spiritual unity." The prospect of a union of mankind founded in love was

a high and glorious ideal, and we have received it from the very foundation of our religion. It ought to float before our eyes as the goal and guiding star of our historical development. Whether mankind will ever attain to it, who can say? but we can and ought to approximate to it, and in these days—otherwise than two or three hundred years ago—we feel a moral obligation in this direction. Those of us who possess more delicate and therefore more prophetic perceptions no longer regard the kingdom of love and peace as a mere Utopia.[5] *

The Ritschlian theology was eminently Protestant in its insistence on returning to the nuclear faith of the primitive Gospel. In Catholic thought during the same period progressivism found expression, more consistently perhaps, in the "Modernist" doctrine of an unfolding revelation of the divine Word in the history of the Catholic Church and the Christian community. But flattering as this may have been to Rome, it involved innovations of belief in such fundamental disharmony with the traditional teachings of the Church that the so-called Modernist movement in Catholicism earned the implacable opposition of the ecclesiastical hierarchy and failed to

* Most German theologians remained uninfluenced by evolutionary thought, unlike their Anglo-Saxon counterparts, but see Karl Beth, *Der Entwicklungsgedanke und das Christentum* (Berlin, 1909), which reaches conclusions not unlike those of the British and American theologians of evolutionary progress discussed below on pp. 92–94 and 96–99.

85

survive as a movement after its condemnation in 1907 by Pius X in his encyclical *Pascendi gregis*.

Modernism had its spiritual center in French Catholicism, although it also attracted important followers in Italy, Germany, and Great Britain. It learned something from Bergson, and the chief philosophical mentor at least of the French Modernists was Léon Ollé-Laprune. In a deeper sense Modernism originated in the characteristically German conception of history, from Lessing on down to Hegel, as the progressive revelation of God's will, as the "education of the human race." Cardinal Newman—in so many other ways a conservative—contributed to it in his *Essay on the Development of Christian Doctrine* (1845), a skillful assault on Protestant "Bibliolatry." In Newman's judgment, the Church, under the inspiration of the Holy Spirit, gradually unveiled the truth, which existed only in germinal form in the Bible. The leading French Modernist, Alfred Loisy, was introduced to Newman's work by his Austro-English friend Baron Friedrich von Hügel as early as 1893; its impact upon him was apparently quite profound.[6]

Loisy was in most respects the pivotal figure in the Modernist movement and not unexpectedly the principal target of *Pascendi gregis*. He was born in 1857 and ordained in 1879. As a seminarian, he experienced serious difficulties in accepting scholastic theology, but he felt that his place was inside the Church. In the 1890s he became well known as a biblical historian and critic, and in 1902 he published *The Gospel and the Church,* which was intended to serve as a reply to Harnack's *What Is Christianity?*. Rarely has the work of a major German theologian been so effectively answered by a non-German. Using all the resources of modern biblical criticism (itself mainly German in origin), Loisy argued that Jesus had been essentially an apocalyptic preacher, warning the world of its imminent end. But the kingdom of God that he prophesied had arrived in a quite different way, not as something absolute and eschatological, but as something relative and time-bound: the historic Catholic Church. Harnack had tried to limit Christianity to the personal message of a historical redeemer, a once-only sudden irruption of the divine into the temporal, whereas, in reality, the Gospel was only the seed from which Christianity, led by the Church, had gradually grown over the centuries.

But Loisy did not confine himself to a critique of Harnack and Protestantism. If the New Testament was not infallible, if the

Gospel was not final, then the dogmas of the Church could not be protected for all time against the fresh inspiration of spiritual progress. Loisy went far enough beyond orthodoxy in *The Gospel and the Church* that the Archbishop of Paris felt compelled to condemn the book as a danger to the Faith, and within a few months of his election to the Holy See, Pius X placed it on the Index. For the next several years Modernism gathered strength. A direct confrontation with the hierarchy seemed unavoidable. These were also the culminating years in the political struggle against the Catholic establishment in France; Pius could not but see a certain connection between the atheistic politicians of the Third Republic and the theologians of Modernism. In both instances the modern world preferred itself to ancient wisdom. *Pascendi gregis,* issued in September of 1907, put the Church clearly on the side of ancient wisdom. Loisy replied with his impenitent *Simple Reflections* in 1908, and Pius excommunicated him. The break proved complete and final. Loisy left the Church to accept a professorship at the Collège de France, where he remained until 1932, writing a long series of books expounding his religious views, which tended to depart more and more from Catholic faith. He died in 1940.

In *The Gospel and the Church,* Loisy was still obviously a child of Rome. Harnack had eliminated the Church from his definition of Christianity, but for Loisy in 1902 the Church and the Gospel were inseparable. The Gospel could not have survived or developed without the Church; in fact, the relationship between the two was somewhat like that between a boy and the man he later becomes.

> The identity of a man is not ensured by making him return to his cradle. The Church, to-day, resembles the community of the first disciples neither more nor less than a grown man resembles the child he was at first. The identity of the Church or of the man is not determined by permanent immobility of external forms, but by continuity of existence and consciousness of life through the perpetual transformations which are life's condition and manifestation.[7]

To affirm the immutability of the Gospel was to petrify it: it had to be a living faith, adapted and re-adapted to the needs and conditions of each age. And each fresh adaptation would represent not only change but also improvement in man's understanding of the eternal and eternally inaccessible absolute.

After his excommunication, Loisy expanded his conception of the

Church to include the total life of man, and in particular his religious life. Guided by religion, civilization was the vessel of divine grace, the means by which man sought to reach the divine kingdom here on earth. Loisy gave up all effort to fathom God in his transcendence. God could be known to man only in his immanence, as he labored through the works of man to improve the world. Christianity was, admittedly, the greatest effort in history to elevate mankind, yet its progress had been brought to a halt by dogmatism and blindness to modern ideals. "Humanity," he wrote in 1923, "bursts through the limits of Christian evangelism and belief. . . . But Christianity has more effectively than any other religion prepared the coming of human religion, the kingdom of humanity." [8] Religions had "a way of dying, which allows them to perpetuate themselves, somewhat altered, in those which replace them." [9] So it was with Christianity. If the Church had faltered and ossified, nonetheless she had helped make possible the nascent religion of humanity. As he wrote to his loyal friend and biographer, Maude Petre, in 1917, Catholicism was essentially a religion of humanity "and could have been the best of all." The "failure" of the Christian churches had driven

> those who can no longer live in them . . . to seek a wider human religion and to promote it alongside Christianity, just as they would have promoted it within Christianity itself if they had remained Christians. . . . Every religion is a religion of humanity, and . . . religious progress—which is identical with human progress—consists in the realization of a higher and more perfect humanity.[10]

Most of the other leading Modernists in France did not join Loisy in his flight from the Church. Their respect for ecclesiastical discipline was too compelling and their immanentism rather less radical. The most engaging figure is no doubt Maurice Blondel, by profession a philosopher, who taught most of his life at the University of Aix-Marseille. Blondel's philosophical and religious position might best be described as denaturalized vitalism, an outlook centered not on "life" but on "action," given direction by the divine grace immanent in the will and spirit of man. The Christian life, he insisted, called man to action, to experience, which could never be wholly captured in the formulas of traditional faith, but had to be renewed time and again in the unceasing motion of man toward

God and of God toward man. Still, God remained transcendent as well as immanent, above as well as within, so that Blondel felt no need to propose a doctrine of divine emergence or finitude.[11]

At the same time, Blondel's dynamic conception of history and truth involved an idea of progress. Whereas pagan thought equated man with nature, Christianity had "placed the human person above the entire cosmic order." From this new valuation of man and his civilization had developed the higher society and the great scientific achievements of medieval and modern times. "The ascendant movement of civilization has been promoted, in the moral, political, and material form that it has taken in our contemporary societies, by the slow and progressive influence of the Christian *sursum.*" This progress was threatened in the twentieth century by our perverse inclination to cut ourselves off from the Christian sources of our civilization, but Blondel remained optimistic. Even if man were not destined to conquer evil once and for all in earthly time, "it is legitimate and profitable to work for the amelioration of this place of trials." [12] *

At least one prominent Modernist, however, chose to remain in the Church even though she rejected him. He was the Irish Jesuit George Tyrrell, the most brilliant figure in the whole history of the Modernist rebellion, and one whose proper place in the history of the idea of progress can be defined only with the greatest difficulty.

Tyrrell was less liberal in some respects than Loisy, but his views on dogma and church government hewed to those of his French colleague. His masterpiece, *Christianity at the Crossroads,* which was not quite finished when he died in his late forties in 1909, ranks as one of the great polemical works in the history of religion, a remarkable fusion of clarity, anger, and faith. With Loisy, he refused to accept the papal condemnation of Modernism in 1907 and was excommunicated; but he could not follow Loisy into the wilderness

---

* *Cf.* Lucien Laberthonnière, *Le Réalisme chrétien et l'idéalisme grec* (Paris, 1904), which contrasts the static, pessimistic idealism of Greece with the dynamic, progressive, unfolding faith of Christianity. Father Laberthonnière followed Loisy in arguing that the Christian faith "progresses in humanity as in individuals, like a child who becomes a man, while continuing to be the same person" (Laberthonnière, p. 80). See also Auguste Sabatier, *Outlines of a Philosophy of Religion Based on Psychology and History,* tr. T. A. Seed (London, 1902). Sabatier was a leading French Protestant theologian whose views both influenced and harmonized with those of his Catholic compatriots.

of a generalized religious humanism or even some of his own fellow Modernists in Britain into the fellowship of the Church of England. He died receiving the last rites of Rome.

Christianity at the Crossroads had much the same starting point as *The Gospel and the Church*: a good Catholic's sincere distaste for Protestant primitivism, and in particular for the theology of Harnack. Liberals like Harnack "wanted to bring Jesus into the nineteenth century as the Incarnation of its ideal of Divine Righteousness, i.e., of all the highest principles and aspirations that ensure the healthy progress of civilisation. They wanted to acquit Him of that exclusive and earth-scorning other-worldliness, which had led men to look on His religion as the foe of progress and energy." [13] Their alleged submission to the findings of "science" reduced, on closer investigation, to something quite different: the use of some of the techniques of scientific criticism to justify a totally anachronistic view of the Christian Gospel.

"Whatever Jesus was, He was in no sense a Liberal Protestant." Tyrrell's Christ proved to be a most unmodern man, superstitious, devoted to ritual and ceremony, convinced of his own divinity, and certain that the end of the world and the coming of the kingdom were imminent. "The necessity," he wrote, "of finding in Jesus a German Liberal Protestant, guided entirely by the light of a sweet, nineteenth-century reasonableness, requires us to ignore everything in the Gospel that suggests the visionary or the ecstatic, even though to do so makes the narrative incoherent and unintelligible." The Christ of the New Testament believed that "the world was as good as finished" and would end in "universal failure and illusion." To find a notion of progress in this Christ was to miss the whole point of the New Testament as a prophetic book. The gospels gave no hint of

> a reign of morality here upon earth to be brought about by the gradual spread of Christ's teaching and example. The parables of the mustard seed and the leaven, adduced in its favour, are irrelevant. They merely contrast the slightness of the cause with the greatness of the effect; man's natural efforts with God's supernatural response. Jesus did not come to reveal a new ethics of this life, but the speedy advent of a new world in which ethics would be superseded.

The early Church shared Christ's vision and passed it on to the Middle Ages. "The notion that Good was to triumph by an immanent process of evolution never entered into the 'idea' of Jesus or of the Church." [14]

In all of this, Tyrrell echoed the most recent developments in New Testament scholarship, which were already beginning to undermine the Ritschlian theology in Germany. Loisy had done the same, without drawing from his apocalyptic interpretation of Jesus conclusions that imperiled his belief in progress. It was quite otherwise with Tyrrell. Jesus had not been a prophet of progress, but Tyrrell contended that even modern man had no reason to believe in it, if by progress he meant the gradual achievement on earth of the kingdom of heaven. Clearly, he could also no longer share the early Christian anticipation of an imminent end of the world. Heaven was only the symbol of an ineffable transcendent spirituality; no modern man could understand it literally. But then it was just as foolish to assume that a literal heaven could be realized on earth. The modern gospel of progress was a fallacy, based on the mistaken assumption that the race and the whole world, "like the individual organism, are inherently predetermined to pass through a series of stages ending in a definite final perfection." Nothing could be further from the truth. The course of history was unplanned and unpredictable, and every material and moral victory brought with it unexpected sorrows, new sources of frustration and misery. Men were no happier for their "progress" and no nearer to transcendence. Neither modern man nor his posterity "shall find goodness, happiness, truth and beauty united in this life." But the illusion hung on stubbornly, reinforced in modern times by the supposed representatives of Christ on earth. "The Churches chatter progress, and the secular and clerical arm are linked together in the interests of a sanctified worldliness." [15]

Up to this point Tyrrell could pass for one of the radically disillusioned theologians of the years after 1914 (who will be discussed in Part III), but he was not able to escape altogether the conceptual frame of the generation in which he lived. He endorsed without question the current belief in the progress of religion, morality, science, and technology. Commenting on his own interpretation of the traditional doctrine of immortality, he found nothing surprising in the fact that it had come "late in the history of religion." The "purer morality and more spiritual religion" of modern times had been "similarly delayed. . . . Man's nature unfolds its potentialities gradually, the deepest and most fundamental being the last to appear."

Religious experience "grows stronger and purer as man rises morally and spiritually." Tyrrell also noted that modern civilization, unlike earlier civilizations, could not fall, but would continue to thrive and grow as long as man inhabited the planet.[16] As he wrote in another book, *Through Scylla and Charybdis,* published in 1907, "Life is essentially progressive, or rather is a progress. . . . It expands in every dimension, upwards as well as outwards. . . . It amplifies in quality as well as in degree." [17]

Yet in the final reckoning, Tyrrell was not an authentic believer in progress. Despite the advances made by the human race in certain areas of its common life, including areas of the highest importance to any theologian, mankind had not progressed in true happiness, and for Tyrrell this was what mattered most. Heavenly bliss would never be known on earth: every victory had to be paid for with a defeat. Tyrrell's intellectual indebtedness to the progressivist tradition is not without interest, but he stands outside the camp of the faithful.

Liberal Protestant thought in Britain followed a more predictable course before the First World War. It produced no one of the stature of Tyrrell, but two representative figures were Henry Drummond and R. J. Campbell. Both men led spiritually adventuresome lives, Drummond as a colleague of the Chicago evangelist Dwight Moody and the lifelong champion of a theological alliance between evangelical Christianity and evolutionism, and Campbell as a liberal Nonconformist transmuted into a conservative Anglican during the First World War.

Drummond's *idée fixe* was the analogy between the natural and the spiritual world, a theme he first treated at length in his *Natural Law in the Spiritual World,* published in 1883. Everything that science found in nature one might also find in Christian faith: law, causation, design, order. Instead of resisting the teachings of natural science, the faithful Christian should rejoice in them. But the most important revelation of modern science was the theory of evolution, which supplied a new and invaluable key to the true meaning of the Christian epic. Creation, history, progress, and the destiny of man were all explicable in terms of a single master-idea, and Drummond set himself to the task of providing a "connected outline of this great drama" in his Lowell Lectures, *The Ascent of Man,* in 1894.

Christian objections to evolutionary theory, Drummond wrote, grew from a misunderstanding of the place of struggle in the evolutionary process. There were two kinds of struggle, which could be traced back to the beginning of time: the struggle for life and the struggle for the life of others, egoism and altruism, appetite and love. Each had its necessary function in the scheme of things, but the share of the latter steadily increased as the share of the former steadily decreased. From the sexless unicellular organism, up the biological scale to the mammals, and finally to man, love inexorably enlarged its dominion, revealing the final purpose of the universe. Drummond even found a rudimentary sort of love in the attraction of atoms to one another, their tendency to combine and form more complex substances, from which life had ultimately developed. But what mattered most was the future.

> Whatever controversy rages as to the factors of Evolution, whatever mystery enshrouds its steps, no doubt exists of its goal. The great landmarks we have passed, and we are not yet half-way up the Ascent, each separately and all together have declared the course of Nature to be a rational course, and its end a moral end. . . . Evolution has ushered a new hope into the world. The supreme message of science to this age is that all Nature is on the side of the man who tries to rise. Evolution, development, progress are not only on her programme, these are her programme. For all things are rising, all worlds, all planets, all stars and suns. An ascending energy is in the universe, and the whole moves on with one mighty idea and anticipation. The aspiration in the human mind and heart is but the evolutionary tendency of the universe becoming conscious.[18]

The goal of evolution, then, was the perfection of life through love. Drummond could see "an undeviating ethical purpose in this material world, a tide, that from eternity has never turned, making for perfectness." At the same time he could not be satisfied with a simple naturalist metaphysic based on evolution. If science pointed the way, so also did Christianity, and each reinforced the other. Science enabled the theologian to see God as immanent, working through evolution, and Christian theology enabled the scientist to see evolution as a fundamentally spiritual and teleological process. Evolution and Christianity, interpreted with the help of the light shed by each, proved to be one and identical. Each was a method of creation, to make more perfect living beings through love. No man could study the evolutionary process without ending in a complete acceptance of Christian truth and faith. "A religion which is Love and a Nature which is Love can never be but one." Echoing the eighteenth-century deist Matthew Tindal, Drummond argued that "Christianity did not begin at the Christian era, it is as old as Nature; did not drop like a bolt from Eternity, came in the fulness of Time." And its work would be consummated in due course, with the coming of an earthly kingdom of love. "The Struggle may be short or long; but by all scientific analogy the result is sure. All the other Kingdoms of Nature culminated; Evolution always attains; always rounds off its work. . . . The Further Evolution must go on, the Higher Kingdom come." [19] *

Drummond died in 1897. His place as a popularizer of liberal Protestantism in Britain was taken during the Edwardian period by R. J. Campbell. A Congregationalist, Campbell served as minister of the City Temple in London from 1903 to 1915, and expounded his view of religion in several widely read books, including *The New Theology* and *Christianity and the Social Order*. Acknowledging his debt to such thinkers as Tyrrell and Sabatier, he called for a modern theology that restated the essence of Christianity in terms of modern science, social consciousness, and the immanence of the divine. "Upon the foundations laid by modern science a vaster and nobler fabric of faith is rising than the world has ever before known. Science is supplying the facts which the New Theology is weaving into the texture of religious experience." We now understood that God was like a child becoming a man, actualizing what before had been only potential. God existed, "but it will take Him to all eternity to live out all that He is. . . . God is getting at something, and we must help Him. We must be His eyes, and hands, and feet; we must be labourers together with Him." [20]

Campbell went on to repudiate the doctrine of the Fall. Science and history showed that man had risen, not fallen. Civilization as a whole had advanced since antiquity to a "higher level of intellectual attainment" and "greater sympathy, a keener sense of justice." "Slowly, very slowly, with every now and then a depressing setback, the race is climbing the steep ascent towards the ideal of uni-

---

* A similar, although rather less elaborate, evolutionary theology embodying a doctrine of progress was expounded by another Scotsman, James McCosh, especially in his book *The Religious Aspect of Evolution* (New York, 1890). McCosh was living in the United States at the time the book appeared, and it had its greatest influence in this country.

versal brotherhood." [21] The socialist ideal of modern enlightened Western man was nothing less than the spirit of Christ working in the hearts of men, building the kingdom of God, which Campbell defined as "a universal brotherhood, a social order in which every individual unit would find his highest happiness in being and doing the utmost for the whole." [22]

But Campbell's liberalism did not survive the First World War. As early as 1911, he began to entertain serious doubts about the health of modern civilization and man's capacity for self-salvation. Albert Schweitzer's *Quest of the Historical Jesus,* which first appeared in an English translation in 1910, made him see the futility, he later reported, of a "modern" and "reasonable" Jesus, and he came under the influence of Eucken and von Hügel, whose works helped to turn him away from the radical immanentism of *The New Theology.* The war that broke out in 1914 finished the conversion of Campbell to orthodoxy. He experienced a profound disillusionment with so-called modern values, left his pulpit at City Temple, purchased the rights to *The New Theology* in order to withdraw it permanently from the market, and was ordained in 1916 as an Anglican priest. His explanation calls to mind the parallel experience of Karl Barth in Switzerland: "I wanted to follow God's way, not my own." [23] If any progress were to be made in human affairs, it could be made only with the help of advenient divine grace. In the end, Campbell became, not unlike Tyrrell, a living witness against the belief in progress of his own generation.

The United States contributed its full share of religious thinkers and theologians to the propagation of progressivism in the post-Darwinian era. Indeed, the characteristic late-nineteenth-century American faith in progress found expression more often in the sermons and writings of liberal religion than in any other way. Systems of "evolutionary" or "scientific" theism involving a doctrine of progress were legion, both inside and outside the churches. Among lay thinkers, theistic philosophies of progress were advanced by Joseph Le Conte of the University of California and by the Chicago publisher and philosopher Paul Carus, carrying on in the tradition of Herbert Spencer's American disciple John Fiske. But the clergy needed no help from their brethren in the laity. Modernist factions erupted in most of the leading Protestant denominations. The movement for a "social gospel," led by Walter Rauschenbusch, stressed the implications for social progress of Christian ethics. So strongly did modernism and liberalism dominate American clergymen that the movement swept on with undiminished fervor into the years after the First World War, encountering little significant resistance until the 1930s.

At the beginning of the Progressive Era, a large percentage of the major liberal voices in American church life were New England Unitarians, who owed more to the deism of the Enlightenment and to German idealism than to evolutionary thought. In any event, they unanimously subscribed to a belief—following the words of a popular Unitarian credo—"in the progress of mankind onward and upward forever." A representative figure was James Freeman Clarke, a minister in Boston for more than forty years, and the author of some thirty books, including a pioneering study in the field of comparative religion, *Ten Great Religions,* which appeared in two volumes between 1871 and 1883. "The progress of the human race," he wrote, "is fixed by laws immutable as the nature of God. The fidelity of man may hasten it; the wilfullness of man may retard it, but Divine Providence has decreed its certain issue." [24] Clarke believed that mankind had progressed morally in his own lifetime, and he predicted that the vices and failings of the human race would be entirely abolished by the year 2000.

Clarke spoke for moderate and conservative Unitarians. Further to the left stood men such as Minot J. Savage, who drew upon neo-Lamarckian thought in *The Religion of Evolution,* published in 1876, and Francis Ellingwood Abbot, best known for his *Scientific Theism,* published in 1885. When he wrote *Scientific Theism,* Abbot was not even a Unitarian, having broken away in 1867 with several other liberals to found the Free Religious Association, in protest against what he considered to be the excessively Christian bias of the Unitarianism of his time. Among the "Fifty Affirmations" of the F.R.A., which Abbot himself drafted, were included the propositions that free religionists believed in man "as a progressive being" and held as their "great ideal end . . . the perfection or complete development of man—the race serving the individual, the individual serving the race." [25] *

* A Unitarian layman, W. H. Carruth, summed up the world-view of evolutionary theism with nice economy in two familiar lines from his poem "Each in His Own Tongue," written at the turn of the century: "Some call it evolution,

The mother church from which Unitarianism had emerged in the early nineteenth century, the Congregational Church, retained much of its liberalism in the latter part of the century, in America as in Great Britain. The equation of evolution with progress, and of both with the will of God, was made almost routinely by Congregationalist clergymen. Lyman Abbott, for example, who collaborated with Henry Ward Beecher and succeeded him as editor of *Christian Union* (later *The Outlook*) and also as pastor of Beecher's church in Brooklyn, wrote several books, each of which, in his own words, "assumes the truth of the principle of evolution as defined by Professor Le Conte, and attempts to apply that principle." The most interesting for our purposes is *The Theology of an Evolutionist,* which Abbott published in 1897. The chapter titles alone expound the author's point of view: "Creation by Evolution," "The Evolution of Revelation," "Redemption by Evolution," and so on. God, Abbott noted, had traditionally been seen as a kind of glorified Roman emperor in his dealings with man and as a mechanic in his manipulation of nature; both analogies were now exposed by modern science as erroneous. God clearly had but "one way of doing things." That way might be described "in one word as the way of growth, or development, or evolution." Natural and divine law were all one, and "all nature and all life is one great theophany. . . . Scientifically this is the affirmation that the forces of nature are one vital force; theologically it is the affirmation that God is an Immanent God." Redemption had, then, to be reinterpreted as the process by which man progressed to Christ-like perfection, and Abbott looked forward to the culminating day "when man shall be presented before his Father without spot or wrinkle or blemish or any such thing." [26]

Abbott's fellow Congregationalist Francis Howe Johnson also espoused an evolutionary theism, fortified by his intimate familiarity with contemporary biological and philosophical thought. His first major work, *What Is Reality?*, presented an immanent, finite, naturalized God who achieved his purposes through the evolutionary process. Man's noblest ideals, peace and love, were in fact

/ And others call it God." The poem "sprang into immediate popularity and has been in continuous circulation all over the world since." *The National Cyclopaedia of American Biography* (New York), XIV (1917), 486. Carruth was professor of German at the University of Kansas.

God's ideals, "God working in us, for the bringing about of the great end toward which the process of creation has been moving from the beginning even until now."[27]

Twenty years later, in his *God in Evolution: A Pragmatic Study of Theology,* Johnson restated his position in the language of pragmatism; the new book was virtually a compendium of the ideas of the pre-1914 generation. It began by noting that "the stream of thought" on which the author's earlier work had been launched "has swollen into a great river." The dreary mechanistic positivism of the nineteenth century had been replaced by biopsychological vitalism and pragmatism, with insights of the greatest significance for theology. James, Dewey, Schiller, Bergson, Driesch, and others were reeducating the human race. In religion the response of modern-minded men was to turn away from "the thought of God as external to the universe, and toward some conception of Him as its living, in-dwelling principle." Like so many other thinkers of his time, Johnson found the new world-view—with its stress on change, action, experience, and will—an exhilarating and inspiring approach to life. "The conviction that the world of man is growing, daily expanding and deepening, revealing new vistas for exploration, new possibilities of realisation—this is the secret, the motive power that generates the energy and the enthusiasm of all modernism. It is this that gives zest even in the midst of weariness, that makes the future glow with expectancy though the present be discouraging." Earlier times had their quiet charm "for us who look back to them from the hurry and changefulness of our day, but, as compared with the present, those ages were only half-alive."[28]

God revealed himself to man in various ways, but for Johnson, the doctrine of evolution was "the greatest of all the revelations that have successively dawned upon the mind of man." It included "all other revelations and immensely augments their value by giving them their proper setting as parts of one great world manifestation." Evolutionary thought disclosed that reality was a single vast system of divinely inspired progressive change, "one grand continuity of becoming, one long, consistent story of successive triumphs pointing still onward to we know not what great consummations." Man, with his intellect, his social organization, and his religious faith, had quite clearly reached the highest point in the evolutionary process yet known, and the significance, therefore, of evolution was contained in him, and in what he might still become. But he owed his ascendancy not to himself alone. Working with him was God, the creative force in evolution. Like man, God struggled, suffered, and erred, but in spite of all obstacles and limitations, he accomplished great things. "He appears as one Who shares the battle with us, Who counts on us as supporters in the world-process." The obsolete doctrine of God's omnipotence had divided him "as by an unfathomable gulf, from us. . . . The God of evolution is, on the contrary, one Whom we can measurably understand, one with Whom we can live in sympathy." He demanded of us not blind faith but enlightened courage and loyalty.[29]

Somewhat better known than the Unitarian and Congregationalist theologians of progress in the United States were the liberal Baptists, such as Shailer Mathews, Walter Rauschenbusch, and Harry Emerson Fosdick. Their emphasis fell more on the social applications of the liberal gospel than on its theological or metaphysical underpinnings, but each put forth an idea of progress firmly supported by a theological position.

Mathews, dean of the Chicago Divinity School from 1908 to 1933, preferred to think of himself as a modernist rather than a liberal, concerned to reassert the truth of Christianity in terms of modern science, rather than to deny traditional doctrines in the reductionist spirit of liberalism. He was hardly any less contentious than, for example, the Unitarians in his dealings with the orthodox. He repeatedly inveighed in his prolific writings against "scholasticism," "ecclesiasticism," "dogmatism," and "doctrinal legalism." The believing Christian of the sixteenth and seventeenth centuries had felt a need for official and binding creeds, but the modern mind, trained by science and historical criticism, could dispense with them. It would be no less Christian for doing so.

Mathews subscribed to most of the tenets of evolutionary theism. He saw God primarily as immanent in the world. Evolution was the divine will in action. Through the whole process, one could observe a growing complexity of structure and a persistent tendency toward personalization, which culminated in the emergence of mankind. In human history, the process of personalization pressed on to ever-higher levels. History was the record of the progress of personal freedom and the progress of religion. Only the highest religion, Christianity, was fully in accord "with the tendency of human progress," and alone could sustain the modern type of civilization that men of all nations were striving to build. Mathews called upon evangelists and missionaries to join "in the adventure of bringing to the world the spiritual blessings heralded in the message and the example of Jesus." [30]

In the light of modern hope, Mathews saw a new relevance for the eschatological prophecies of the biblical authors. A kingdom of God on earth was no longer unthinkable, if man persevered in drawing on divine power. "We look forward," Mathews wrote in 1924, "with a hope that is more than a desire to a day when, because men are embodying the attitudes and convictions of Jesus Christ in their individual and social lives, the coöperation and help of the God of law and love will make their world a social order in which love and justice will be supreme." [31]

The use of the telling phrase "social order," which Mathews equated with the kingdom of God, calls to mind that he was also one of the earliest exponents in America of the "social gospel." His book *The Social Teaching of Jesus* appeared in 1897, ten years before the publication of Walter Rauschenbusch's first major work, *Christianity and the Social Crisis*. But it was Rauschenbusch who eventually became the chief spokesman for the social gospel movement in American theology. He was born in 1861, and served for eleven years as the Baptist pastor of a working-class congregation on New York's West Side before accepting a post at Rochester Theological Seminary, where he remained until his death in 1918. Nearly all his writings focused on the relevance of Christian teaching to the problems of the working class in the early twentieth century: his central theme was the expectation of an equalitarian kingdom of God, to be established progressively on earth as the result of divinely aided human effort.

Rauschenbusch grounded his optimism in his reading of the Bible, and particularly in the prophetic books of the Old Testament and the gospels of the New. Although the great prophets of the Old Testament had limited their hopes to a basically national restoration, Rauschenbusch found in their "magnificent" optimism a clear foreshadowing of the revolutionary social program of Jesus. They were "the moving spirits in the religious progress of their nation" and "the men to whose personality and teaching Jesus felt most kinship." Rauschenbusch's Christ was, predictably, an outgoing, vigorous champion of the oppressed, who expanded the Jewish national dream into the vision of a universal reign of justice, love, and brotherhood. Neither a mystic, nor an ascetic, nor a pessimist, Jesus had turned away from the apocalyptic hopes and fears of his age to preach an "organic" view of the coming of God's kingdom on earth. "Jesus had the scientific insight which comes to most men only by training, but to the elect few by divine gift. He grasped the substance of that law of organic development in nature and history which our own day at last has begun to elaborate systematically." The kingdom would arrive not as a result of force or some sudden cataclysm, but through the steady growth to maturity of the divinely implanted mustard seed. And it would be above all a social order. "Jesus worked on individuals and through individuals, but his real end was not individualistic, but social, and in his method he employed strong social forces. . . . His end was not the new soul, but the new society; not man, but Man." [32]

But as Harnack had shown in meticulous detail, the early Church managed to evade the revolutionary message of its founder, and Christ's gospel had been tragically adulterated by Hellenic ideas alien to its very essence. Rauschenbusch applauded the modern scholars in Germany, France, and elsewhere who were purging Christian faith at last of all such foreign elements, including the erroneous dogma of the fall of man. Modern history was a record of progressive emancipation. Intellectual life had begun to free itself from enslavement to authority during the Renaissance. Then followed the Reformation, with its liberation of religious life; the Puritan Revolution, with its liberation of political life; and the Industrial Revolution, with its attendant social movements promising the liberation of man from want and injustice. None of these movements had yet gone through to completion, but humanity steadily gained in knowledge and capacity for progressive change, as witnessed in particular by the United States. "The swiftness of evolution in our own country," Rauschenbusch observed, "proves the immense latent perfectibility in human nature." [33]

In his last book, *A Theology for the Social Gospel,* he defined the coming kingdom as "humanity organized according to the will of God." It would guarantee to all persons their highest possible free development, institute the reign of love in human affairs, abolish war, end the exploitation of man by man, and progressively unify all mankind. Rauschenbusch exhorted his readers to revive the millennial hope, "which the Catholic Church dropped out of eschatology." The millennium could be achieved, not in the sense of a static utopia, but as an always improving earthly social order, which came to us in an infinite series of installments. Some distant day the planet would no doubt become uninhabitable, but "meanwhile we are on the march toward the Kingdom of God, and getting our reward by every fractional realization of it which makes us hungry for more. A stationary humanity would be a dead humanity. The life of the race is in its growth." [34]

The hopes of American theological liberals and social gospellers were summed up with exuberance by another Baptist, Harry Emerson Fosdick, in his Cole Lectures for 1922 at Vanderbilt University, *Christianity and Progress*. Fosdick became somewhat disenchanted with liberalism in later years, but in this book he was very much the prophet of progress. Christianity could no longer afford to ignore the idea of progress, he warned. The danger was that modern man might grow so confident that he would try to work out his own salvation, forgetting God and the Christian gospel. Since true progress was possible for humanity only with divine help, it behooved Christians everywhere to fashion a Christian idea of progress in harmony with their faith. The choice lay not between the old, static theology and the new, progressive one, but between the new and none at all. To keep Christian thought free of the influence of the belief in progress would mean

> the death of vital faith. . . . Far from being hostile to religion, our modern categories furnish the noblest mental formulae in which the religious spirit ever had opportunity to find expression. . . . To take this modern, progressive world into one's mind and then to achieve an idea of God great enough to encompass it . . . is alike the duty and the privilege of Christian leadership to-day. In a world which out of lowly beginnings has climbed so far and seems intended to go on to heights unimagined, God is our hope and in his name we will set up our banners.[35]

If Western Christianity could be made over again in the image of the gospel of progress, so could other religious traditions. Reform Judaism and Russian Orthodoxy had their advocates of progress during our period. Confucian scholars, such as K'ang Yu-wei and T'an Ssu-t'ung, fused the progressivist interpretation of history with ancient Chinese thought. Closer to our own time, Aurobindo Ghose and Sarvepalli Radhakrishnan took similar liberties with traditional Hinduism. The Baha'i Faith founded in Persia by Baha'u'llah in the 1860s has remained throughout its history a proclamation of the unity and progressiveness of mankind. Although we have chosen to limit our discussion to the Christian West, clearly the idea of progress penetrated religious thought in every part of the world in the generation after Darwin.

7

Progress and Politics

For any political historian not obsessed with wars and revolutions, *la belle époque* was an eventful era. Political power flowed rapidly in several directions: from the old elites to the masses, from Asia and Africa to Europe and North America, from courts and counties to the institutions of the national service-state. It was also the last prolific period in the history of normative political theory. We could reasonably expect to find an abundance of politically oriented doctrines of progress, and indeed we do.

Two facts, however, prevent the present chapter from growing unmanageably long. The first is that we have already discussed some of the most interesting political minds of the age. Many of them were not primarily political thinkers at all, but scientists, sociologists, philosophers, and even theologians, who happened to contribute significant ideas about the political process or who took up ideological positions hinged, so they believed, on their theoretical work in other fields. Friedrich Nietzsche, John Dewey, and Walter Rauschenbusch meet this description, and each has been studied above, although each would deserve mention in any history of recent political thought.

The nature of politics in *la belle époque* also simplifies our task somewhat. The nation-state was triumphant throughout the Western world during these years, and the attention of political thinkers often centered on national institutions and problems, rather than on mankind as a whole. Much of the best political thinking of the age was done entirely within a given national context, and showed little interest in anything so abstruse as the problem of human progress. The same could be said, even more strongly, of historiography. Most historians wrote nothing but national political history. The study of universal history, which had fascinated scholars since the time of Eusebius and Augustine, was unceremoniously abandoned.

All the same, many political thinkers and even a few historians could not resist occasional flights into the stratosphere of world-historical speculative thought, if only to demonstrate how the history and politics of their nation had contributed to human progress. Political thinkers were also influenced by the main currents of scientific and philosophical thought, which tended to drive them willingly or otherwise out of their national nests. The positivist faith in the saving powers of science, the Darwinist conception of progress through conflict, and the philosophies of idealism, vitalism, and pragmatism affected political thought to a greater or lesser degree, with much random borrowing by ideologues in search of extra-political rationalizations for their already formulated political programs. But ideas have a power of their own, and sometimes an ideology became the helpless prisoner of its own rationalizations.

Political theories of progress during *la belle époque* typically belong to one of two camps. Despite an astonishing promiscuity in styles and loyalties, most political prophets of progress may be classified either as national liberals or as socialists. Both made extensive use of other political ideals, including democracy and internationalism, but for different reasons and with different ultimate objectives. In the one camp, the ultimate objectives were freedom and national unity; in the other, social equality. The common denominator was a strong conviction that only the achievement of these goals would ensure the general progress of mankind.

National-liberal politics had not yet fully arrived at the beginning of *la belle époque,* except in Germany. In France, Britain, and the United States, many liberals still clung to a more traditional point of view, blending the inspirations of classical economics, John Stuart Mill, T. H. Green, and Charles Renouvier. For them, progress was more or less identical with the rise of individual freedom in Western civilization, and many expressed alarm in the 1880s and 1890s that Western freedom was endangered by the advancing collective power of the masses. Such eminent Victorian liberal prophets as W. E. H. Lecky and Sir Henry Maine filled their last books with intimations of disaster. In France, the elderly editor-in-chief of the *Journal des Economistes,* Gustave de Molinari, warned his readers that the survival of militarism and the advent of socialism in modern times demonstrated "the lagging of moral progress behind material progress." [1]

Of this purer and older liberalism, which the new national liberalism in time supplanted, the spokesman most important for students of the idea of progress was the English Catholic historian, Lord Acton. Gertrude Himmelfarb has stressed the more pessimistic aspects of Acton's thought and even argued that "he invites identification with none of the dominant currents of thought in Victorian England." [2] But this is to take an unduly narrow view of the Victorian mind. Acton was a severe moralist, almost a prig; he had no use for those who sought to ignore or excuse evil, and he saw history as a titanic battle between the forces of light and the forces of darkness. But it was his very willingness to find a moral meaning in history that allowed him, in the final analysis, to enlist in the ranks of the believers in progress, and to appear in retrospect as a fairly representative Victorian type, his reputation for eccentricity during his own lifetime notwithstanding.

Born in 1834, Acton was refused admission to Cambridge because of his religion and decided to continue his education at the University of Munich, where he studied history and theology under Döllinger. Germany infected him with a durable passion for the new scientific historiography of Ranke. At the same time his religious faith kept him safe from the relativistic implications of *Historismus.* After a varied career in politics and journalism, Acton accepted the Regius Professorship of Modern History at Cambridge in 1895 and in his inaugural lecture exhorted his future students and colleagues "never to debase the moral currency or to lower the standard of rectitude, but to try others by the final maxim that governs your own lives, and to suffer no man and no cause to escape the undying penalty which history has the power to inflict on wrong." [3] The scholar had the manifest obligation to judge all historical figures according to whether or not they conformed to the "eternal" moral law.

Guided by his own interpretation of that law, Acton maintained that "the establishment of liberty for the realization of moral duties" was the purpose of civil society, defining liberty as "the assurance that every man shall be protected in doing what he believes his duty

against the influence of authority and majorities, custom and opinion." [4] The vicissitudes of liberty comprised the central theme of
history. As he told his audience at Cambridge in 1895, many modern
thinkers doubted that "the world is making progress in aught but
intellect." Yet for him, the "constancy of progress . . . in the direction of organised and assured freedom is the characteristic fact of
modern history, and its tribute to the theory of Providence." Ab-
solutism had been overthrown, infallible authorities controverted,
liberty of conscience secured, and "a fair level of general morality,
education, courage, and self-restraint" reached in the Western world.
With Maine, Acton rejoiced in the rejection of the rule of stability
and continuity, under which ancient man had languished for thousands of years. "In this epoch of full-grown history men have not
acquiesced in the given conditions of their lives." He anticipated the
spread of Western liberty and prosperity to other parts of the world
and complimented the historians of his own epoch for their power
"to learn from undisguised and genuine records to look with re-
morse upon the past, and to the future with assured hope of better
things." [5]

But Acton shared the apprehensiveness of the liberals of his generation in regard to both democracy and nationalism. Although in
his later writings he came to feel that full popular participation in
government would not necessarily lead to chaos or despotism, he
warned repeatedly against totalitarian democracy. The "true democratic principle" was that "none shall have power over the people,"
as distinguished from the false principle that under democracy "none
shall be able to restrain or to elude its power." When democracy
claimed "to be not only supreme, without authority above, but
absolute, without independence below," the liberal had to raise his
voice in protest.[6] Absolute power in the hands of the people was no
less dangerous than absolute power in any hands: to quote Acton's
own most celebrated saying, absolute power corrupted absolutely.

As for nationalism, Acton was too much the cosmopolitan to find
excitement in its slogans and appeals. He thought well of both the
Austrian and the British empires and predicted that "the national
theory" would in time collapse, once it had served its historical purpose of overthrowing the power of absolute monarchy and bringing
the spirit of revolution to its highest and final stage. Although it
functioned, therefore, as a necessary evil in the world economy, it
was at the same time "absurd" and "criminal" and a "retrograde

step in history." Its course would be marked "with material as well as moral ruin." [7]

Acton died in 1902, by which time a new school of liberal thought had appeared all over the Western world, which showed little fear of nationalism and democracy, and even borrowed some of the ideas of socialism. This national-liberal movement expressed perfectly the needs and aspirations of the middle class in a democratic age. It was still unmistakably liberal. Even in Central Europe its dedication to the ideals of civil liberty, free enterprise, and the rule of law cannot be questioned. But it saw in the institutions of the nation-state great opportunities to help safeguard and expand private wealth, and to heal the ills in the body social evidenced by the growing discontent of the working class A concern for the general welfare of mankind—or, if nothing else, enlightened self-interest—dictated that the state be enlisted actively in the solution of the "social question."

National-liberal thought developed earliest and most vigorously in Germany, where governments had always been compelled to play a major role in economic development, and where the struggle for national unification had absorbed much of the energy and idealism of liberals from the 1840s to the proclamation of the Reich in 1871. The most popular theorist of the new order in its first quarter-century was the historian Heinrich von Treitschke, often libeled in Western accounts as the archetype of Teutonic barbarism and a forerunner of Adolf Hitler. His *magnum opus* as a political thinker was a course of lectures delivered at the University of Berlin every year from 1874 to 1896 and published in two volumes shortly after his death. In his introductory remarks, he dissociated himself from the doctrine of progress in the broadest sense. Mankind developed along many different routes, he observed, and on some the highest peaks had already been climbed. Original sin ensured that the individual man was no better disposed to behave morally than in the past, although ethical theorists might argue that progress had taken place in mankind's recognition of the abstract good. For the race as a whole, Christianity had long ago scaled the heights by its teaching of love. In sexual relationships early societies had "discovered very soon the absolute ethical standard of marriage," beyond which no progress was conceivable. In sculpture and oratory, the classical Greeks had achieved the ultimate. The only area in which mankind had apparently made fairly continuous progress through the millennia was in "culture," by which Treitschke meant ways of communal life—economic, social, and political.[8]

But as a nationalist he insisted that the progress of human culture had resulted primarily from the activities of states, which struggled, rose, and fell according to iron laws, and in the vicissitudes of their corporate life made all such progress possible. The state was a moral being, firmly founded on power, which it had a moral duty to preserve. It was "called to positive labours for the improvement of the human race," and "bound to take its appointed part in the education of the human race."[9] The individual man served humanity best by unswerving allegiance to his state, by obeying its laws and fighting its wars. There was a growing tendency all through the history of political culture for states to increase their power, and at the same time to increase the liberty of their subjects. The best states, therefore, were strong and large, expressing the natural will of racially unified nations and securing to their citizens the blessings of both liberty and order.

Treitschke dismissed as an immoral pipedream the hope of universal peace among nations. He found the just wars of such modern states as Bismarck's Germany entirely obedient to the highest moral ideals of humanity, and made out a similar case for the overseas expansion of Europe's power. "The world beyond Europe," he noted, "is bulking larger and larger upon Europe's horizon, and there is no doubt that the European nations must go out to it and subdue it directly or indirectly to themselves." He labored under no illusion, however, that the inferior races of Asia and Africa could be treated like Europeans.

> It is clear that the political methods of dealing with races upon a lower level of civilization must be adapted to their capacity for feeling and understanding. . . . We must not blame the English who in the imminent peril of the Indian Mutiny bound Hindus to the cannon's mouth, and blew their bodies to the winds. It is evident that the situation demanded such measures, and we cannot condemn them if we accept the English contention that England's rule in India is beneficial and necessary.[10]

As the modern national state educated European humanity, so it would accept the charge of educating all men, modifying its methods whenever circumstances dictated.

But Treitschke was not immune to the appeal of the liberal in-

ternationalist tradition of the German *Aufklärung*. He accepted
Herder's idea that "the rays of the Divine light are manifested,
broken by countless facets among the separate peoples, each one
exhibiting another picture and another idea of the whole." Each
people was rightfully proud of its particularity and had to "assert
its rank in the world's hierarchy and in its measure participate in
the great civilizing mission of mankind." Although wars even
among the most civilized great powers would no doubt be fought
in future years, such wars would tend with the progress of culture
to become rarer and shorter, and Treitschke at one point admitted
that "the ideal towards which we strive is a harmonious comity of
nations, who, concluding treaties of their own free will, admit restrictions upon their sovereignty without abrogating it." [11] If this
conception is far from the ideal of a world republic, it is equally far
from the views sometimes attributed to Treitschke by his Western
critics.

The new national liberalism in Germany found another popular
apologist and an unremitting believer in progress in Friedrich
Naumann, whose political career began shortly after Treitschke's
death. Born in 1860, the son of a Lutheran minister, Naumann became a Lutheran clergyman himself in the 1880s, and from the beginning of his pastoral work evinced an unusual interest in the
social question. In 1896, having resigned from the ministry, he
founded the short-lived National Social party. From 1907 to 1918
he sat in the Reichstag as a Progressive, and in the year of his death
(1919) he helped to organize one of the major liberal parties of the
Weimar Republic, the Democratic party. Naumann's books were
more the efforts of a political journalist than of a scholar, but with
all their shortcomings, they provided the German public of his
generation with an eloquent amalgam of liberalism, nationalism,
socialism, monarchism, democracy, and social Christianity. Naumann made all the disparate elements of Wilhelmian Germany
seem to fit miraculously together, from the Kaiser and his newly
acquired empire to the demands of the masses for public assistance
and the mushroom progress of German industry.

His first book, *Democracy and Emperorship,* published in 1900,
was his most successful. He agreed with classical liberalism that the
fulfillment of the individual was the ultimate objective of politics.
But the individual by himself could do very little. His personal en-

richment depended on the wealth and power of the social organism in which he lived. In the case of Germany, the individual was thrice blessed: he enjoyed the many rights embodied in the traditional German *Rechtsstaat;* he was represented by a democratically elected imperial parliament; and he had an emperor, whose authority unified the German states, and who was capable of providing strong executive leadership at a time when Germany had finally earned the opportunity to establish herself not only as a great power but also as the greatest of all the powers. If the classes and the masses of the new Germany fell in behind their emperor, nothing on earth could prevent German world-ascendancy in the twentieth century.

Despite his preoccupation with German problems, in his first book and in those that followed Naumann occasionally addressed himself to the larger question of the progress of mankind. Human advancement always depended, he wrote, on the individual's will to advance himself. "The progress of human culture is a consequence of the progress of individual men." [12] Progressive laws were of no avail if individual men did not bind themselves to obey them. Nor was there anything automatic about freedom itself. All good things in life came through struggle. "Therein lay the mistake of the old fighters for freedom: they believed in eternal, innate rights. There is no such thing. The only rights are those earned and won by struggle. No human right has originated without struggle, and each one will collapse and be lost as soon as the will to fight for it no longer exists." [13]

Naumann carried his faith in the efficacy of struggle into the larger sphere of international politics. Christian ethical values did not apply in the political struggles of nations: here natural and human values prevailed. But this did not mean that mankind was doomed to perpetual fratricide. The progressive nations, disciplined by reason, were hard at work instituting the rule of law in their domestic affairs and forging a new world order based on industry and commerce.

> Neither the prophets nor the agitators are creating human unity; rather, it develops without their help and will in many ways be best promoted by precisely those people who in theory most strenuously resist it. . . . Just as the socialist regulation of production will in the main be achieved by anti-socialist circles, so internationalism will go forward under national flags. . . . Population growth and machines

are the moving forces in the emergence of mankind. It is not China, not India which realize the idea of mankind, but in capitalist Western Europe that the new condition of man has arisen and spread from there to all the world.

The more the world became unified by commerce, the more each new war would seem an intrusion into the important business of life. "Let this logic of interests speak! It is the best peace propaganda." [14]

Looking into the immediate future, Naumann anticipated that the vanguard of the Western powers would be Germany, already the country in Europe with the highest culture and the greatest industrial base. Because human unity was still far from achieved, war among the powers, chiefly for economic reasons, could be expected in the new century. Germany could not carry out her work for human progress without a formidable army and fleet and without extensive imperial possessions; and she could count on any future wars to be wars between whole peoples, involving the total commitment of their industrial might. Naumann was especially fascinated by the possibility of a war between Russia and the Central Powers, which would benefit progress by liberating Eastern Europe from the tyranny of the tsars.

In 1915, after the war he predicted had already begun, he published a volume on war aims entitled *Middle Europe,* which envisaged the establishment after the war of a great central European federation to include Germany, Austria-Hungary, Poland, the Balkans, and perhaps Scandinavia and the Netherlands. Such a federation, dominated by the German-speaking peoples, would take its place with the other three surviving world powers, the British, North American, and Russian, as the organizers of the future life of mankind. In the process many of the smaller nations of Europe would be submerged. "This is a harsh necessity, a heavy fate, but it is the overpowering tendency of the age, the categorical imperative of human evolution. . . . People may submit to necessity earlier or later, freely or from compulsion, but the universal watchword is spoken and must be complied with." Sobered by Germany's inability to win the war in its opening months, Naumann pointed out that the same fate could overtake the Germans if they did not bestir themselves. His earlier hopes for German world-ascendancy notwithstanding, he could now imagine Germany falling helplessly into a satellite relationship with the British or Russian empires. Germany's only chance for a place in the sun in the postwar world lay in joining forces with the Dual Monarchy and in raising her population and industry to the highest possible levels. The decline of France was attributable to the failure of her will to grow. "Now the urgent, heartfelt, beseeching call must resound, mingled with the ringing of the peace bells, calling to men and women in town and country: beget children!" [15]

Naumann warned that even after the transition to the age of the superpowers, wars would still have to be waged.

> Before the organisation of humanity, the "United States of the World," can come into existence, there will probably be a very long period during which groups of humanity, reaching beyond the dimensions of a nation, will struggle to direct the fates of mankind and to secure the product of its labour. Mid-Europe comes forward as one such group, and that indeed a small one: vigorous but lean! [16]

Nonetheless, for whatever comfort it might give Naumann's liberal readers, he did foresee the unification of man in the far distant future. As the superpowers evolved from the nations, so the world state would ultimately evolve from the superpowers.

France, in the days of the Third Republic, after seven major revolutionary changes of regime in eighty years, was more than ever a house divided against itself—half radical, half reactionary, and suspicious of moderate politics. Although the French had given the world Montesquieu, Voltaire, and de Tocqueville, they seemed at times to deserve Emile Faguet's bitter *mot* that "liberalism is not French." But the national-liberal movement did not entirely pass them by. In the middle of the political spectrum, between the xenophobic integral nationalists and nostalgic monarchists at one extreme and the Marxists and anarcho-syndicalists at the other, stood men such as Léon Bourgeois and his doctrine of progress through "solidarity."

Bourgeois was premier of France for six months in 1895–96 and head of the French delegation to the Hague Peace Conferences of 1899 and 1907. He saw value in both liberalism and socialism, and deplored their unyielding hostility to one another. Society, he declared, made all its progress through the free activity of individuals, but the individual could do little unless he enjoyed full use of his social heritage. It followed that liberalism and socialism were both right, and that a new social liberalism had to be devised, which could bring into dynamic rapport the insights of both ideologies.

Bourgeois's reading of the evolution of mankind from savagery to civilization recalls Comte's. In the beginning progress had resulted only from "the brutal struggle for existence." Later, societies evolved, which at first flourished through warfare and violence and in time learned the arts of voluntary cooperation. Peaceful, contractual relationships largely supplanted those based on force. Such relationships were possible, however, only when individuals enjoyed the freedom to develop their faculties to the maximum extent compatible with the equal needs of others. "Every political or social arrangement which seeks to determine in some other way the limits of men's liberties will be contrary to the natural laws of the evolution of society." [17]

On the other hand, the classical economists wrongly assumed that the individual could or should act entirely on his own account. Men were not isolated beings, but associated one with another: "At the point of contact, these liberties, mutually limiting one another, ought not to clash, to block and to destroy one another, but on the contrary, like forces brought to bear on a common point, they should have the result of accelerating the movement of the whole system." The atomic individual of classical economic theory was a figment of the imagination. "Man living in society, and not being able to live without it, is at all times its debtor. That is the basis of his obligations, the cost of his liberty." Bourgeois hinged his whole political philosophy on the principle of indebtedness. In entering into association with others, the individual contracted an obligation not only to those with whom he did business but also to his ancestors, to everyone who had ever lived and contributed something to the progress of civilization. He was in the debt of all those who had transformed the earth from a wilderness into a land of plenty, in the debt of artists and musicians, of scientists and scholars, and of those who had led the human race "from the state of violence and hatred, and little by little directed it toward the state of peace and harmony." [18]

This debt could be repaid in only one way: by working for posterity as our ancestors had worked for us.

> Thus every man, on the morrow of his birth, as he enters into possession of that improved state of mankind prepared for him by his ancestors, contracts, unless he breaks the law of evolution which is the

same for his personal life as it is for the life of the species, the obligation to contribute by his own effort not only to the maintenance of the civilization in which he is about to take his place, but also to the further development of that civilization. His liberty is encumbered with a double debt: in the assessment of the charges fixed naturally and morally by the law of society, he owes, besides his part in the exchange of services, what one may call his part in the contribution to progress.[19]

Bourgeois did much to crystallize the ideological position of the Radical and the Radical Socialist parties in France in the early years of the twentieth century. In practical terms, the principle of solidarity demanded a modification of traditional liberalism to permit the equalization of opportunity by public action. Guarantee of employment and a minimum standard of living, ample free education, and a full program of social insurance, to be financed with the help of a progressive income tax, were among the aims of his policy, although he could do little during the short time he held the office of premier.

The national-liberal movement in the Anglo-Saxon world took a rich variety of forms, from radical programs that denounced Western imperialism and incorporated generous amounts of social-democratic thinking, to doctrines of Anglo-Saxon supremacy suggestive of Continental European racism. But national-liberal thinkers of all persuasions in Britain and America agreed on the paramount importance of human freedom and the need to promote social solidarity in the modern nation-state as the foremost condition of future progress.

Typical of the left-leaning national-liberals were L. T. Hobhouse and J. A. Hobson in Britain and Herbert Croly and Walter Weyl in the United States, contemporaries of Naumann and Bourgeois whose best work was done in the years immediately before the First World War. All four evinced a strong interest in progress, both as a fact of history and as a duty of modern states to promote by active and humane social policies.

The most sophisticated and—at the same time—the most tender-minded exposition of liberal thought in the first fifteen years of the new century was L. T. Hobhouse's book *Liberalism,* which translated into political terms the author's philosophical sociology and retained its emphasis on progress.* Developing the ideas of Mill and

* See above, pp. 50–54.

Green, he argued that legal equality and legal freedom were not, in themselves, enough. Society owed its progress to the progress of the individual, who could grow and fulfill himself only in organic interaction with the body social. It was a matter, then, of justice and not simply of charity that all individuals should have the fullest possible opportunity to achieve maximum self-development in the society to which they belonged. Unemployment, exploitative wage scales, and public neglect of the special needs of mothers, children, the incapacitated, and the aged, were not compatible with the liberal ideal. Hobhouse's formula was "Liberal Socialism," democratic and personalist, which preserved freedom by equalizing opportunity. Opposed to liberal socialism, he warned, stood mechanical socialism, which sought to impose a logically developed system of state control, ignoring the needs and inner resources of the human person. "Every constructive social doctrine," he concluded, "rests on the conception of human progress. The heart of Liberalism is the understanding that progress is not a matter of mechanical contrivance, but of the liberation of living spiritual energy." [20]

Hobhouse's humanism strongly influenced another liberal of his generation, the journalist, economist, and political thinker, J. A. Hobson.[21] More than twenty-five books issued from Hobson's pen, but he is best remembered for a magisterial indictment of late nineteenth-century imperialism, first published in 1902, which also contained a reasoned defense of the belief in progress. Hobson despised imperialism both for its exploitation of non-European races and for its degenerative effects on the Western peoples. It led the exploiters themselves, he contended, down the road to parasitism, decay, and eventual collapse. The same thing, for all practical purposes, had happened many times before in ancient history. One could almost see an organic destiny at work in the history of nations. Citing Brooks Adams, Hobson warned that "nature is not mocked: the laws that, operative throughout nature, doom the parasite to atrophy, decay, and final extinction, are not evaded by nations any more than by individual organisms." [22]

But civilized states were not under any natural compulsion to play out the life history of the parasite. Through the deliberate exercise of will and reason, they could cease to feed parasitically on their victims and break free to a new way of life in harmony with the upward movement of evolution. Here Hobson took on two arguments simultaneously: the idea that imperialism is a predestined higher stage of national development and the idea that man is by nature an aggressive animal, who survives and progresses only through war. Both were demonstrably false. Progress did occur through struggle and selection, but Karl Pearson and other supposedly scientific apologists for imperialism missed the decisive point that the instruments of progress as civilization evolved were chiefly reason and cooperation, the social application of human intelligence to the conquest of nature, rendering the cruder, lower forms of struggle obsolete. Pearson saw this with respect to the internal development of nations, but failed inexplicably to realize that it held good in international relations as well. If class war was inefficient and unprogressive, why not also warfare among nations? "As nations advance towards civilization," Hobson argued, "it becomes less needful for them to contend with one another for land and food to support their increasing numbers, because their increased control of the industrial arts enables them to gain what they want by conquering nature instead of conquering their fellow-men." Piracy, in effect, was replaced by industry; theft by production. "The struggle has become more rational in mode and purpose and result, and reason is only a higher form of nature." [23] *

Despite the evils of imperialism, Hobson saw many hopeful indications that modern civilization was marching forward. Although imperialism grew, so also did internationalism. The nations engaged in more cooperation one with another; men, ideas, and goods moved ever more freely across national frontiers, promoting progress through intermarriage and peaceful competition. In due course the nations might well elect to try "experimental and progressive federation, which, proceeding on the line of greatest common experience, shall weave formal bonds of political attachment between the most 'like-minded' nations, extending them to others as common experience grows wider, until an effective political federation is established, comprising the whole of 'the civilized world'." [24]

But such a federation would fail in its ultimate aims if it turned out to be merely a sophisticated device for maintaining class rule in the West and imperial rule overseas. Hobson also insisted on the need to continue the progress already made in recent times toward

---

* In *Problems of a New World* (London, 1921), pp. 259–60, Hobson added the point that human nature as such was not changing appreciably: historical progress occurred chiefly through the improvement of man's environment and the social transmission of knowledge.

the establishment of a liberal, democratic and humane political order in the civilized nations. Imperialism, in his analysis, was a disease resulting from the persistence of *laissez-faire* capitalism. It benefited at least temporarily a small class of acquisitive capitalists, while the mass of the people suffered. Only if power fell into the hands of the whole people could the grip of imperialism and unreformed capitalism be broken. In a true democracy the citizenry would concentrate on developing their own markets for the prosperity of all, instead of spending "men and money in fighting for the chance of inferior and less stable foreign markets." The great obstacle to progress was clearly the usurpation of power by "certain commercial and financial interests. Depose these interests, and the deep, true, underlying harmonies of interest between peoples, which the prophets of Free Trade dimly perceived, will manifest themselves." [25]

Hobson's ideal society, in fact, closely resembled the democratic welfare state of twentieth-century northern Europe: liberal, preserving civil liberty and freedom of enterprise; democratic, installing the people in full political power; and socialist, guaranteeing equality of opportunity and public protection against the hazards of life. Such a national society, Hobson felt, would be unwarlike and prepared to enter into ultimate political federation with all similarly advantaged nations. It would represent the triumph of reason and cooperation over animal greed and aggression.

But what of the so-called lower races? Hobson did not ignore them; they, too, had their role in the progress of humanity, and he sketched out the rudiments of a scheme for world economic development that was far ahead of its time in 1902. To begin, he doubted that most of the "backward" peoples of the world, except perhaps those of tropical Africa, were actually inferior to the white race. The civilizations of Asia were older than those of the West: culturally different, less developed technologically and industrially, but not, on balance, inferior. The Western powers would do humanity a disservice by trying to make them over in the Western image. At the same time, it was both impractical and unnecessary that all the riches of the non-Western world should lie unused. Hobson called upon the Western nations to collaborate in the planned utilization of the world's resources for the common good of all mankind. Such utilization would have to be managed without unduly altering local ways of life and without exploitation of any kind.

The progress of world-civilization is the only valid moral ground for political interference with "lower races.". . . That this process of development may be so conducted as to yield a gain to world-civilization, instead of some terrible débâcle in which revolted slave races may trample down their parasitic and degenerate white masters, should be the supreme aim of far-sighted scientific statecraft.[26]

Hobson's final goal was a unified world. The various nations, both Western and Eastern, would advance each in its own way, with international assistance where needed, until all their peoples at last enjoyed prosperity and political well-being.

The First World War shook Hobson's confidence in human rationality, but in his intellectual autobiography, published in his eightieth year, he reaffirmed his prophecy of a coming world order. Humanity, he wrote, was not alone, but "the highest present product of powers which permeate the universe . . . the corporate part of a system inspired and moulded by some evolving process that may be realized as purpose or even spirit." It was understandable that after 1914 the older generation of liberals, whose hopes had risen so high, should feel "a sudden panic in our aged breasts. But if our lifelong sense of progress has found support in so many centuries of intellectual and moral advancement, it seems inherently unlikely that a few years' debauch of folly and of hate can permanently reverse the course of human history and plunge us back to barbarism." [27]

The transformation of liberalism into the philosophy of the national-liberal service-state also occurred in the United States. Nothing is more characteristic of the Progressive Era in American politics than the abandonment by liberals of their insistence on minimal interference by government in society: almost suddenly, the Jeffersonians became Hamiltonians. The leading theorists of Progressivism, Croly and Weyl, both subscribed to a doctrine of progress through the efforts of the liberal and democratic nation-state. Croly's first book, *The Promise of American Life,* Weyl's *The New Democracy,* and their articles and editorials for *The New Republic,* founded by Croly in 1914, provide invaluable indices to the American political climate in the early twentieth century.[28]

Croly came well by his faith in progress. His father, David, was an Irish-born American disciple of Auguste Comte, who had had young Herbert "baptised" in the Positivist religion of humanity. *The Promise of American Life,* which first appeared in 1909, gave its many readers a vision of the United States as the hope of the world. American patriotism, wrote Croly, was unlike the patriotisms of Europe. The European tended to look to the past, whereas the American, however much he revered his country's past glories, felt an even greater commitment to the future. His was the "land of promise," whose destiny was to realize the democratic ideal, and serve as a beacon and shining example to all the peoples of mankind.

> That the American political and economic system has accomplished so much on behalf of the ordinary man does constitute the fairest hope that men have been justified in entertaining of a better worldly order. . . . [It] has given to its citizens the benefits of material prosperity, political liberty, and a wholesome natural equality; and this achievement is a gain, not only to Americans, but to the world and to civilization.[29]

In Croly's thinking, for which he gave due credit to the Hamiltonian tradition in American political theory, democracy was not enough: men also needed the warmth and strength of the modern nation-state, "the best machinery as yet developed for raising the level of human association." It was, in fact, a school for progress, and "everybody in the schoolhouse . . . must feel one to another an indestructible loyalty. . . . As a worldly body they must all live or die and conquer or fail together." [30] But the most highly developed nations were the democracies, led by the United States, and now including a fair number of European powers, with whom America would be well-advised to establish cordial relations, although not at the price of entanglement in European power politics. The European countries themselves had to recognize that they could not accelerate "the march of Christian civilization" unless they worked together in relative harmony. Unfortunately some powers in Europe insisted on living by the sword, which required their neighbors to hold extensive armed forces in readiness; even the unification of Europe, when it finally came, would in all likelihood involve a certain amount of bloodshed, after the example of the process by which Germany had been unified. Each nation, meanwhile, had to keep its powder dry and maintain the highest possible level of national efficiency.

Yet Croly did not equate progress and power. In the final pages of *The Promise of American Life,* he made clear the difference between means and ends in his philosophy. The nation, and democracy itself, were means to a higher end: the production of heroes and saints. "For better or worse, democracy cannot be disentangled from an aspiration toward human perfectibility." Since the common man could perfect himself only by imitating men of transcendent virtue, his success would "depend upon the ability of his exceptional fellow-countrymen to offer him acceptable examples of heroism and saintliness." [31]

Walter Weyl, writing four years later in *The New Democracy,* shared Croly's view that democracy had a purely instrumental value. A "socialized" and "plenary" democracy, safeguarding liberty and solving the social question, would almost certainly meet the needs of progress for many centuries to come. "If, however, for any reason democracy becomes incompatible with progress and happiness, it will simply cease." Nor should one expect democracy to transport mankind to utopia. In tune with the pragmatists and the Bergsonians, Weyl argued that "society does not strive towards fulfillment, but only towards striving. It seeks not a goal, but a higher starting point from which to seek a goal. . . . Our present ideal of a socialized democratic civilization is dynamic. It is not an idyllic state in which all men are good and wise and insufferably contented. It is not a state at all, but a mere direction." New dangers would undoubtedly arise to confront mankind when the present dangers had been eliminated. Prosperity would test man, as poverty once had. But sufficient unto each day was the evil thereof. "For this century we need but take this century's forward step. If we can extirpate misery, that will be progress enough." [32]

Weyl also echoed Croly's warnings against the nineteenth-century American's "illimitable, supreme, categorical optimism" and his "belief in the inevitableness of progress." [33] This simple faith in the future was one of the sources of America's early greatness, but the time had now come for a fundamental overhauling of the American system, in order that the American dream could be realized for the many, as well as for the few. Reconstruction demanded a vast outpouring of will. To cling to a belief in automatic progress, unaccompanied by fresh thought and creative response to new challenges, might prove fatal to the Republic.*

* Perhaps one good illustration of what Weyl was warning against is provided by the leaders of the feminist movement. "Woman Suffrage is not a receding wave, it is a mighty incoming tide which is sweeping all before it; . . . no *human* power,

Following in the tradition of Bancroft, liberal American historians also rejoiced in democracy, and especially American democracy, seeing in its progress an index to the progress of humanity. As the Pennsylvania historian Edward P. Cheyney told the American Historical Association at its annual meeting in 1923, moral progress through the growth of freedom and democracy was a "law" of history.[34] The founder of the "New History," James Harvey Robinson of Columbia University, made his contribution to the literature of progress in the form of an intellectual history of mankind, *The Mind in the Making,* a eulogy of the free human mind and the limitless powers of science.[35] His vision of progress inspired Robinson to leave Columbia in 1919, after twenty-seven years, to found the New School for Social Research in New York, where historians and social scientists could collaborate freely across disciplinary boundaries to help solve the nation's and the world's problems.

Other national-liberal thinkers in Britain and the United States leaned "rightward," in the direction of racism and concepts of national or imperial mission, which stressed collectivist goals somewhat at the expense of liberal ones. In Britain, for example, it was not at all difficult for apologists such as Viscount Milner to see in the unprecedentedly vast British Empire of late Victorian and Edwardian times a mandate from history to carry Western civilization to the "darkest" corners of the world and to surpass all past states in power and service to mankind. "I believe," Milner told an audience in the East End of London in 1912, "in development on national lines, and I believe in the mission of my country, of the British race—that it stands for something distinctive and priceless in the onward march of humanity." National social reform was necessary, he added, to end cleavage along class lines in order that the nation might be able to do its best in the world struggle. The people most successful in reducing class antagonism would "take the lead in the rivalry, not necessarily a hostile rivalry, of nations, which, with all its deplorable excesses, is one of the greatest factors in human progress." [36]

Theories of racial supremacy were plentiful in America during these same years, often explicitly linked to a doctrine of progress.

no university professor, no parliament, no government, can stay its coming. It is a step in the evolution of society and the eternal verities are behind it." Carrie Chapman Catt, in her presidential address to the International Woman Suffrage Alliance in 1911. Carrie Chapman Catt papers, A-68, folder 16, Radcliffe College Woman's Archives. I am grateful to Mrs. Helen Lefkowitz Horowitz for this citation.

In a textbook of political science published in 1890, John W. Burgess of Columbia University explained the political capacity of nations almost entirely in terms of racial characteristics. Slavs and Celts, he showed, had no gift for political organization beyond the tribal or communal level; the Latin peoples had given birth to the imperial state-form; and the Teutons had created the highest of all polities, the national state. Empires tended to stagnate and turn to despotism, but the national state was the strongest and the freest of all state-forms known to political history. The fact that the national state was "the creation of Teutonic political genius stamps the Teutonic nations as the political nations *par excellence,* and authorizes them, in the economy of the world, to assume the leadership in the establishment and administration of states." In view of the ascendancy of national states since the final collapse of the Roman imperial idea, Burgess concluded that they were designed by providence to serve "as the prime organs of human development." When they defended their interests on the field of battle, they did God's own work.[37]

It was also incumbent upon them to take up the task of the political education of the other races. The greater part of the world was "inhabited by populations which have not succeeded in establishing civilized states; which have, in fact, no capacity to accomplish such a work; and which must, therefore, remain in a state of barbarism or semi-barbarism, unless the political nations undertake the work of state organization for them." But the continuance of barbarism anywhere on the globe could not be tolerated. The political nations possessed a mandate from divine providence to force organization upon the barbarian peoples by any means judged necessary; and if they met with implacable resistance, they might well be justified in "clearing" the territory in question and making it "the abode of civilized man." The rights of barbarian populations were "petty and trifling" in comparison with the right of civilized states "to establish political and legal order everywhere." [38]

Sooner or later, then, all mankind would enjoy the same advanced political life presently enjoyed by the Teutonic peoples. But the state had higher objectives than the mere maintenance of order. In the Hegelian politics of Burgess, the ultimate purpose of the state was "the perfection of humanity; the civilization of the world; the perfect development of the human reason, and its attainment to universal command over individualism; the apotheosis of man. This end is wholly spiritual; and in it mankind, as spirit, triumphs over all fleshly weakness, error, and sin." In due time, after many centuries or even "cycles," the national polities would inevitably coalesce into a true world order, but such a step could not be safely taken until the whole race was fully prepared for it. Nature and history prescribed a fixed sequence, which peoples defied at their peril. Constitutional government came first, followed by the maturation of the national state, and then, ultimately, by the world state. Any effort to bypass these necessary stages of development would result in "dissolution and anarchy" and a return to barbarism.[39]

The most widely read apostle of racism in America was not a political scientist or a biologist or a historian, but a Congregational clergyman, the Rev. Josiah Strong, whose tract for the American Home Missionary Society, *Our Country,* was published in 1885 and went on to sell 175,000 copies. Most of the book addressed itself to the perils facing the young Republic—immigration, Roman Catholicism, Mormonism, alcohol, socialism, materialism, and "our rabble-ruled cities." But with all its failings, America remained for Strong the land of hope, and the nineteenth century had witnessed more progress than all past ages. Machinery, science, liberty, womankind, and respect for life had all made unprecedented advances in America in the first eighty years of the century, and since the rate of progress was constantly increasing, the last twenty years might "outmeasure a millennium of olden time." [40]

Strong called special attention in his forecast of the future progress of mankind to two closely related developments: the growth of home and foreign missions, chiefly through the enterprise of the English-speaking peoples; and the steady expansion of the Anglo-Saxon race throughout the world, above all its settlement of the American West. Providence had decreed that the Anglo-Saxon peoples should be the principal instruments of progress in modern times. They had grown from fewer than six million souls in 1700 to almost one hundred million in 1880, and now ruled more than one-third of the earth's surface and one-fourth of its people. More significant still, they were the greatest living representatives of the forces that had in the past always contributed most to the improvement of mankind: pure Christianity and civil liberty. "It follows, then, that the Anglo-Saxon, as . . . the depositary of these two greatest blessings, sustains peculiar relations to the world's future, is divinely commissioned to be, in a peculiar sense, his brother's keeper." It also followed that North America was destined to serve as "the great home of the Anglo-

Saxon, the principal seat of his power, the center of his life and influence," and within North America, it was the Western states that would in time wield the greatest authority.[41] From a dedicated, vigorous, Christian American West, all mankind would receive light and guidance in the next century.

But Strong was too much captivated, at least in 1885, by Darwinist imagery to confine himself to prophecies of Anglo-Saxon and North American leadership. The Anglo-Saxon race would not only guide the rest of mankind: it would also supplant much of the rest of humanity through success "in the final competition of races." Long before the world's population had reached its maximum,

> this race of unequaled energy, with all the majesty of numbers and the might of wealth behind it—the representative, let us hope, of the largest liberty, the purest Christianity, the highest civilization—having developed peculiarly aggressive traits calculated to impress its institutions upon mankind, will spread itself over the earth. If I read not amiss, this powerful race will move down upon Mexico, down upon Central and South America, out upon the islands of the sea, over upon Africa and beyond. And can any one doubt that the result of this competition of races will be the "survival of the fittest"?

Quoting a colleague, Strong asked if it were not, perhaps, God's plan to repopulate the world with better human stock. The Anglo-Saxon had, in particular, shown himself to have extraordinary powers of adaptation to strange climates. Wherever he went, he prospered and bred prolifically. "Thus, in what Dr. Bushnell calls 'the outpopulating power of the Christian stock,' may be found God's final and complete solution of the dark problem of heathenism among many inferior peoples." [42]

As for those races able to resist the Anglo-Saxon tide, they would no doubt survive, "but, in order to compete with the Anglo-Saxon, they will probably be forced to adopt his methods and instruments, his civilization and his religion." Strong could find no reason to doubt that the Anglo-Saxon race "unless devitalized by alcohol and tobacco, is destined to dispossess many weaker races, assimilate others, and mold the remainder, until, in a very true and important sense, it has Anglo-Saxonized mankind." When all this had taken place, "the coming of Christ's kingdom in the world" would be assured.[43] *

* In Strong's later works, however, this identification of American world hegemony with the arrival of the Christian millennium was more or less discarded in

It was in the socialist camp that the greatest confidence in the progress of mankind found voice in the political literature of *la belle époque*. Although few socialists of the period could add significantly to the perfervid expectations of the socialist prophets of the mid-nineteenth century, they all believed in the coming victory of socialist virtue over capitalist evil. If they were divided among themselves, it was only on questions of tactics and strategy, on the troublesome matter of "revolution" versus "evolution." In their view of progress and in their vision of the socialist future, they were all essentially of one mind.

Even the problem of grand strategy divided socialists less than they imagined. Apart from such mavericks as the French syndicalists and Lenin's faction of the Russian Social Democrats, socialists after the death of Marx reached a fair amount of consensus on the question of whether a violent seizure of power, on the model of 1848 or the Paris Commune, would be necessary to convert modern society from capitalism to socialism. The "orthodox" talked revolution, the "revisionists" talked evolution, but both agreed that social democracy could achieve its aims in the modern world without violence. In effect, the classical revolutionism of early Marxist socialism had ceased to be a vital factor in socialist thinking or politics by the end of the nineteenth century.

The heartland of socialism between the death of Marx and the Bolshevik Revolution continued to be Germany, and among German socialists Karl Kautsky served as the leading theorist and philosopher of progress. His textbook of Marxism, *The Economic Doctrines of Karl Marx* (1887), introduced Marxist thought to a whole generation of socialists throughout the world. From 1883 to 1917, he edited *Die neue Zeit,* the foremost Marxist journal of its time. When the challenge to Marxist orthodoxy came from the "right," in the form of Eduard Bernstein's revisionist movement, Kautsky led the doctrinal struggle against revisionism. When the challenge came from the "left," in the form of a triumphant Bolshevism in the years after 1917, he led the doctrinal struggle against Bolshevism. His greatest theoretical work, *The Materialist Conception of History* (1927), expounded the Marxist approach to history in two large volumes.

favor of a more conventional interpretation of the social gospel, including a plea for world federal government and a world brotherhood of races on the principle of "unity in diversity." See *Our World* (Garden City, N.Y., 1913), chs. 1 and 4.

Kautsky's earlier writings contained little original philosophical thought. His duel with Bernstein was a contest between an almost literally expounded orthodoxy and a deliberate deviation from orthodoxy. But when Lenin emerged as the leader of the first successful Marxist revolution, and claimed to represent Marxist thought more faithfully than Kautsky himself, Kautsky was forced to strike out boldly on his own, in explicit defense of his own interpretation of the Marxist gospel. No great gulf separates the earlier and the later Kautsky, but Lenin's challenge helped to stimulate an ampler development of his world-view.

Kautsky pointed repeatedly in *The Materialist Conception of History* to the continuity that he found existing between the laws of evolution and the laws of history. If one spoke of progress in either context, one could mean only increasing complexity and diversity, not increasing fitness, since the so-called lowest creatures were just as well-adapted to their environment as the highest. The laws of social evolution, or history, constituted a "natural continuation" of the laws of biological evolution, so that what passed among bourgeois meliorists for "progress" was not the qualitative improvement of mankind but only the growing complexity of the conditions of human life, to which mankind responded by becoming more complex and diverse itself. Men were not better or happier or more humane as a result of the historical process. Nonetheless, it could be argued fairly that they were achieving steadily "higher" levels of development, and for Kautsky this was good in itself. In short, although men were not individually better, they lived in a society that was better, and would one day live in a still better society, according to Kautsky's own conception of the good.

The force initiating change and development in history, in Kautsky's view, was man's inventive spirit, which led him to make tools for the production of desired goods and services. "With the production of the means of production," he wrote in 1906, "the animal man begins to become the human man." [44] But as soon as a rudimentary technology existed, it set in motion a material process independent of human volition, by which man's environment became steadily more complex, new relations of production were necessarily contrived, and the dialectical rhythm of history described by Marx and Engels began to function. The further evolution of man "to ever higher, that is to say, more diverse, more efficient forms" was mechanical in the sense that "the problems which arise from the new inventions and which give the impetus to further evolution, are not foreseen and willed, but rather constitute a force that works independently of the will and knowledge of man and, indeed, points out to him the course he must pursue." [45] The freedom enjoyed by men consisted "only in the voluntary execution of what they have recognized as necessary." [46]

Human history, then, recorded the progress of civilization from the simplicity of primitive communism through slavery and serfdom to the industrial society of modern times, which contained the seeds of its own negation and would lead inevitably to the still higher communist society of the future. Although class hostility could be expected to increase as the day of the proletarian revolution drew nearer, Kautsky opposed the notion of a violent revolution achieved through force and conspiracy. The victory of socialism had to be won democratically, through the ballot box; Kautsky held this view as strongly before the Bolshevik seizure of power in Russia as after it. War and civil war were equally abhorrent to his thoroughly civilian mind. He also attempted, with typical Marxist "realism," to resist the pleasures of utopography. In many respects, in ethical behavior, in justice, in humanity, in freedom, in health and happiness, the communist society of the future would do well to equal the standards already achieved by prehistoric man, which had so often been sacrificed to greed or to the cause of so-called efficiency in the course of historic time by the exploiting classes. The individual man would be no better morally than men of other periods; no "supermen" would emerge from the masses; and progress would continue to be limited to improvement in social relations, technology, and the sciences.

At the same time Kautsky's coming age had many of the features of the nineteenth-century socialist millennium. Conforming to his law of progress, its society would be more heterogeneous than any past society since the progress of science and technics and the liberation of all individuals from social bondage would enable every man to make his own life as he saw fit.

In this sense socialism will offer a hitherto unheard of possibility for the free development of personality. . . . The fullest development of the abilities of individual personalities and the greatest freedom for the exercise of these abilities in a society without class distinctions . . . must enormously increase the diversity and efficiency of this society. . . . In the life of the spirit, in technics, in the development of the personality, diversity will grow more than enough to counterbalance the effects of the levelling of classes and races and the impoverishment of the wild part of organic nature. And for progress in this direction no limit is visible for a long time to come.[47]

Only the cooling of the sun would bring the human adventure to a close.[48]

Kautsky's most formidable rival for the ideological leadership of German social democracy, Eduard Bernstein, had no fundamental quarrel with Kautsky on the question of socialist political strategy. Both were pacifists and democrats. But Bernstein could not accept much of the philosophical substructure of Marxism, and arrived at a position that might best be described as socialist pragmatism. What most disturbed Bernstein about Marxist methodology was its failure to predict the future. Marxism claimed to be a science with predictive power, but it had failed to foresee the relationship between the classes that had developed by the end of the nineteenth century. The class struggle was not intensifying, the proletariat was not becoming more desperate and impoverished, capital was not concentrating in a few hands, the bourgeois state was not turning a deaf ear to the pleas and needs of the working class.

In his controversial book, *Evolutionary Socialism,* first published in 1899 in German, under the title *Die Voraussetzungen des Sozialismus und die Aufgaben der Sozialdemokratie,* Bernstein therefore proposed to retain only some of the moral content and the topical interests of Marxism, moving beyond dialectical materialism to a commonsense, pragmatic, social evolutionism. The dialectic was abandoned, as a Hegelian irrelevance; history was to be seen not as a Calvinistically predetermined material process, but as a gradual development, coming under human control by degrees, and less and less subjected to the influence of purely economic forces. Such forces, he wrote, "create, first of all, only a disposition for the reception of certain ideas, but how these then arise and spread and what form they take, depend on the co-operation of a whole series of influences." The more we knew of how the material world affected us, the less power it could wield over us. "The economic natural force, like the physical, changes from the ruler of mankind to its servant according as its nature is recognised." [49] The notion that socialism could ever be a science was entirely wrong-headed, the result of listening to Hegel when one should have listened to Kant. "No *ism,*" Bernstein asserted, "is a science." [50] The socialist was a maker of values, leading mankind in a direction that he deemed desirable, a creative force, and not a gazer into scientific crystal balls.

Bernstein also deplored the romantic utopianism in Marx. In common with the American pragmatists, he felt that the movement was all-important, the ends unfathomable. There would be no final, catastrophic revolutionary abolition of history as we knew it, no achievement of a socialist Eden, but only a steady growth upwards toward justice and freedom, without drama or dialogue of sharply opposed forces. As he told his British readers in 1909, his interest centered on

> that work in the furrows of the field which by many is regarded as mere stop-gap work compared with the great coming upheaval. . . . Unable to believe in finalities at all, I cannot believe in a final aim of socialism. But I strongly believe in the socialist movement, in the march forward of the working classes, who step by step must work out their emancipation by changing society from the domain of a commercial land-holding oligarchy to a real democracy.[51]

For the mysteries of the dialectic, as Peter Gay suggests, Bernstein had substituted a straightforward theory of unilinear progress "closely akin to the view of the nineteenth-century Positivists."[52]

Still more heretical from the Marxist point of view was the idealist socialism of Jean Jaurès, the leader of the revisionist camp in French social democracy. No socialist of his time believed more deeply in progress and humanity. By temperament and education he was quite incapable of the tough-mindedness of orthodox Marxism: he looked upon the world and, in spite of everything, he found it good.

Jaurès was born in 1859 in the south of France. A classmate of Bergson's and Durkheim's at the École Normale Supérieure in Paris, he served as a radical deputy in the Chamber between 1885 and 1889, and also lectured in philosophy at the University of Toulouse. His philosophical training inclined him to the idealist tradition, and when he joined the socialist movement in 1893, he immediately challenged the validity of historical materialism as a world-view for socialists. History, he wrote, had two aspects, the one material, the other spiritual. The first was regulated by mechanical laws, which had to be understood; but man was fundamentally a spiritual being, fulfilling the aspiration of the universe toward harmony and order. Without the help of creative spiritual energy, the material dimension of reality could make no progress. Without spirit, there could be no grounds for hope or reason for living. Jaurès found it inconceivable that "human progress, which has brought our race from primitive brutality and savagery to a beginning of order, liberty, and equity, results from the clash of blind and mechanical forces"—and nothing more. Technical and economic change had their place in the record of man's progress, but their origins could not be discovered and their human uses could not be fathomed unless one also took into account the role of will and conscious aspiration. It was impossible, he wrote, "to dissociate so-called material, mechanical, and economic progress from moral progress, the progress of spirit and conscience." [53] In history, as in nature, matter and spirit evolved together, and neither could be understood in isolation from the other. [54]

What one called socialism, therefore, was only a name for the next step in the continuing upward movement of the human race, a fuller realization of man's perennial search for the good life. No sharp break with the past was required. In one of his most important essays, written in 1901, Jaurès took the Marxists to task for their insistence on the historical necessity of the total pauperization of the proletariat and the catastrophic downfall of capitalism, ridiculing Marx's vision of things to come as "a Hegelian transposition of Christianity." The proletariat, according to Marx, would have to play the role of a crucified savior-deity. Just as the "infinite abasement of God" in the Christian myth led to "the infinite elevation of man, so, in the dialectic of Marx, the proletariat, the modern Saviour, had to be stripped of all guaranties, deprived of every right, degraded to the depth of social and historic annihilation, in order that by raising itself it might raise all humanity." Marx had not surprisingly experienced "a sort of joy" in cataloguing the woes of his proletarians. [55]

But Marx was a mystic and a romantic, and Jaurès feared that he had very much missed the mark. The working classes of Europe were far removed from the depths described by Marx, if only because, like all modern men, they shared in mankind's victory over savagery and barbarism, and over slavery and serfdom. Every reform democratically achieved brought the still higher justice of socialism a little closer. "It is not by an unexpected counter-stroke of political agitation that the proletariat will gain supreme power, but by the methodical and legal organisation of its own forces under the law of the democracy and universal suffrage." To revert to the slogans of *The Communist Manifesto*, "so obviously superannuated by the course of events," was to condemn oneself "to a life of chaos." The need of the hour was not "revolution" but "revolutionary evolution." [56]

In Great Britain, the cradle of modern industrialism and also of parliamentary democracy, the political climate was still less favorable than in republican France to the production of elaborate ideological programs or revolutionary schemes for the apocalyptic achievement of socialism. In their separate ways, William Morris's utopian socialist writings and H. M. Hyndman's quasi-Marxist Social Democratic Federation proved too exotic to flourish for long on British soil, and from the 1890s until the First World War the development of socialist thought rested chiefly with the men and women of the Fabian Society, an organization remarkable neither for the rigor nor the clarity of its doctrinal position, but far more finely tuned to the realities of British politics than Marx, Morris, or Hyndman. As their name indicated, the Fabians deeply disbelieved in rash and theatrical assaults on the British power structure. Even more emphatically than did the "revisionists" of European Marxism, their program called for gradual and evolutionary development through education and permeation of national political life, rather than conspiratorial revolutionary politics.

The Fabian Society was organized in 1884 by a small group of young men still in their twenties. Gradualism was not, at the outset, part of their political philosophy, but they were led to it by one of their first recruits, Sidney Webb, then a young clerk in the Colonial Office. He soon became the dominant intellectual force in the movement and, together with his wife, Beatrice, retained the leadership of British socialism until the 1930s. We have already dealt with some of the other luminaries of Fabianism, such as Shaw and Wells,* but the central figure in the movement and the master draftsman of its plans for the winning of socialism in Britain was unquestionably Webb.

The first book published by the Fabians appeared in 1889, a collection of essays by seven different authors, including Webb and Shaw. It brought the Fabian Society to the attention of the general public for the first time, and is still the best-known Fabian publication, although hundreds have appeared since. As the Society's official historian pointed out, these early *Fabian Essays in Socialism* "based Socialism, not on the speculations of a German philosopher, but on

* See above, pp. 49–50 and 74–76.

the obvious evolution of society as we see it around us . . . ; it proved that Socialism was but the next step in the development of society, rendered inevitable by the changes which followed from the industrial revolution of the eighteenth century." [57] In short, the Marxist insistence on the dialectic, on revolution and class warfare, was replaced in Fabianism, somewhat as in the thought of Bernstein and Jaurès, by a faith in rectilinear progress from less to more social justice, achieved through political reform and general enlightenment. In the words of Sydney Olivier, who contributed the chapter in *Fabian Essays* on socialist ethics, socialism was "but a stage in the unending progression out of the weakness and the ignorance in which society and the individual alike are born, towards the strength and the enlightenment in which they can see and choose their own way forward." The moral ideas of socialism could be seen just as much "in the increasing philanthropic activity of members of the propertied class" as in the aspirations of the proletariat.[58]

Sidney Webb's essay on the historic basis of socialism developed at much greater length this cardinal Fabian belief in the catholicity of the socialist movement. All the burgeoning new life and thought of the nineteenth century contributed to its progress: democracy, industrialism, intervention in the economy by local and national government, even the thought of Comte, Darwin, and Spencer, who had persuaded modern men to think no longer "of the ideal society as an unchanging state." No philosopher, Webb added, "now looks for anything but the gradual evolution of the new order from the old, without breach of continuity or abrupt change of the entire social tissue at any point during the process." The earlier socialism of utopian and revolutionary romance was finished.[59]

Webb paid special attention in his essay to the multiplication of government services and controls in nineteenth-century Britain. Orthodox political economy had forbidden it, the two leading parties had denounced socialism, the voices of committed socialists were few and weak, and yet, little by little, starting often at the municipal level, government had steadily enlarged its powers at the expense of private enterprise and in the service of the public weal. The only explanation of this astonishing development lay in the sovereign authority of history: the times were ripe, the *Zeitgeist* was favorable, and the task of socialists was not so much to take over this historical process as to help it along through education of both the general public and the governing classes. It should be added that Webb did not have in mind the necessary appearance of a socialist political party. Even when a Labour party was created, relations between it and the Fabian Society down to 1914 were often quite strained, and Fabians on occasion preferred Liberal to Labour candidates, or even ran against Labour candidates themselves.

American socialism during the late nineteenth and early twentieth centuries, although more vigorous than it was ever to be in later years, contributed few thinkers of international rank to the socialist movement. Its representative Marxist, Daniel De Leon, was an ideological purist, who preached a rigid economic determinism and looked forward to a single, total, political revolution, at which time socialism would be achieved once and for all. Gradualism and reformism, bitterly condemned by De Leon, flourished in the Socialist party founded and led for many years by Eugene Debs, although Debs himself stood on the far left of his own party. Henry George, the one thinker who had a considerable impact on world socialism, and Fabianism in particular, was not in the strictest sense a socialist at all, but a radical economist. His *Progress and Poverty* first appeared in 1879, and by the turn of the century had sold two million copies.

Progress and Poverty is one of the few books by a thinker in or close to the socialist movement in the period under consideration to focus directly on the problem of human progress. A self-educated printer and journalist, George knew the plight of the poor from hard personal experience, and his point of departure was the contrast between the affluence made possible by modern science and technology and the stifling poverty in which the majority of men still lived, even in the most prosperous cities of the New World. Progress and poverty, presumably incompatible, nevertheless appeared to go hand in hand. To resolve the paradox one had to consult history.

George's analysis of history has led some critics to see him as a forerunner of Spengler and Toynbee, but it would be just as appropriate to call him an American Karl Marx. Like Marx and his more orthodox followers, George rejected the bourgeois idea of the gradual ascent of civilization. As currently expounded by the Darwinists, the progress of civilization was "a development or evolution, in the course of which man's powers are increased and his qualities improved by the operation of causes similar to those which are relied upon as explaining the genesis of species—viz., the survival of the fittest and the hereditary transmission of acquired qualities."

War and misery were reputedly "the impelling causes which drive man on, by eliminating poorer types and extending the higher." [60] But such a definition of progress not only outraged common morality: it was also contradicted by history. What about the civilizations of Asia, which had changed scarcely at all for thousands of years, to which George applied the pre-Toynbeean label of "arrested civilizations"? And what about those other civilizations that had grown, prospered, declined, and fallen? If the "laws" of progress were fixed, why should mankind not have gone steadily onward and upward, throughout the world?

For George, the causes of stagnation and decline in the history of civilizations were quite clear. He had no tolerance of theories of racial differences or organic destiny. True progress consisted of the devotion of "mental power . . . to the extension of knowledge, the improvement of methods, and the betterment of social conditions," a process facilitated by the tendency of men to associate in ever-greater numbers under civilized conditons of life. But since in-creased association also brought with it the risk of increased conflict, and conflict meant the waste of mental power, advanced civiliza-tions were always in danger of losing their forward momentum. Society could be likened to a boat. "Her progress through the water will not depend upon the exertion of her crew, but upon the ex-ertion devoted to propelling her. This will be lessened by any expenditure of force required for bailing, or any expenditure of force in fighting among themselves, or in pulling in different directions." Conflict, in turn, resulted most often from a denial of equality of rights on the part of some men in the body social to other men in the same body. The progress of association was eventually canceled out by the progress of exploitation, and when this happened the civilization lost its social health and ultimately petrified or perished. To maintain continuous progress the only correct formula was association in equality. "Association frees mental power for expenditure in improvement, and equality, or justice, or freedom—for the terms here signify the same thing, the recognition of the moral law—prevents the dissipation of this power in fruitless struggles." [61]

Nothing guaranteed that modern Western civilization would not fail just as its predecessors had failed. All the familiar signs of imminent collapse were at hand. "Wages and interest tend con-stantly to fall," George warned, "rent to rise, the rich to become very much richer, the poor to become more helpless and hopeless, and the middle class to be swept away." Unless this tendency were reversed at once, modern civilization would "decline to barbarism." [62] But although he did not share Marx's faith in the inevitability of the socialist revolution, George entertained great hopes that the downward tendency could be reversed. He was, in spite of everything, a prophet of progress.

Before turning to his scheme for salvation, we should note that George did not subscribe to a Spenglerian relativism regarding his own civilization. It might go under, but even if it did, it would still have achieved more than any other in history. He found no reason to doubt "that our own civilization has a broader base, is of a more advanced type, moves quicker and soars higher than any preceding civilization." [63] It had surpassed Greco-Roman civilization to the same degree as the latter had surpassed Asian civilization. George's cyclical theory of history, on closer examination, proves to be a spiraliform theory.

The West, he continued, owed its superiority to its unprecedented success—until recently—in promoting equality as it also expanded the scope of association, a success made possible by the moral influence of Christianity and by the political fragmentation of the West, thanks to Teutonic traditions of self-government, which had foiled every effort to establish a single Western empire on the fatal model of Rome and Byzantium. All through the earlier pages of Western history one saw the steady advance of equality—the abolition of slavery and hereditary privileges, the substitution of parliamentary for arbitrary government, the rise of civil and religious liberty. "This tendency," he observed, "has reached its full expression in the American Republic . . . the most advanced of all the great nations." Nevertheless, political equality and freedom had not saved Americans or any other advanced people from "the tendency to inequality involved in the private ownership of land, and it is further evident that political equality, co-existing with an increasing tendency to the unequal distribution of wealth, must ultimately beget either the despotism of organized tyranny or the worse despotism of anarchy." [64] Rome, too, had been a great republic at one time, and the Caesars had waited for centuries before discarding all the trappings of her republican constitution.

The West, which had soared so high, could expect, therefore, to go on to still greater heights, or fall farther than any civilization had ever had a chance to fall. The present moment in history was unprecedentedly fateful. "The civilized world is trembling on the verge of a great movement. Either it must be a leap upward, which will open the way to advances yet undreamed of, or it must be a plunge downward, which will carry us back toward barbarism." [65]

Steeply up, or steeply down—that was the choice. And George had the remedy for the social sickness besetting modern man, a remedy he spent the rest of his life tirelessly preaching on both sides of the Atlantic. It was not a political revolution or a general strike or Fabian permeation, but the device of a single tax on land, which would have the effect of confiscating all rents. His panacea had all the strengths and weaknesses of simplistic solutions. It was easily understood and converted into slogans, but most socialist intellectuals preferred the diagnosis to the cure.

George left no doubt that after the abolition of rent the millennium would come, and progress could continue uninterruptedly forever. With want exchanged for abundance, fear for fraternity, greed for nobility, "who shall measure the heights to which our civilization may soar? Words fail the thought! It is the Golden Age of which poets have sung and high-raised seers have told in metaphor! . . . It is the culmination of Christianity—the City of God on earth, with its walls of jasper and its gates of pearl! It is the reign of the Prince of Peace!" [66]

Land and Labor Clubs sprang up in many parts of the United States to bring George's gospel to the people, and virtually every early leader of the Fabian Society recorded his debt to *Progress and Poverty*. American and British socialism also owed much in the 1880s to two utopian books, Laurence Gronlund's *Co-operative Commonwealth* and Edward Bellamy's *Looking Backward,* both published in that decade.

Although George's forecast of the coming society is more than a little apocalyptic, he was no teacher of violent revolution. As we have already noted, the steady progress of democratic reform and labor politics in the late nineteenth and early twentieth centuries eroded the appeal of old-fashioned conspiratorial revolutionism in leftist circles. The authentic revolutionary impulse flourished in only a few odd corners of the socialist world during *la belle époque.*

In one of these, the syndicalist movement in France, the belief in progress was bitterly denounced as another example of bourgeois chicanery. "The grandeur of the country, the domination of the forces of nature by science, the march of humanity toward the light," wrote Georges Sorel in *The Illusions of Progress,* "this is the humbug we encounter at every turn in the speeches of democratic orators." [67] Real progress could begin only after the achievement of the syndicalist millennium. [68]

But a doctrine of progress linked with a sober acceptance of the need for revolutionary politics did thrive in Russian socialist thought. The Russian political situation in the final decades of Romanov despotism was wholly unlike that of Germany or France or any other Western country. Revolution appeared inescapable to many observers even outside the socialist movement. Here, if nowhere else, it was entirely possible for pragmatic political thinkers to espouse both Marxist progressivism and Marxist revolutionism.

Of the two principal theorists of scientific Marxism in Russia during our period, G. V. Plekhanov and V. I. Lenin, Plekhanov was clearly the more cerebral, if not necessarily the more brilliant. His place in the history of Russian socialism closely parallels that of Kautsky in German socialism. Both men considered themselves orthodox Marxists. Both were primarily theorists rather than politicians. Both were born in the mid-1850s and had emerged in the 1880s as men of the extreme left, committed to revolution, living in exile. Both eventually found themselves occupying a position to the right of Lenin, unable to accept his cynical attitude toward democracy and his tough-minded authoritarian conception of socialist party politics. In Plekhanov's case the break came in 1903, when he quarreled with Lenin over the question of readmitting members of the Menshevik faction to the editorial board of the socialist journal *Iskra.*

By 1903 Plekhanov had also written most of his major theoretical works. In general they are characterized by a remarkable fidelity to the thought of Marx and Engels. In discussing the place of the individual in the historical process, for example, he reached the not surprising conclusion that great men could delay or advance the coming of inevitable social upheavals, or in various ways modify the exact manner in which such upheavals occurred, but that the broad outlines of historical change were fixed. The individual made history by learning its laws and doing its will. Such obedience to history was free and conscious, yet one should never imagine that the inevitable would not happen if he failed to act or if he set himself in opposition to the future. Human nature "can no longer be regarded as the final and most general cause of historical progress," Plekhanov maintained, rather "we must regard the development of productive forces as the final and most general cause of the historical progress of mankind." It followed that "a great man is great not because his personal qualities give individual features to great historical events, but because he possesses qualities which make him most capable of serving the great social needs of his time." [69] In the same way the aspirations of men helped to produce revolutions, but such aspirations took a necessary course, determined by socio-economic laws, and sociology could therefore become a science only when "sociologists are able to understand the appearance of specific aims in social man (social 'teleology') as a necessary consequence of the social process, determined in the last analysis by the march of economic evolution." [70]

Plekhanov also insisted on the fundamental distinction between the gradualist, evolutionary view of progress taken by the heirs of the Enlightenment and the dialectical, revolutionary historical outlook of Marxism, grounded in the wisdom of Hegel. Gradual changes in history only prepared the way for sudden, dramatic changes. This had been the scandal of Rousseau's theory of history, in the eyes of the *philosophes*. For them progress resulted from the slow ascent of reason. For Rousseau, who anticipated Hegel and Marx, the progress of intellect and civilization had led to moral retrogression, a decline in harmony and equality since primitive times, which afflicted the whole history of civil society. In civil society were to be found the seeds of its own negation, and this the men of the Enlightenment could not grasp. Engels had been well-advised to rank Rousseau among the great dialectical thinkers in modern history.[71]

We need not concern ourselves here with Plekhanov's Menshevik associations and his activity in 1917. Like so many others, he put his faith in the democratic process after the March Revolution, but by 1918 he was dead, and Russia had moved into a new era under the leadership of his former disciple, Lenin. Plekhanov's influence among Russian Social Democrats fell off sharply in that last year of his life. Although he had been a skillful exegete and apologist of Marxism, he lacked Lenin's sense of timing and political sagacity. But the same could be said of most Russian socialists in 1917!

Including Lenin among the prophets of progress raises one very serious question at the outset. "Leninism," writes Alfred G. Meyer, "is Marxism beset with many inner doubts—a distrust of history, of the masses, even of the conscious leadership. This leads to the preoccupation with manipulation, organization, and coercion."[72] Anyone who follows Lenin's career carefully or reads between the lines of his writings soon comes to sense his instinctive reluctance to rely on the promises of Marxist historical analysis, or on anything except his own political judgment. He understood better than most of his Western European comrades the enormous strength of the capitalist system, which had increased rather than decreased in the years since Marx's death. He did not feel sure enough of its "inevitable" collapse to wait patiently for the Great Day. The best way to ensure the revolution was to form a tightly disciplined and fanatical conspiracy prepared to seize any opportunity that offered itself.

But although this ingrained pessimism may help to explain Lenin's revolutionary strategy, it did not prevent him from adhering in principle to the general outlines of the Marxist idea of progress. Mankind, he agreed, had reached the present by a series of dialectical transformations necessary to its progress from savagery, and communism was destined to win—sooner or later. Lenin also offered three important clarifications of the Marxist doctrine of progress, all of which, at least on the surface, betrayed no lack of faith in history. The first was his analysis of imperialism as the highest and final stage of capitalism, the second his conception of the dictatorship of the proletariat, and the third his discussion of the transition to pure communism.

Lenin's theory of imperialism, as presented in his pamphlet of 1916, is too well known to require detailed treatment here. It extends in a logical and plausible way Marx's discussion of monopoly capitalism. Lenin interpreted imperialism as a last-ditch effort on the part of capital to save itself from its internal contradictions; but imperialism in turn was riddled with internal contradictions of its own, which had the same origins as those of domestic capitalism. In resorting to imperialist adventures, capitalism was therefore sealing its fate. Again, whether Lenin really believed this, or whether he conceived of his theory as a propaganda device to feed the confidence of his followers, is impossible to determine. Many thinkers no less acute than Lenin allowed themselves to believe all kinds of things in those difficult years. In any event, the doom of the capitalist era was spelled out in Lenin's essay, and at the same time the faithful Marxist obtained an explanation of the temporary reprieve that capitalism had apparently won for itself, enabling it to keep its strength beyond the limits indicated by Marx.

In good time the revolution would come, and Lenin assumed that it would have to be a violent revolution, a forcible seizure of power that no exploiting class would yield in any other way. But whereas many socialists of his generation, including the Russian socialists, expected that after the acquisition of power the processes of liberal democracy could be trusted to establish a new socialist order, Lenin interpreted Marx's "dictatorship of the proletariat" in its strictest sense, to mean "the organisation of the vanguard of the oppressed as the ruling class for the purpose of crushing the oppressors."[73] The state could not, in Engels' often misquoted phrase, "wither away" until the last remnants of the exploiting bourgeoisie and of bourgeois psychology had been extirpated. Only a workers' state, with a government composed of the leaders of the workers' party, could do this hygienic work of purifying society, and it was therefore idiotic to prattle about the preservation of "parliamentary" institutions and the safeguarding of the "rights" of minorities, *i.e.,* of exploiting classes.

Lenin anticipated that communism would be achieved in two stages. Developing the terminology used by Marx in his critique of the "Gotha Program" of the German socialists in 1875, Lenin described a preliminary stage, "socialism," in which the proletarian state would address itself to the building of an order of society free of exploitation, private capital, and private enterprise. Such an order would require forcible oppression of class enemies and rigorous "factory" discipline. Prices, rents, and wages would still be paid, and each man would receive a share of the public wealth in proportion to the amount of work done by him. Exploitation by means of private ownership would be abolished, but not—for some time—the state of mind associated with private ownership, the state of mind that led a man to care more about getting what was "due" him than about his work or the common good.

Once socialism was achieved, however, the new order could begin building true communism, and here Lenin abandoned his cynical anthropology to preach with a certain eloquence the utopian vision of Marx and Engels. "We set ourselves the ultimate aim," he wrote in 1917, "of abolishing the state, *i.e.,* all organized and systematic violence, all use of violence against man in general." Violence would no longer be needed after the last vestiges of bourgeois psychology had been eliminated, "since people will *become accustomed* to observing the elementary conditions of social life *without force* and *without subordination.*" Men would learn, free of the fear and shame of exploitation, to work together happily. They would enjoy their work, instead of hating it. The rule, "From each according to his ability, to each according to his needs," could at last apply "when people have become so accustomed to observing the fundamental rules of social life and when their labour is so productive that they will voluntarily work *according to their ability.* . . . There will then be no need for society to make an exact calculation of the quantity of products to be distributed to each of its members; each will take freely 'according to his needs'." [74] The state would dissolve into public administration. The lesser peace of socialism would give way to the greater peace of communism.

Lenin's ideas are not in themselves extraordinary for a man of his time, but he was alone among all the socialists whose doctrines of progress appear in this chapter in having had the opportunity to lead his party through a successful revolution. His view of the transition from socialism to communism remains the official position of the Soviet government.

PART THREE

Progress on Trial

Both professor and prophet depress,
For vision and longer view
Agree in predicting a day
Of convulsion and vast evil,
When the Cold Societies clash
Or the mosses are set in motion
To overrun the earth,
And the great brain which began
With lucid dialectics
Ends in a horrid madness.

—W. H. Auden (*1947*)

8

The Decline of Hope

BELIEF IN PROGRESS did not perish and disbelief in progress did not first take root "in Flanders fields." But the summer of 1914 is a convenient point of vantage from which to survey the spiritual condition of modern Western man. By the common judgment of historians, it marks the end of the nineteenth century conceived as an epoch in the history of politics and ideas. Up to 1914 meliorism prevailed, despite much uneasiness in certain segments of the Western intelligentsia. Since 1914 the tendency has been toward despair, despite the persistence in certain other segments of honest hopefulness. "August 1914 is the axial date," writes William Barrett, "in modern Western history, and once past it we are directly confronted with the present-day world." The sense of man's power has been replaced by "a sense of weakness and dereliction before the whirlwind that man is able to unleash but not to control."[1]

The First World War rarely deprived firm believers in progress of their faith, but it acted as a powerful stimulant to doubt and despair among those previously sceptical. Carl Gustav Jung, not yet forty when the war began, and already in rebellion against the rationalism of Freud, complained in a characteristic post-war lecture that modern man was "the disappointment of the hopes and expectations of the ages." He challenged his listeners to contemplate the contrast between two thousand years of Christian idealism and their dénouement in contemporary history, "the World War among Christian nations with its barbed wire and poison gas. What a catastrophe in heaven and on earth!"[2] Raised in a liberal home during *la belle époque,* the Norwegian novelist Sigrid Undset wrote near the close of her life that "the war and the years afterwards confirmed the doubts I always had had about the ideas I was brought up on." The progressivist ideologies of the prewar era had assumed that human nature could be changed, but 1914 showed her conclusively that it could not.[3] Undset's renunciation of her childhood culminated in 1924 with her conversion to Roman Catholicism.*

Even when the war in and of itself had little to do with the despair of seminal thinkers, it provided the best possible spiritual situation for the favorable reception of despairing books. Consider Oswald Spengler's *The Decline of the West.* Its first draft was complete when the guns began to sound in 1914, but it was not published until the summer of 1918, and its phenomenal public success owed much to the gloom engendered by the war. The arrival of peace did not reduce its appeal; it became one of the most widely discussed books of the Twenties.

And it continues to be read. What makes 1914 the most important turning point in modern history, of course, is not the fact of the First World War alone, or the pessimism to which it gave rise, but the disasters that have overtaken the Western world without interruption or relief in the half-century since Versailles. *The Decline of the West* might seem almost as curious a production today as Spencer's *Synthetic Philosophy* if some sort of tranquillity had returned to the Western world by the 1930s. Instead, one must agree with H. Stuart Hughes that the expectation of catastrophe "has become the characteristic attitude of social observers and the general public alike, both in Europe and, more recently, in the United States."[4] The First World War, which cost ten million lives, has been followed by a second, which cost fifty million. The likelihood of a third, which could destroy most of the human race in a few hours, haunts every thinking man. The threat to human freedom implied in the rise of mass society and the new demagogy of the nineteenth century has been carried out with unprecedented ruthlessness in the regimes of Mussolini, Hitler, Stalin, and Mao; and even nominally free societies sway under the impact of economic depression, monopoly capitalism, omniscient bureaucracy, racial hatred, depersonalizing technicism, and megalopolitan sprawl. Pollution, overpopulation, and the waste and depletion of natural

* See the similar case of R. J. Campbell, pp. 94–95.

resources jeopardize the economic progress of the century, which in any event has been restricted mainly to the Western world. The non-Western peoples have at last extricated themselves from Western political domination, but it becomes steadily more apparent that they may be able to equal the prosperity of the West, if at all, only by aping the worst forms of twentieth-century Occidental despotism. Meanwhile, the loss of empire in itself has been a severe shock to many Western minds, further confirming fears of the "decline of the West."

Nor is the predicament of modern man confined to social, economic, and political catastrophe. The history of thought in the twentieth century reveals problems of another order altogether: the growth in most of the knowledge-seeking disciplines of a radical relativism that seems to invalidate man's traditional search for truth, goodness, and beauty. Orthodox religious belief continues, as in the foregoing century, to lose ground, but the new scepticism does not stop at challenging theological faith: it undermines the rational foundations of faith itself, leaving man stranded in a meaningless cosmos, man who is meaningless even to himself except as a creature compelled to live and condemned to die.

The response of sensitive thinking people to the political, ecological, and spiritual crisis of the twentieth century has been wholly predictable. Just as the art, music, and literature of the century mirror the disintegration of Western man's self-confidence, so the major systems of thought are, on the whole, systems of rebellion against modernism and its gospel of progress. Disillusionment with the ways of man and a longing for the lost anchorage of religious faith have driven many typical Western minds of the twentieth century back toward a theological orientation of some kind, although few have found the peace and certainty of the old orthodoxies. Contemporary theology, nonetheless, has taken full advantage of its opportunity to denounce the anthropolatry of the nineteenth century, and some of the most telling attacks on the belief in progress have been made by theologians. There has also been a lively interest in the soteriological virtues of mysticism and the religions of Asia. Franco-German existentialism, the best known philosophical movement of the 1930s and 1940s, offers a tragic humanism not necessarily incompatible with theism, which focuses on man's finitude and estrangement. The dominant trends in psychology have pointed to the power of the irrational in human existence. Horror provoked by the savagery of world war and modern totalitarianism and by the impersonality of technicism and mass democracy has served as the inspiration for scores of counter-utopian novels and tracts that convey a searing conviction of the futility of civilization. Cyclical theories of history predict civilization's imminent total collapse.

Although a warming trend has developed in Western thought since about 1960, we must agree, therefore, with Emil Brunner's judgment that just as the nineteenth century marked "the climax of . . . belief in progress, so clearly is the twentieth century the time of its rapid decline." [5] But two caveats must be issued before proceeding further. The engines of publicity in the scholarly world and the popular press have made the most of the literature of despair in the twentieth century. It is widely read, abundant, and influential. Still, a clear distinction must be kept in mind between fear, anxiety, and alarm on the one hand and the renunciation of hope on the other. The belief in inevitable progress, for example, has largely evaporated in the twentieth century, but this is not the same thing as a belief in possible progress, and there are many thinkers in our century who denounce the former with as much warmth as they endorse the latter. Before 1914 the strongly voluntarist emphasis of pragmatism, vitalism, and other systems of thought centered on action as opposed to science or reason had already struck determinist theories of progress a heavy blow. This emphasis has become almost compulsory for believers in progress today, but they remain believers all the same. It is also interesting that many thinkers who vehemently oppose the "modern" view of progress still retain vital fragments of the progressivist tradition in their historical outlook. They may feel, let us say, that "man" or "science" cannot bring about progress, but that God acting in history, with or without man's cooperation, can do so; the result is optimism with regard both to past and future, even if the reasons given would not satisfy a Comte, a Spencer, or a Marx.

There is one other caveat. We must take care not to overlook the nineteenth-century roots of contemporary despair. The exaggeration of the significance of anomalous figures such as Kierkegaard or Dostoyevsky is an unfortunate practice of many recent cultural and intellectual historians, but it is just as misleading to write off the entire nineteenth century as an age of unqualified confidence in progress. That such confidence dominated even the last third of the century has been shown above. But not all the best minds were progressivists, and some of those who contributed most usefully to the diffusion of the belief in progress were not fully committed believers themselves, if we adhere faithfully to our own definition of the idea of progress. Among thinkers studied above, the identification of Nietzsche, James, and Tyrrell as believers in progress is subject to grave reservations, and one might also have his doubts about Huxley, Shaw, and several more.

Casting our nets more widely, we can find many other thinkers who much less ambiguously foreshadowed contemporary pessimism, whether or not they had an appreciable influence on it. In the chapter that follows, we shall explore this nineteenth-century despair by way of background to the historical pessimism of the years since the First World War.

9

Romanticism, Positivism, and Despair

T HE CULTURAL LIFE of Europe and America in the nineteenth century followed a cyclical pattern, not unlike its commercial life. As depression alternated with prosperity, so romanticism alternated with positivism. The struggle was carried out in every field of art and thought, in every country, and quite often in individual minds, from Goethe at one end of the century to Zola at the other. If by romanticism we mean a value commitment to such organismic concepts as "intuition," "spirit," and "will," and if by positivism a value commitment to "reason" and "science," then clearly the romantic impulse tended to prevail in the first third of the century, from 1815 to 1848. Positivism (or naturalism) dominated the middle decades, down to 1880. During *la belle époque,* as we have seen, the younger generation inclined toward some form of romanticism, although positivism continued in favor in many quarters.

To resort to such sweeping terms as "romanticism" or "positivism" is always dangerous, and they do not bear directly on the problem of the belief in progress, but as a device for organizing a survey of pessimism in the nineteenth century, the opposition of "romanticism" and "positivism" has some heuristic value. Most prophets of decadence, doom, and despair in the nineteenth-century context are readily distinguishable either as romantics or as positivists, just as there are romantic and positivist strategies for attacking the belief in progress, founded on different concerns and expressing different types of anxiety.[1]

Despair came more easily to the romantic, perhaps, than to the devotee of science and critical reason, if only because the romantic movement was in its very essence the expression of an acute sense of cultural failure. The romantic was a man convinced that the old world of courts, salons, dynastic wars, rational religion, and enlightenment stood in ruins, particularly after the events of 1789 to 1815. He felt himself called to create a new order, whether based upon an intuitive understanding of nature, on the principle of historical continuity scorned by the *philosophes,* on sentiment and passion, or on a new spiritual world outlook that would have been incomprehensible to the sceptics and materialists of the eighteenth century.[2]

Although most romantics rose to their task with enthusiasm and imagination, their vocation involved them in certain inevitable risks. Some could see no way forward except by retreating in fact or in fantasy into a happier and more "natural" or "historical" past. A tendency to nostalgic primitivism is obvious in many of the most representative romantics, and nothing could be deadlier to the faith in progress. The same sort of cultural despair, for all practical purposes, had infected the best minds in Europe between 1300 and 1600, during "the waning of the Middle Ages," the Renaissance, and the Protestant Reformation, when it was universally the fashion to regard the modern world as old and exhausted, and writers and artists turned to the past for inspiration—to Greece and Rome in classical antiquity, to the biblical ages, or perhaps to both at the same time. Living in a period when all beliefs had been exposed to the acids of scepticism, including the rationalism of the Enlightenment, and when the fall of the *ancien régime* in most of Europe and the American hemisphere had led to unprecedented political chaos and class warfare, the romantic had at least as much reason to despair of the present and future as the man of the difficult years between the Avignonese Captivity and the Peace of Westphalia.

But it was not just the times or the challenge of history that sometimes brought the romantic mind to despair. Romanticism glorified the creative power of man's will: the romantic was a man eternally striving to transcend his human limitations, to attain in some sense a Platonically ideal world, and the strain might prove too much for mere man to bear. Romantic enthusiasm called forth its dialectical opposite: romantic world-weariness, mystical retreat or denial, a sense of decadence or overstimulation, a disgust with the material world, but also a morbid obsession with death, lust, and crime, the "romantic agony." If the romantic gave free rein to his enthusiasm, he might rejoice in history and foresee even greater triumphs of the human spirit in years yet to come. If he fell prey to world-weariness he could just as easily reject all temporal effort to improve humanity as futile, and sink into sensualism, mysticism, or fatalism. This oscillation between spirit and flesh, hope and despair, excitement and ennui seems wholly characteristic of romantic psychology, and it is not surprising that many individual minds should have found the shadow side of romanticism more attractive as a life-option than romantic exuberance. For every Beethoven who could mount the ladder of the passions in his Ninth Symphony from strife to celestial joy, there was a Berlioz whose tormented hero, in the *Symphonie Fantastique,* ended his life on the gallows.

Early in the nineteenth century one commonly encountered form of disbelief in progress focused on the threat to society posed by the French Revolution and its sequels. In the eyes of conservatives, both of the *ancien régime* and of the romantic movement, the Revolution epitomized the *hubris* of rationalism, the *hubris* that leads to *nemesis.*

The most remarkable thinker in the anti-revolutionary camp was the émigré aristocrat Joseph de Maistre, who rejected the modern world almost in its entirety, from the Protestant rebellion against Rome to the rebellion of the masses in contemporary France (incited, so he thought, by the sophistical abstractions of the *philosophes*). Even modern science was only a grotesque mechanical simulacrum of the higher intuitive learning of antiquity. Although a passionate Christian, Maistre contrived to incorporate into his thought the classical idea of a golden age existing before the biblical Flood, for which he found evidence in "reason, revelation, and all human traditions." But in spite of his affection for the Middle Ages and for remote antiquity, he insisted on the ineradicable sinfulness of humankind in all ages, and in all ages he found God quick to anger and swift in his terrible judgments against his erring creatures. Sinful nations were regularly punished with war, civil disorder, and natural calamities. None escaped for long.

> The whole earth, continually steeped in blood, is nothing but an immense altar on which every living thing must be sacrificed without end, without restraint, without respite until the consummation of the world, the extinction of evil, the death of death. But the curse must be

aimed most directly and obviously at man: the avenging angel circles like the sun around this unhappy globe and lets one nation breathe only to strike at others.[3]

No Gothic novelist or *fin-de-siècle* diabolist could have said it better.

Maistre's despair was not peculiar to French aristocrats of the age of Robespierre and Napoleon. The evils of revolution, egalitarianism, and rationalism were denounced in every country, although seldom to the accompaniment of such a pessimistic view of history as Maistre's. In Britain, Edmund Burke shared much of Maistre's political romanticism and his abhorrence of revolution, but he expressed a reverent faith in the wisdom of history and civilization that looked forward rather to Hegel. His successors as prophets of Toryism, Samuel Taylor Coleridge and Thomas Carlyle, approached more closely the theocratic outlook of Maistre, but again, despite their condemnation of modern "decadence," occasional presentiments of disaster, and much sympathy for medieval society, they hoped for a resumption of human progress, once the wounds inflicted on the body social by the Revolution had healed.* The same might be said of nearly all the politically minded exponents of cultural despair in the latter part of the nineteenth century, men like Julius Langbehn in Germany, the Russian Pan-Slavists, and Charles Maurras in France. However much they despised the present age, their political faith had its roots in history, and they entertained chiliastic hopes for the future. Nationalism and Christian millennialism, in varying proportions, saved them from an ultimate pessimism.

Here and there, however, the conservative's poor opinion of man combined with a suppression or denial of the utopian impulse to produce views of history quite similar to Maistre's, in substance if not in style. The leading conservative statesman of Europe between 1815 and 1848, Prince Metternich, although not a romantic, espoused some of the characteristic political ideas of the romantic era, reaching conclusions hardly more hopeful than Maistre's. As he wrote in a memorandum to Tsar Alexander I in 1820, the course of the

---

* The early Victorian cult of medievalism was aptly parodied by Charles Dickens in lines spoken by the painted "Cleopatra" of *Dombey and Son,* Mrs. Skewton: "Those darling byegone times . . . with their delicious fortresses, and their dear old dungeons, and their delightful places of torture, and their romantic vengeances, and their picturesque assaults and sieges, and everything that makes life truly charming! How dreadfully we have degenerated!" *Dombey and Son* (London, 1848), pp. 274–75.

world's history was cyclical. Institutions, like everything else, "pass through periods of development and perfection, to arrive in time at their decadence; and, conforming to the laws of man's nature, they have, like him, their infancy, their youth, their age of strength and reason, and their age of decay." The pages of history had been repeatedly stained with blood because of human presumption. Puffed up by power and knowledge, man had ever and again dared to defy the immutable moral law and become his own providence, although Metternich sought "in vain for an epoch when an evil of this nature has extended its ravages over such a vast area as it has done at the present time." [4] As man's pride had grown fat on the material accomplishments of modern civilization, so had his presumption grown, leading inexorably to all the horrors of the era since 1789. Metternich recommended policies to the Tsar that could "save" Europe, and he worked all his life to the same end, but he had no illusions about the real hopelessness of the task. "The existing society is on the decline. Nothing ever stands still . . . and society has reached its zenith. Under such conditions to advance means to descend." [5]

Another conservative statesman at the opposite end of Europe, Arthur J. Balfour, took the trouble to make the belief in progress the subject of a lengthy inaugural address that he delivered on the occasion of his installation as Lord Rector of the University of Glasgow in 1891. Most thinkers, he noted, assumed "that there exists a natural law or tendency governing human affairs by which, on the whole, and in the long run, the general progress of our race is ensured." History, however, gave no such promises. On the contrary, civilizations rose and fell again, and those which for a time contributed to human betterment were plants "of tender habit, difficult to propagate, not difficult to destroy." [6]

One by one Balfour demolished every source of hope in man's perfectibility. Natural selection operated, if at all, as a dysgenic force in modern civilization. Science was still advancing, but Balfour doubted that it would be able to do so much longer; the limits were obviously being reached. Schemes for the amelioration of humanity through the application of the principles of political science in the work of the state overlooked the important point that reason was far from sovereign in politics. Society was founded, and would always be founded, on emotions, beliefs, and customs. It could not be rationally regulated nor could its future be predicted. "The future of the race," he concluded, "is thus encompassed with darkness; no faculty of calculation that we possess, no instrument that we are likely to invent, will enable us to map out its course, or penetrate the secret of its destiny." Yet, as Balfour had argued in an earlier address, whatever the future might hold, one thing at least was clear. The millennium would never arrive. Such mortal ills as "separation, decay, weariness, death" were beyond man's power to remedy, and "nothing that humanity can enjoy in the future will make up for what it has suffered in the past." [7]

Despair and world-weariness also pervade much of the romantic imaginative literature of the nineteenth century, more markedly at the end of the century than at the beginning. The introspective egoism of the early romantics and their predilection for the wild, the macabre, and the horrific often deepened, as the century advanced, into a programmatic aestheticism that scorned the world or rejoiced, masochistically, in its decay. The belief in progress was sometimes discarded only by implication, but few of the writers catalogued in Mario Praz's *The Romantic Agony,* from Sade to Huysmans, have any standing as disciples of progress, to say the very least.[8]

One of the first major literary manifestoes against the doctrine of progress appeared in 1835 with the publication of Théophile Gautier's romantic novel, *Mademoiselle de Maupin.* In a substantial preface Gautier ridiculed the idea that literature had to be "useful." A novel was not a pair of boots, nor a sonnet an automatic syringe, nor a drama a railway. The most useful place in a man's house was, in fact, his water closet, but beauty belonged to another realm altogether, and nothing beautiful could be of the slightest "use" whatsoever. As for progress, only the ravings of a Fourier could be counted as an adequate guide to true progress, and they suffered from the inconvenient defect of being impossible to translate into reality. All the blessings of modern civilization were as tinsel and rubbish by contrast with the accomplishments of antiquity. The ancients, Gautier quipped, had "three or four thousand gods in whom they believed, and we have only one, in whom we scarcely believe at all. This is progress of a strange sort. Is not Jupiter much greater than Don Juan, and is he not a better seducer? In truth, I do not know what we have invented, or even perfected." [9]

Gautier was also strongly drawn to the decadence of imperial

Rome, as witnessed by the tastes of the Chevalier d'Albert in *Mademoiselle de Maupin*. The sense of decadence among many of the later romantics was heightened by a morbid attraction to "evil," so powerful that not a few felt compelled to seek refuge in the bosom of the Church. Whatever personal solution they sought for their neuroses, they were nearly all ardent disbelievers in progress. Charles Baudelaire, Gautier's successor as the high priest of the aesthetic revolt in French letters, excelled him in his disdain for the modern world, although he never renounced it, since he found artistic inspiration in its very decadence. But the belief in progress was "a most fashionable error, from which I want to guard myself as I would from hell . . . a grotesque idea which has flowered in the rotten soil of modern self-conceit." [10]

The decadent movement reached its culminating point in J.-K. Huysmans's *Against the Grain,* published in 1884. Its hero, the duc Des Esseintes, having exhausted himself in a round of solitary and super-refined pleasures, returned at the end of the novel to Paris, lamenting the triumph in his century of the despised bourgeoisie, who had transported "the vast, foul bagnio of America" to Europe. " 'Well, crumble then, society! perish, old world!' cried Des Esseintes. . . . He could deceive himself no more, there was nothing, nothing left for it, everything was over." Perhaps the world would be destroyed in cataclysms of fire, as in biblical times, or perhaps it would drown in the flood of its own filth. In any event, the consolations of pessimism gave him no peace, and the duc called upon God to "take pity on the Christian who doubts, on the sceptic who would fain believe, on the galley-slave of life who puts out to sea alone, in the darkness of night, beneath a firmament illumined no longer by the consoling beacon-fires of the ancient hope." [11] Much the same apocalyptic vision appeared in Elémir Bourges's novel *The Twilight of the Gods,* also published in 1884.*

---

* Without sharing in the decadent movement, the Catholic apocalypticism of Léon Bloy and the *mélange* of eschatological nationalism, socialism, and Christianity in the thought of Charles Péguy further contributed to the sense of cultural disaster in France just before the First World War. For a sharply critical study of Péguy, emphasizing the anti-progressivist despair and proto-fascism of his later years, see Hans A. Schmitt, *Charles Péguy: The Decline of an Idealist* (Baton Rouge, La.), 1967. A good introduction to Péguy as a prophet of "hope and joy" is Emmanuel Mounier, "La Vision des hommes et du monde," in Mounier, Marcel Péguy, and Georges Izard, *La Pensée de Charles Péguy* (Paris, 1931), especially pp. 143–208.

Romantic despair made fewer conquests in the German and Anglo-American literary worlds, but similar tendencies are readily discernible in both. Under the influence of the French decadents, notably Mallarmé, Stefan George brought the gospel of aestheticism to Germany in the 1890s. He rejected the crassness of Wilhelmian bourgeois society and directed his followers back to the purer light of Hellenic antiquity and the Middle Ages. Shortly after the turn of the century, however, George and his circle abandoned their earlier cosmopolitan cultural pessimism in favor of a mystical and messianic German nationalism not unlike Péguy's vision of the divinely appointed destiny of France. Another gifted poet of the decadent movement, the Austrian Hugo von Hofmannsthal, after a similar career in the 1890s, dedicated to melancholy and thoughts of death, became a respectable family man, and wrote libretti for Richard Strauss. At the same time, although he parted company with the decadents, Hofmannsthal remained a man whose deepest loyalties lay in the past, rather than in the future. In his mature years he took up the cry for a "conservative revolution" to save a warring, mobocratic world from chaos and to restore the Europe of the Hapsburgs.[12] *Buddenbrooks,* Thomas Mann's early masterpiece, can also be read as an exercise in *fin-de-siècle* pessimism, with its relentless narrative of the decline and fall of a great North German merchant family, and its portrait of the arrogant Thomas Budden-brook, who sought consolation for his failures in the maxims of Schopenhauer.

The breach between the romantic artist and the progressive world of commerce and industry appeared early in the nineteenth century in Britain. Attention has already been called to the medievalist utopism of Coleridge and Carlyle. A wild, despairing sense of estrangement and meaninglessness was powerfully developed in the verse of Lord Byron. Aestheticism emerged in mid-century at Oxford in the thought of Walter Pater, and passed on from him to Oscar Wilde, Aubrey Beardsley, and Ernest Dowson, all men of the 1890s who were dead by 1900. The decadent movement in English letters lacked the lugubrious quality of Continental decadence, except perhaps for the poetry of Dowson, but it clearly belonged to the same artistic climate. Nostalgia, melancholy, and cultural despair, not unmixed with hope, found more profound utterance in the prose and poetry of a Victorian romantic of the older generation, Matthew Arnold. In the United States, too, some of the most vivid expressions of romantic anxiety and horror appeared in mid-century, in the work of Edgar Allan Poe, who influenced Baudelaire, and in the novels of Herman Melville.

Philosophy and historical scholarship in the romantic spirit also contributed to despair, especially in central Europe. Indeed, from the contemporary point of view, the richest vein of nineteenth-century romantic pessimism can be found in the writings of Schopenhauer, Kierkegaard, Burckhardt, Nietzsche, and (to include one figure of the early twentieth century) Spengler, all philosophers or philosophically inclined historians, who preached a despair as deep as any expounded by more recent writers, and whose impact upon thought has generally been greater in our century than in the nineteenth.

Of the five, only Arthur Schopenhauer aroused strong interest in the nineteenth century itself, and even he had to wait until the last decade of his life for recognition. His fame continued to grow after his death in 1860; he became the symbol, despised or revered, of all the forces in European culture that turned away from modern life and civilization, much as Machiavelli had served for centuries as the archetype of the immoralist or Spinoza of the atheist. Today he can be numbered among the founders of existentialism. He was one of the first modern thinkers to reject absolutely the analogy between the human mind as a creator of value and the natural and social world outside the mind. To follow Morse Peckham's argument, he "sundered order and meaning from value," by positing a universe ruled by blind will, in which all passions were useless, all sensory gratifications empty, all striving fruitless.[13]

Schopenhauer did not deny that the world had its own order and meaning, and he defined the world in romantic terms as the activity of a cosmic will, below and beyond reason. At the same time he denied human value to this cosmic order. The more man strove to fulfill himself within it, the more miserable he became. History, therefore, appeared to Schopenhauer as a tragedy that repeated itself endlessly. "The motto of history in general should run: *Eadem, sed aliter*."[14] The same things happened over and over again, albeit in different ways. To know one page of history was to know them all. The route to salvation from life's unhappiness could not be found by charting a course for the Middle Ages or Athens or the New Reich: the self could be saved only from within, by escape into the inner world of art and, still further, into the emptiness of Nirvana, the absolute negation of will taught by the sages of India.

In the words of an American admirer, written twenty-five years after his death, Schopenhauer had been "the first to detect and logically explain that universal nausea which, circulating from one end of Europe to the other, presents those symptoms of melancholy and disillusion which, patent to every observer, are indubitably born of the insufficiencies of modern civilization." [15] But few readers of Schopenhauer were able to follow him all the way. His most brilliant posthumous disciple, Eduard von Hartmann, attempted in his *Philosophy of the Unconscious* (1869) to give Schopenhauer's thought the historical dimension that the master's chronophobia had compelled him to reject. It was as if Hegel and Schopenhauer, thesis and antithesis, had finally been resolved in a higher synthesis after both were dead.

For Hartmann, the cosmic will and the spirit of man possessed a common destiny after all. Man was morally obliged to live out history, despite its pains and sorrows, to the bittersweet end; he who chose for himself the path of mystic world denial only delayed the inevitable. The end, however, was not the millennium, or the victory of mankind in any ordinary sense. Hartmann proposed that the end of the world would arrive when, with the progress of consciousness and pessimistic philosophy, all men had become linked in a common loathing of life and would use their collective racial will, stronger in its unity than nature's will, to annihilate the entire fabric of being. The aristocratic nihilism of Schopenhauer yielded to democratic nihilism, somewhat as Nietzsche's elitist philosophy of the Superman was transformed by Shaw into a doctrine of universal superhumanity.[16]

Hartmann's debt to the philosophy of Hegel is clear, despite the chasm that separates their world-views. No such debt was contracted by Schopenhauer's younger Danish contemporary, Sören Kierkegaard, who devoted his career as a philosopher to the demolition of Hegelianism. Hegel had attempted to derive vital existence from reason; Kierkegaard sought to recall man to his existence through the concrete experience of self-encounter. Man's existence was, moreover, personal rather than collective, and characterized not by robot obedience to collective or supra-personal forces, but by acts of choice in specific situations. The only serious questions, then, were either/or questions, answered by individuals in their day-to-day existence. For Kierkegaard, as a Christian, all such questions in turn sprang ultimately from the overwhelming fact of man's estrangement, as a free and finite being, from God; the life of religion consisted of the effort, always perilous and difficult, to bridge the rationally unbridgeable gulf between the two.

It followed that Kierkegaard's ethico-religious thought fastened obsessively on such themes as anxiety, doubt, suffering, sickness, and death; and that, as a social critic, he savagely denounced the attempts of liberals, Hegelians, and other distinctively "modern" minds to save mankind through the gradual world-historical "improvement" of civilization and the subordination of the individual to the state, the greatest happiness of the greatest number, or the demands of social and industrial "progress." As he complained in his *Concluding Unscientific Postscript* (1846), to allow oneself to be directed, in the Hegelian manner, by the alleged great forces in world history was a "topsy-turvy notion." The reconstruction of past lives was possible only under the condition that one learned by living how men lived. "But it is certainly a topsy-turvy notion . . . to go and try to learn from the dead, apprehended as if they had never lived, how one should . . . live—as if one were already dead." [17] Yet even at its best, the study of history was a luxury; history, like nature, had nothing serious to tell man. Because the nineteenth century could not accept this proposition, it could not accept Kierkegaard.

Much of what Kierkegaard had proclaimed intuitively the Swiss historian Jacob Burckhardt discovered in his own way, with the help of the careful study of history itself, following and enlarging upon the latest methods of the Rankean school of historiography. In his own day he became well known as a specialist in Renaissance studies. Between 1868 and 1871 he delivered several lectures in his native Basel on the meaning of history (they were not published until 1905, eight years after his death), in which he attacked the Hegelian notion that man's reason could construct a model of history, true for all time. Only a divine being could see its whole structure and purpose, and of gods, Burckhardt, unlike Kierkegaard, refused to speak. To the historian, history revealed itself as a flux, or continuum, and nothing more. Of course the actors in history had other ideas; they felt that they were living in ages of good fortune or bad, of progress or decline; and they read into history various linear and cyclical schemes of development, all demonstrably relative to the mental climate in which they happened to live. In the nineteenth century, the doctrine of progress prevailed, as in earlier times cyclical theories had held the center of the stage. But progressivism, together with the rationalism from which it had emerged, were "the deadly enemies of true historical insight." [18] Instead of trying to understand the past, the progressivist saw it only as the scaffolding for modern civilization and an opportunity for self-congratulation.

The relativistic tendencies implicit in historicism, which will be discussed at greater length in Chapter 10, were not, however, the only forces at work in Burckhardt's mind. Beyond the relative, he identified certain transcendent universal values and certain universal characteristics of the human condition that imparted to the restless flow of temporal events a meaning that any Greek or Christian philosopher might also have found in history. He assumed, above all, the unchangeableness of human nature. Men were born with a capacity for both good and evil, which disclosed itself in time of peace as well as in time of war. Struggle, suffering, and evil played just as inevitable a part in the economy of world history as goodness and happiness, since all change in history, all growth and achievement, stemmed from man's perpetual discontent with the world as it is, and his willingness to fight to improve his lot. Burckhardt denied even the theoretical possibility of utopia; he could conceive of nothing more horrible than the hypocrisy of a world in which all men acted well, although their hearts remained black with evil intentions.[19]

He was also quite unable to confine himself to a value-free relativism when he came to consider the state of Europe in his own century. As Kierkegaard had warned, the leveling tendencies of modern democracy, commercialism, and rationalism threatened an end to all culture, all faith, all self-discipline. It was the age of the herd. The churches had declined to the status of purveyors of a tawdry optimism. All things and all men were for sale. Burckhardt foresaw the likely further degeneration of the Western countries into omnipotent military-industrial states engaging in great national wars as a way of keeping the masses in order.

Bismarck's Prussia and Louis Napoleon's France and their historic clash in 1870–71 suggested the shape of things to come. Although "good and evil, perhaps even fortune and misfortune, may have kept a roughly even balance throughout all the various epochs and cultures," what passed for moral progress in the modern era might better be described as moral decay. Individuality had been "domesticated," partly as a result of "the vast increase in the power of the State over the individual, which may even lead to the complete abdication of the individual, more especially where money-making predominates to the exclusion of everything else, ultimately absorbing all initiative." Burckhardt condemned nostalgia for the Middle Ages, and yet he had to confess that whereas modern lives were dissipated in "business," the lives of medieval men had been taken up in "living." Nor could modern man afford to boast of intellectual progress, "since, as civilization advanced, the division of labor may have steadily narrowed the consciousness of the individual." [20] What did the accumulation of knowledge profit mankind, if each individual became steadily more specialized in his learning and more ignorant of the world as a whole? Clearly, Burckhardt regarded modern civilization as having suffered a net loss in the values he held most precious, and further losses were predictably in store. The historian who, in another place, had warned equally against "senseless despair" and "fatuous hope," [21] became the prophet of something very near despair when he confronted his own times, deploring the materialism of an age that spurned its heritage and converted its men into insects.

Schopenhauer, Kierkegaard, and Burckhardt all figure, then, as romantic rebels against their civilization, against reason, science, and the life of technique and organization, on behalf of time-transcending systems of value. Schopenhauer offered salvation through an Occidentalized Buddhism, Kierkegaard through a proto-existentialist Christianity; Burckhardt's credo might be defined as a Stoicism colored by *Historismus.*

Elements of the thought of all three recurred in the work of Friedrich Nietzsche. Schopenhauer gave Nietzsche his conception of life as will. He came to a kind of Kierkegaardian existentialism independently of Kierkegaard. Burckhardt, an older colleague at the University of Basel for a decade, contributed to his historical education and encouraged his sceptical proclivities. He shared the negative attitude of all three toward modern civilization, and their rejection of the Enlightenment doctrine of progress. He saw the immediate future of the West in much the same mournful terms as did Kierkegaard and Burckhardt, and with them he has done much to stimulate twentieth-century historical pessimism. But Nietzsche was also, as we have argued above,* a great yea-sayer in the Faustian spirit. Although any formal analysis of his thought must lead to the exclusion of his name from the believers in historical progress, if only because of the great importance he attached to his theory of the "eternal recurrence," in his own way he remained always the meliorist, challenging men to self-transcendence and proclaiming the heroic coming age of the *Übermensch*.

Of another order entirely was the philosophy of Oswald Spengler, whose first book, *The Decline of the West,* did more than any other literary production of our century to awaken in contemporary Western minds a sense of the total and irrevocable failure of their civilization. Not that Spengler was ignorant of Nietzsche's thought. On the contrary, he knew it well and acknowledged its decisive influence, along with Goethe's, on his own. His affinities with the main tendencies of German romanticism are obvious.[22] But whereas Nietzsche extracted joy and hope from his insight into life as the will to power, Spengler's discovery that cultures rise and fall according to a cyclical rhythm imposed on them by an organic destiny beyond their power to evade gave its discoverer little pleasure. His idea of "destiny" is much like Schopenhauer's "will": inscrutable, fatal, fraught with suffering and woe. Unlike Schopenhauer, Spengler urged men to a life of action, but it was not Nietzsche's life of creative self-transcendence. For Spengler individual wills were powerless to change the predestined course of history. If a man were wise, he would seek only to bring his private will into harmony with the larger will of destiny—back to Hegel!—and find fulfillment in conforming to the spirit of his age.

There is no need to recapitulate here Spengler's elaborate cyclical theory of history.[23] Suffice it to say that the "Faustian" culture of the West had in Spengler's judgment reached its winter season. His descriptions of life in the earlier stages of cultural history, contrasted with his view of the final stage and the future of the West, leave no doubt where his sympathies lay. All the familiar clichés of nineteenth-century cultural despair recur: life was once simple, heroic, and full of promise; as time unrolled, each culture grew and ripened, realizing all its unique potentialities for self-fulfillment; at the end, nothing remained but the barren life of technology and trade, and the new barbarism of imperial conquest, as blood triumphed over gold, and the homogenized mass of humanity waited

* See above, pp. 57–62.

to fall under the rule of a single world dictator. Spengler urged his readers to adjust their aspirations to "the hard cold facts of a *late* life, to which the parallel is to be found not in Pericles's Athens but in Caesar's Rome." Western culture had exhausted its possibilities in art and thought; young men of the twentieth century were admonished to devote themselves "to technics instead of lyrics, the sea instead of the paint-brush, and politics instead of epistemology." In any case, "a task that historic necessity has set will be accomplished with the individual or against him. *Ducunt Fata volentem, nolentem trahunt.*" [24]

With Spengler nineteenth-century romantic pessimism reached its culmination. He himself claimed not to be a pessimist in the truest sense, since even in "late life" a civilization had much work for its people to do; after the publication of *The Decline of the West* in 1918–22, he spent the rest of his life trying to persuade his native Germany to set its house in order and play the leading role in the coming world empire. But the service done by his *chef-d'oeuvre* was the same in all parts of the Western intellectual world: the transmission to the twentieth-century mind of the most fatalistic and despairing intuitions of nineteenth-century romanticism.

Although it could be argued that every thinker active in the nineteenth century fell under the sirenic influence of romanticism in one way or another, some varieties of pessimism in the century proceed more from positivistic assumptions than from romantic ones. Most of Spengler's immediate forerunners as exponents of a cyclical theory of history fit this description. None had his success with the general public, and none in all likelihood contributed to the shaping of Spengler's mind, but each is interesting in its own right. In Russia, for example, there was Nikolai Danilevsky, who earned a degree in botany at the University of St. Petersburg in 1849 and later became a government expert on fisheries. His books include studies of Darwinism, economics, and linguistics; *Russia and Europe,* which contains his theory of history, was first published in 1869. Contrary to the views of the "Westernizing" party among the Russian intelligentsia, Danilevsky contended that the Slavic peoples had created a civilization of their own, quite distinct from the "Germano-Romanic" civilization of the West and some five hundred years younger. Holding that all civilizations pass through four periods from birth to death, he found that Europe had already entered its fourth epoch, the age of decadence, whereas the Slavic civilization led by Russia had only just reached its third epoch, the age of cultural maturation and florescence. For the next several centuries the Slavs could expect to exercise world leadership, before in due course they also declined and fell by the wayside. In the nineteenth-century Russian context, *Russia and Europe* should be described as an optimistic book, larded with messianic nationalism, but it was also a refutation of the doctrine of progress.[25]

Although Danilevsky's book sold slowly at first in Russia, two editions published in the late 1880s did well enough to warrant the appearance of a French translation in Paris in 1890, through which his ideas reached an appreciable number of readers in Western Europe. Five years later the first edition of another contribution to cyclicalism appeared in London, *The Law of Civilization and Decay,* by the American lawyer and gentleman historian, Brooks Adams. The American edition followed in 1896; a French edition, considerably augmented, in 1899; and a German one in 1907. Adams interpreted history primarily from the point of view of economics; the immediate stimulus for writing *The Law of Civilization and Decay* came from the business panic in the United States in 1893 (Spengler found similar inspiration in the second Moroccan crisis of 1911). He was also attracted, like his older brother, the Harvard historian Henry Adams, to the possibility of a correlation between physics and history. As he wrote in the preface to the American edition, his thought was "based upon the accepted scientific principle that the law of force and energy is of universal application in nature, and that animal life is one of the outlets through which solar energy is dissipated." [26] Since the activities of man represented one type of animal energy, the rise and fall of human societies had, therefore, to conform to the laws governing energy in the cosmos.

Adams proposed that every society oscillated between barbarism and civilization, or, what amounted to the same thing, between the dispersion of energy and its concentration. In the early phase of civilizational growth the dominant motive was fear, which led to religion and militarism. Wealth was accumulated and stored, but "however large may be the store of energy accumulated by conquest, a race must, sooner or later, reach the limits of its martial energy, when it must enter on the phase of economic competition." [27] The possessors of the wealth so acquired became the controlling force in society, and greed replaced fear as the dominant motive in life, leading to exploitation, waste, and the dissipation of energy, and at length to the disintegration of the exhausted society. What the man of fear produced, the man of greed spent, and in the end humanity found itself once more at the starting point, with no energy left but that of brain and muscle.

Applying his theory to modern Western civilization, Adams discovered that it had reached the same stage attained by late imperial Rome, when the life of faith, imagination, and heroism was no longer possible, and all men were subject to the rule of greed. "As consolidation apparently nears its climax, art seems to presage approaching disintegration. The architecture, the sculpture, and the coinage of London at the close of the nineteenth century, when compared with those of the Paris of Saint Louis, recall the Rome of Caracalla as contrasted with the Athens of Pericles." [28] The doom of the West stood plainly revealed, and Adams could not hold out hope for a miracle as the shadows inexorably fell. Years later he recalled that in early middle life he had learned, "as a lawyer and a student of history and of economics, to look on man, in the light of the evidence of unnumbered centuries, as a pure automaton, who is moved along the paths of least resistance by forces over which he has no control." As Calvin had shown, fallen man was fatally disposed at any cost to pursue his own selfish interests. "Christ taught that we should love our enemies. To compete successfully the flesh decrees that we must kill them. And the flesh prevails." [29] The flesh, then, would insist on dragging the masses and the plutocrats of the new Rome down to disaster; they could not help themselves.

Despite the physicalist imagery, and the strong admixture of New England Calvinism, Brooks Adams' philosophy of history clearly anticipates Spengler's in its sense of aristocratic despair over the failure of modern democracy and industry to build a just, harmonious, stable social order. Both men grew to manhood in countries that had just entered the industrial age, joining the fray late but with great success, and both were sickened by the rapacious materialism of their times. The real force of both arguments, Spengler's in *The Decline of the West* and Adams' in *The Law of Civilization and Decay,* lay in their exploitation of the analogy between the decline and fall of Rome and the alleged failure of the modern Western world. After the publication of the books that brought them fame, both men also prophesied that their own nation was destined to rule the world in its final years unless the colored races made good their threat to overwhelm white civilization—Spengler in such books as *The Hour of Decision* (1933) and Adams in *The New Empire* (1902).

For some late nineteenth-century French observers, the imagination of disaster was kindled by the defeat of France at the hands of Prussia in 1870–71 and by the subsequent decay of her influence in world politics, as well as by her falling birth rate and her steadily worsening position in world industrial competition. One little-known set of four volumes by a Belgian military officer, Ernest Millard, published between 1903 and 1908 under the title, *An Historical Law,* appears to have been—at least in part—a response of just this sort to the fact of French decline in the nineteenth century.

Less insightful than Danilevsky or Adams, Millard's work is considerably bulkier, rivaling even Spengler in scope and length. He explained the destiny of peoples in purely physicalist terms, as the effect of "magnetic currents" set in motion by solar phenomena. Every nation passed through five epochs over and over again, each of approximately 250 years in length, which he labeled ages of formation (or reorganization), growth, sickness, splendor (*grand éclat*), and decadence. The Chinese, for example, had run through this cycle no fewer than five times since the year 3468 B.C., and in 1894 had entered the last epoch of their sixth cycle. The French nation, including its predecessors in ancient Gaul, had experienced three complete cycles, and entered the "decadent" stage of the fourth in 1870. The English had reached their third age of "splendor" in 1815, and could expect to remain in it for at least another century, but meanwhile the Germans were rapidly approaching their fourth age of "sickness," after more than two hundred years of steady "growth" beginning in 1657. The difference between an age of "splendor" and an age of "decadence" was of special interest to Millard, as a military man living in a country that belonged to the French-speaking world. "The great law of history will teach us," he wrote, with obvious bitterness, "that love of country and the instinct of war are paroxysmal among peoples arrived at their apogee, while they count for nothing among peoples in a state of decadence. And it is quite simply as the result of the decadence into which the leading people of our time (the French) have fallen, that the ideas of cosmopolitanism and universal pacification have won numerous supporters today." [30]

Not every positivist opponent of progressivism in the nineteenth century chose, of course, to convey his fears in the form of a cyclical theory of history. More representative was the thinker for whom nature herself became, in a sense, the enemy of human aspiration, the barrier to human perfection. On the whole, nineteenth-century minds were able to cope with new scientific theories as they appeared, and often to see in them further confirmation of the gospel of progress, but even when followed with devotion, science and reason did not inevitably lead to a world-view that nourished hope.[31] The disenchanted rationalist is a familiar figure as early as the last part of the eighteenth century.

Thomas Malthus, for example, published his essay on population, which challenged the meliorism of Adam Smith, in 1798. But the archetypal prophet of naturalistic despair was his contemporary the Marquis de Sade, whose novels—drawing on premises not unlike those of Malthus—disclosed a universe satanically hostile to the moral ideals of the Enlightenment. Sade's anti-heroes, for all their success as libertines and criminals, repeatedly confessed their hatred of the natural order, which had ordained perpetual war, famine, pestilence, and crime in order to maintain its grim balance. Their rule of life—to enjoy oneself at no matter whose expense—was a transparent counsel of despair, in a world bereft of all purpose or goodness.

To be sure, the reader must look closely. On the surface, lust carried the day, and procured for its devotees endless pleasures. But Sade never went so far as to redefine good in terms of evil. Although he recommended the evil life, it remained always evil, and, in the final analysis, bitter. A representative text is the great discourse on theology by the statesman Saint-Fond in *Juliette* (1797). Saint-Fond urged upon his atheist colleagues in crime the need to believe in God, but a God who was the abhorrent master of evil, who governed the universe according to the principles of eternal and universal evil, and who could best be served by a life dedicated to crime. The statesman's friends refused, however, to grant him even this much. As one of them pointed out, it was better not to believe in God at all "than forge one in order to hate it." [32] If Sade had been able to effect a Nietzschean transvaluation of all values, his spokesmen would have had no cause to hate the supreme author of evil, whether mythical or not, but the truth was that he found the death of God and the savagery of nature desolating. In the end, man stood defenseless, a rational being contemplating the total depravity of the universe and driven by his very rationality to join in the universal slaughter. "This most sublime life of men is to nature of no greater importance than that of an oyster, and she has abandoned us all equally." [33]

Sade's eccentric image of nature won little acceptance in the first half of the nineteenth century, but in the second half, in the form of Darwin's theory of natural selection, it haunted thousands of sensitive thinking people, including many of those who were most prepared to accept it intellectually. We have already discussed the case of T. H. Huxley, who salvaged a belief in progress only at the price of urging his fellow men to disobey the laws of nature and carry on an eternal struggle against the "cosmic process." * Others could not salvage it at all, especially when they took into consideration the latest teachings of physics and astronomy, which conjured up the vision of a cosmos inconceivably vast and inaccessible, gradually dissipating its heat in accordance with the second law of thermodynamics, and threatening, in the case of our own solar system, to make all life on earth impossible in the geologically near future because of the cooling of the sun.[34]

Several writers inclined to scientific pessimism in the late nineteenth century painted lurid prophetic pictures of the "last men on earth." Huxley's student H. G. Wells supplied one of the best in his first scientific romance, *The Time Machine* (1895), taking advantage of his time traveler's remarkable vehicle to visit a decadent society of the remote future and then, still further along in time, to witness the extinction of all terrestrial life under the baleful eye of a cold, blood-red sun. The traveler "thought but cheerlessly of the Advancement of Mankind, and saw in the growing pile of civilization only a foolish heaping that must inevitably fall back upon and destroy its makers in the end." [35] Wells's early work also included *The War of the Worlds* (1898), which depicted the near annihilation of mankind by invading Martians, and *When the Sleeper Wakes* (1899), a counter-utopian novel of the twenty-second century that foreshadowed *Brave New World*.

Through most of the twentieth century Wells repressed his earlier

* See above, pp. 47–49. We have also noted the thought of the sociologist Ludwig Gumplowicz, who found the belief in progress untenable for strictly scientific reasons. See above, p. 42.

tendency toward pessimism, becoming an evangelist of progress through science and socialism,* but in his old age, during the Second World War, despair returned to claim him once more. His wartime writings warned almost monotonously of the possibility that man would never learn the lessons of evolution and would be struck down by nature's impartial hand. In his last book, *Mind at the End of Its Tether* (1945), the Huxleyan cosmic process made its appearance as "The Antagonist," an unknowable alien presence in the universe, which had for a time tolerated human life, "and has now turned against it so implacably to wipe it out." Man had imagined that he lived in a comprehensible and reasonable cosmos, which gave him a fair chance to win his struggle. But man had been proved wrong. Civilization was not merely bankrupt: "There remains no dividend at all; it has not simply liquidated; it is going clean out of existence, leaving not a wrack behind. The attempt to trace a pattern of any sort is absolutely futile." [36] What Gilbert Murray once called "the Friend behind phenomena," whether conceived as a deity or as a benevolent natural order, had become transformed for Wells into a hostile force, something like the anti-God of Sade's Saint-Fond, bent on destruction rather than progress, and contemptuous of man's hopes.

A few months before the first appearance of *The Time Machine,* Anatole France published one of his most characteristic books, *The Garden of Epicurus* (1894), which expounded its author's pessimism in a series of sceptical parables, including one that offered a vision of the end of the world. A rationalist, a moderate socialist, and an enemy of obscurantism much like Wells, France nonetheless found life hard, a constant struggle against adverse conditions. The whole universe was "one vast gehenna, where animal life is born only to suffer and to die." Although some progress might be traced in the history of man's civilization, it occurred very slowly, and did not make men less inclined to folly and wickedness by nature. In the end, the earth would become too cold and dry to support life. "The last inhabitants of earth will be as destitute and ignorant, as feeble and dull-witted, as the first. They will have forgotten all the arts and all the sciences. They will huddle wretchedly in caves alongside the glaciers that will then roll their transparent masses over the half-obliterated ruins of the cities where now men think and love, suffer and hope." Their successors as masters of the world might well

* See above pp. 49–50.

prove to be the social insects. "Who knows," France asked, "if in their time and season they too may not praise God?"[37] Like Wells, he had his periods of optimism and confidence, and on the whole he might even be included among the believers in progress, but he was too much the sceptic and the Epicurean to embrace the progressivist gospel with anything like warmth.[38]

Henry Adams, inspired by the efforts of his brother Brooks to refute the doctrine of progress, surpassed the most despondent of the European positivists in 1909 with an essay on "The Rule of Phase Applied to History," which proposed a law of natural acceleration based on physics that predicted the end of history in the year 1921, or, at the latest, 2025.[39] No doubt Henry was not entirely serious, but his disenchantment with the idea of progress is indisputable. Another essay, dating from 1910, matched the "Evolutionist" against the "Degradationist," the believer in Darwin against the believer in Clausius. The former rejoiced in progress, the latter "proclaimed the steady and fated enfeeblement and extinction of all nature's energies." The Darwinists had done well for a time, but now the theory of degradation prevailed in every scholarly field except history, and Adams, the historian, urged its triumph there.[40]

Pessimism originating in a basically naturalistic world-view also informed the novels of Thomas Hardy, above all *Jude the Obscure* (1895), and of Mark Twain, whose thoughts on the "meaninglessness" of history are analyzed in a recent monograph by Roger B. Salomon.[41] But we should not bring our discussion of nineteenth-century despair to a close without citing one broad theme in the political literature of *la belle époque:* the theme of racial failure. Fears of the "passing" or the "decay" of the nobler races, often coupled with warnings of the yellow or the colored "peril," filled many late nineteenth-century minds at the very point in history when white world hegemony had become almost complete.

As early as the 1850s, Count Arthur de Gobineau doubted that France could continue to exist for more than thirty years because of the declining influence in modern times of Teutonic blood, relative to Celtic and Mediterranean, in her national life. In France at the end of the century, Joséphin Péladan devoted a cycle of fifteen novels, *The Latin Decadence*, to the idea that the "Latin" race was finished; and the anthropologist Georges Vacher de Lapouge warned that natural selection had ceased to operate in man's history, with dysgenic effects that would result in the eventual disappearance of the Aryan master-race from the world if measures were not taken to halt its decline. Madison Grant's *The Passing of the Great Race* similarly sang the praises of Nordic man and advised Americans that their "altruistic ideals" and "maudlin sentimentalism" were in danger of "sweeping the nation toward a racial abyss. If the Melting Pot is allowed to boil without control, and we continue to follow our national motto and deliberately blind ourselves to all 'distinctions of race, creed, or color,' the type of native American of Colonial descent will become as extinct as the Athenian of the age of Pericles, and the Viking of the days of Rollo." [42]

The most perceptive British prophet of racial failure, and one of the most astute prognosticians of his generation anywhere in the Western world, was the historian and educator Charles H. Pearson. His *National Life and Character: A Forecast* (1893) must take its place with the works of Burckhardt, Spengler, and Brooks Adams as a prescient, although not infallible, guide to twentieth-century life. Drawing on his experience as minister of education in Victoria, Australia, Pearson looked forward to the expansion of the power of the state in the coming century. As the frontiers closed in America and Australasia, and populations steadily grew, the national societies of the future would turn increasingly to some form of state socialism. The great age of individual enterprise would yield to an age of collectivism, security, equality, urbanization, and militarism. The white race, Pearson affirmed, had reached its peak of expansion and would lose much ground in years to come. The "lower" races could be expected to learn the arts of white civilization and establish powerful independent states in their turn. In a hundred years they would outnumber the white peoples by three to one; their willingness to work harder with less personal comfort would give them a great competitive advantage over the white man on the world market, which they would not fail to exploit. But with no new worlds to conquer, all men would feel a sense of depression and hopelessness, quite alien to the spirit of adventure that had driven the white race out upon the oceans of the globe in modern times.

Although Pearson predicted that national life in the future would be more stable and democratic than in the nineteenth century, it would not be a progressive life. Science had already done "her greatest and most suggestive work. There is nothing now left for her but to fill in details." Literature would be supplanted by journalism. The world, despite its prosperity, would be "left without deep convictions or enthusiasms." Wars would continue to be fought, waged by vast standing armies administered by the apparatus of the all-powerful state. Every passing year brought closer the time "when the lower races will predominate in the world, when the higher races will lose their noblest elements, when we shall ask nothing from the day but to live, nor from the future but that we may not deteriorate." Still, life would continue. "Simply to do our work in life, and to abide the issue, if we stand erect before the eternal calm as cheerfully as our fathers faced the eternal unrest, may be nobler training for our souls than the faith in progress." [43]

The nineteenth-century gospel of progress, then, did not lack criticism in its own time. The last word belongs to the great realist, sceptic, and anti-romantic Gustave Flaubert. The battle between the advocates and enemies of progressivism wearied him, as all of man's works and thoughts tended to weary him. For his unfinished satirical novel, *Bouvard and Pécuchet,* he left in his notes the outline of a debate that his two bourgeois anti-heroes were to have conducted on the subject of the future. Pécuchet would see it "in dark colours." Modern man had been "whittled down and become a machine." Wars would rage, religion and morality would perish, vulgarity would become universal, and the world would finally end "through cessation of heat." Bouvard, on the other hand, was scheduled to see the future "in a cheerful light." East and West would be fused in a higher civilization. Science would make the world pleasant, and with the disappearance of want would come the disappearance of evil. "People will visit other earths—and when this globe is used up, Humanity will migrate to the stars." [44] The debate was never written, and Flaubert died in 1880, but from the rest of the novel, it is not difficult to imagine how it would have progressed, and how little importance Flaubert would have wished us to attach to it. Monkeys chatter; thus Bouvard and Pécuchet; and thus humanity.

IO

The Relativity of Values

"Among all the labels that have been applied to the present moment in history," writes Charles Frankel, "none, I think, has been more generally accepted than the phrase 'The Age of Anxiety'." [1] Frankel even suspects that many twentieth-century minds enjoy their anxiety, on the same principle that the masochist finds delight in pain. Be this as it may, our *Angst* manifestly has something in common with that of the preceding century, and like the nineteenth-century variety, it often culminates in a radical historical pessimism.

In the next chapter contemporary modes of disbelief in progress that originate in anxiety and despair will receive our close attention. But there is another variety of disenchantment abroad in our century that deserves treatment first—if only because it is so deeply rooted in the thought of the nineteenth century. I refer to the relativistic spirit and methodology of modern scholarship. Academicians have given little attention to relativism except as it affects their own particular disciplines, but there is perhaps no more universal phenomenon in the intellectual life of the present age. Its share in modern disbelief in the idea of progress is incalculable, if only because the twentieth century is the era *par excellence* of the academic specialist. The professionalization of disciplines that began in the middle decades of the nineteenth century has now been carried to its ultimate limits, the universities have grown into colossal structures disposing of vast funds, and practically all the intellectual activity now to be observed in Western civilization takes place in the minds of professional academicians, or in the minds of their students, past and present.

Academic relativism has many origins and expresses itself in various ways. One could argue that relativity with respect to values is well-nigh unavoidable once a high degree of professionalization has been attained in any academic field. The teachers and professors of the nineteenth century, for all their incipient professionalism, often had less training, fewer academic guidelines to follow, and more freedom to cross disciplinary boundaries than their counterparts have today. On the whole, editors and publishers of scholarly work were less demanding. But highly professionalized scientific and scholarly work, by its very nature, tends to avoid wholesale generalizations based on judgments of value. It is dispassionate, narrow, and detached. Even nineteenth-century scholarship was substantially more objective than the scholarship of any preceding century. Today the purely spectatorial attitude has become inevitable for nearly all productive scholars.

The professional is also the specialist, and intensive specialization in itself produces a contraction or distortion of perspective that may lead to relativism. The twentieth-century scholar is the learned barbarian of Ortega y Gasset's trenchant analysis, who knows a great deal about very little, and consequently falls into one of two practices. Either he recognizes his ignorance of the world outside his special field of competence, and declares himself incapable of evaluating the world; or he acquires the habit of seeing all things in terms of his own discipline, supposing man (for example) to be "purely" biological, or historical, or psychological, or political, and therefore beyond the reach of normative judgment.

But it is not the spectatorial attitude or specialization alone that accounts for academic relativism. As the scholarly disciplines have grown and expanded, they have added greatly to Western man's knowledge of pre- and non-Western cultures; the more we know of them the more difficult it becomes to sustain the notion that Western culture provides the standards by which all the others should automatically be measured. Physics has advanced beyond the frontiers of the Newtonian universe to formulate a theory of cosmic relativity; and psychology has arrived at its own variety of relativism by explaining behavior in terms of subconscious impulses and traumatic childhood experiences, and by proposing typologies of personality, which attack the Christian, the Enlightened, and the romantic concepts of the ideal man.

Nor can one ignore the part of philosophy and religion in the promotion of academic relativism in our time. As we have already suggested, the whole thrust of philosophy in this century, from Viennese logical positivism and Anglo-American linguistic analysis to French and German existentialism, has been in the direction of an ultimate scepticism with respect to the classical problems of truth, goodness, and beauty. Knowledge is judged relative to the observer, and never equivalent to truth in any final sense; beauty is a matter of culture and personal taste; and goodness becomes whatever we choose to believe good, a matter of decision or preference on the part of the willing self without any universal cognitive significance at all. In short, neither the true, nor the good, nor the beautiful can be known. At the same time, belief in the Judeo-Christian God and in traditional religious dogmas has declined precipitously among intellectuals since the middle of the nineteenth century, especially in Europe, removing yet another possible landmark from the horizon of thought. Developments in philosophy and religion both stimulate and are in turn stimulated by academic relativism; a cybernetic relationship obtains. A further factor is the West's sense of its own declining relative importance in the wake of two world wars, the collapse of most overseas Western empires, and the partially successful efforts of several non-Western powers to challenge Western ascendancy in politics, technology, and industry.

Although the rise of relativist attitudes is itself a source of anxiety in the Western world, creating fear that ours is an exhausted civilization, it seems safe to say that if there were no other reason for us to feel anxiety, no wars or rumors of wars, no tyrannies, no dehumanizing technics, contemporary Western man would still discover ample cause for doubting the truth of the idea of progress in the mere fact of his inescapable commitment to relativism. Three academic disciplines serve as particularly good illustrations of the way in which this doctrine of value-relativity has undermined faith in progress: scientific historiography, anthropology, and sociology.

Of the three, scientific historiography succumbed first, and is most strongly fortified from within against progressivism. We had only a few occasions to refer to historians in Part Two, and one of those cited, Lord Acton, was among the nineteenth century's most intemperate and outspoken enemies of academic relativism. Acton, as a Christian and a liberal, and other historians of the age who had similar or alternative commitments were able to subordinate their role as academic historians to the service of value-systems that transcended relativism, but the new scientific historiography of the nineteenth century, nurtured in the seminars of Leopold von Ranke at the University of Berlin, proceeded from a point of view ultimately hostile to the belief in progress. The new history was deeply suffused with an attitude known in German as *Historismus,* and variously rendered in English as "historism," "historicism," or "historical-mindedness."

Historismus, as Friedrich Meinecke has shown, originated in the eighteenth century; Herder was among its first exponents. It confined itself initially to the suggestion that peoples, institutions, and ideas could be understood only in their historicity, in their uniqueness and mutability as historical phenomena subject to the vicissitudes of time. To think historically, moreover, involved the abandonment of contemporary standards of value or judgment and the examination of historical subject matter in its own terms and its own context. Such a point of view did not attack the idea of progress frontally; many of the historically-minded philosophers, theologians, philologists, and historians who cultivated the historicist attitude in the nineteenth century in Germany and elsewhere clung to some form of belief in progress, since it remained possible, once the past had been studied in its own terms, to make the final results of such studies conform to the thinker's own trans-historical understanding of the meaning of history.

But historicism is a jealous mistress. It tends to invalidate all other types of thought and judgment: in the end, it may lead to complete nihilism and relativism. This is precisely what has happened in the twentieth century. Historicism has suffered what Erich Kahler terms "a hypertrophical degeneration." Or, in Georg G. Iggers' analysis, it has passed from the view that nothing can be understood except historically to "the recognition that all values are historical and that 'historical is identical with relative'." [2] In other words, the possibility of a trans-historical valuation of history disappears, providence and the Hegelian cunning of reason disappear, and the belief in progress disappears, too: for how can it be said that one historical period or one historically created value is "better" than any other, and that progress, therefore, has occurred?

Even to certain nineteenth-century minds, the implications of historicism for the belief in progress were far from unclear. Ranke, the most influential and prolific of all the new scientific historians, could not bring himself to endorse the nineteenth-century idea of general human progress. As he explained in a lecture given before King Maximilian of Bavaria in 1854, only the Western nations could be said to have progressed at all, and even in the West, except in purely "material interests," it was a question not of general progress but only of occasional periods of exceptional achievement in one or two fields of endeavor, paid for by decline in other fields. To argue that some later generation wholly surpassed any of its predecessors would be to accuse God of injustice. The allegedly inferior generation "would not have a significance in and for itself, since it would be the stepping stone of the following generation and would not stand in an immediate relation to the divine. But I assert: every epoch is immediate to God and its worth does not reside at all in what emanates from it but rather in its own existence, its own identity." It followed that "every epoch must be seen as something having its own special merit" and deserving of study for its own sake. At the same time, despite his adherence to Christian belief, Ranke rejected the notion of a divine providence acting in history and moving men toward some divinely appointed goal, since this would virtually abolish "human freedom and write off human beings as tools without a will of their own." History had no goal, and advanced not in a straight line, but like a meandering stream, whose every turn had its special fascination to the historical mind.[3] In effect, God did nothing at all within the confines of historical time, and Ranke's rudimentary philosophy of history was very close to pure historicism.

But Ranke was in no sense a professional philosopher, and the question of the relationship between historicism and value-systems such as the belief in progress remained largely unexplored until the philosophers themselves began to turn their attention to it late in the nineteenth century and early in the twentieth. Significant contributions were made by Wilhelm Windelband, Georg Simmel, Heinrich Rickert, Wilhelm Dilthey, Benedetto Croce, and José Ortega y Gasset. Their thought, important in itself, also helped to shape the views of history of many others, including Ernst Troeltsch, Max Weber, Martin Heidegger, Karl Jaspers, Karl Mannheim, R. G. Collingwood, Raymond Aron, and Carl Becker.[4] Although in most instances these thinkers have chosen to represent themselves as "overcoming" or "transcending" historical relativism, they have ordinarily done so more in the language than in the substance of their arguments. The belief in progress has tended to encounter either explicit rejection or indifference, with certain noteworthy exceptions that will claim our attention in Part Four.

The central figure is perhaps Dilthey, whose mature thought belongs entirely to *la belle époque*—he died in 1911—but whose greatest influence came in the years immediately thereafter, and in the English-speaking world only following the Second World War. Dilthey was among the first historians of ideas to propose an organic continuity between Christian *Heilsgeschichte* and the modern belief in progress, but as a philosopher he insisted on the relativity of all world outlooks. Just as every gesture, every word, every action bore the hallmark of historicity, so a man's world outlook was inescapably finite, the product of his unique experience in the historical world. It had value and meaning for him, but not a universal value and meaning. If a world outlook sought to interpret history, its theory of history, too, was finite and relative.

Dilthey saw nothing tragic in this, and no reason for a man to abandon his beliefs. On the contrary, "the historical consciousness of the finitude of every historical phenomenon, of every human or social condition and of the relativity of every kind of faith, is the last step towards the liberation of man. . . . The mind becomes sovereign over the cobwebs of dogmatic thought." Man was now free to "surrender" his being to life here and now, secure against the fear that any past system of faith could entrap him by its claims to final or timeless truth.[5] To quote one of Dilthey's recent admirers, H. P. Rickman, the idea of progress rests on a belief in absolute values and collapses, with the overthrow of absolute values, into "the empty platitude that the past has produced the present." Dilthey's reading of history as the record of man's creative freedom, Rickman argues, transcended both the false hope of progressivism and the despair of pure relativism.[6]

The task of popularizing the term *Historismus* in Germany fell to Ernst Troeltsch, who was both a liberal theologian in the Ritschlian tradition and an intellectual historian influenced by Dilthey. Troeltsch saw more clearly than Dilthey the dangers of historicism for man's spiritual life, but he was no better able to escape the relativistic undertow of the historicist approach; in the end, he was carried out to sea in the same way. As he wrote shortly before his death, all things had their being in time: "State, law, morality, religion, art, are all dissolved in the stream of history's becoming and are intelligible to us only as constituent parts of historical developments." [7] No possibility of absolute truth or of absolute knowledge of the good existed.

Troeltsch first applied his historicist approach to value in the area of Chrstian theology. Although a former student of Ritschl, he could not accept the idea of a progressive Christianization of the world. Since all religious doctrines, including non-Christian doctrines, belonged to the stream of historical becoming, all were finite, and all were relative to the needs, conditions, and thought-forms of given eras in history. Christianity, therefore, had no basis for its claims to exclusive truth, and no Christian teaching was immune to the effects of time. In the last pages of his *Social Teaching of the Christian Churches* (1911), he concluded that the chief historic forms of Christian social doctrine were now obsolete and had little relevance to twentieth-century social problems. If Christianity were to have any bearing at all on the solution of these problems, "thoughts will be necessary which have not yet been thought, and which will correspond to this new situation as the older forms met the need of the social situation in earlier ages." Even if such new ideas did make their appearance, "they will meet the fate which always awaits every fresh creation of religious and ethical thought: they will render indispensible services and they will develop profound energies, but they will never fully realize their actual ideal intention within the sphere of our earthly struggle and conflict." Life "remains a battle which is continually renewed upon ever new fronts. For every threatening abyss which is closed, another yawning gulf appears." [8]

After the First World War Troeltsch realized that all systems of value stood in mortal jeopardy, and that relativism, for which the diffusion of historicism had to accept some of the blame, threatened to paralyze the will of Western man in his hour of deepest need. He devoted two books to the matter: *The Problem of Historicism* (1922) and a posthumously published collection of lectures intended for English audiences, which appeared in English under the title *Christian Thought* and in German as *Der Historismus und seine Überwindung* (1923).

The absolutization of historicism, the propagation of an historicism "without limits" had unnerved modern man, Troeltsch contended. The solution was for Westerners to return to their faith in a truth above time. Although this truth manifested itself in different ways at different times and in different cultures, we were not thereby absolved from the duty of making our own effort here and now to restore our lost links with the divine life outside of history. Christianity was "final and unconditional for us, because we have nothing else." [9] Other peoples could seek the eternal in their own ways, which might be just as valid—for them. But as H. Stuart Hughes remarks, Troeltsch remained "as much at sea as ever. . . . His last work was, rather, a confession of failure—an admission that it had proved impossible to discover stable values *within* the flow of the historical process." [10] In any case, knowing that a realm of absolute truth existed could scarcely be of much practical value if all efforts to bring it down to earth were doomed to a merely "relative" value.

The views of Troeltsch's close friend, the historian Friedrich Meinecke, are subject to the same strictures. Spiritual values, wrote Meinecke, arose within the historical process and commanded the service of men, freeing them from bondage to the blind mechanism of nature. But such values could not be measured by any single universal standard. The "infinite fascination of the historical world" lay in its constant creation of "new spiritual entities," which it refrained from arranging "in a progression of ascending rank." It was enough to acknowledge the existence of an "unknown absolute" standing behind history, which man would never be able to unveil. "Only weak souls of little faith could despair and quit under the burden of this relativizing historicism." [11]

Whatever else might be said, Meinecke's own spirit was not broken by the experience of living through the Nazi era, which he saw as a catastrophe for Germany.[12] He remained personally capable of adhering to trans-historical values, above all to those celebrated by Herder and Goethe, the heroes of his study of *The Origins of Historicism* (1936). In much the same way, his contemporary Benedetto Croce survived the Fascist epoch in Italy, refusing illogically to abandon either the basic premises of historicism or his intuitive apprehension of timeless categories of spiritual life. Croce's philosophy of history, expounded in *History: Its Theory and Practice* (1915) and *History as the Story of Liberty* (1938), illustrates in depth the dilemmas of historicism and its fundamental hostility to the belief in progress. By temperament, he was a typical nineteenth-century liberal optimist and idealist, indebted to Hegel, drawn to both Marx and Bergson without accepting either, and a warm admirer of the Risorgimento. But he was also powerfully influenced by German historicism and the thought of his countryman Vico. Although the three-cornered struggle in Croce's mind between liberal progressivism, idealism, and historicism was never logically resolved, historicism always tended to get the upper hand.

Croce went further than his German predecessors in defining the province of historical study and methodology. He was also more acutely ware of the relativity of historiographical judgments. Historical study, he insisted, embraced the whole life of man; all knowledge of man was necessarily historical knowledge. Even philosophy dissolved, for Croce, into historical thought. "It is a curious fate," he observed, "that history should for a long time have been considered and treated as the most humble form of knowledge, while philosophy was considered as the highest, and that now it not only is superior to philosophy but annihilates it." [13] But in another sense, history dissolved into philosophy, for all true history was "contemporary history," current thought in living minds, filled with concern for their own times and studying the documents of the past from the perspective of their own present. To men alive today, not one instant of the past was available, except as they could make that instant vibrate in their own beings. It followed that no historian could conceive or write "universal" history, a history of all ages for all men. "Such a history disappears in the world of illusions, together with similar Utopias, such, for instance, as the art that should serve as model for all times, or universal justice valid for all time." [14]

The so-called philosophy of progress was invalid in Croce's eyes because it posited an end extrinsic to history, "conceiving of it either as that which can be reached in time (*progressus ad finitum*), or as that which can never be attained, but only indefinitely approximated (*progressus ad infinitum*)." But the meaning of history could legitimately be sought only within history itself, since there was no point outside of space and time from which the human mind could discern ends or meanings. The end of history, therefore, "is attained at every instant, and at the same time not attained, because every attainment is the formation of a new prospect, whence we have at every moment the satisfaction of possession, and arising from this the dissatisfaction which drives us to seek a new possession." In the same way, it was not possible for the historian to speak of "good" and "evil" facts, or of "progressive" and "regressive" ages. "Since all facts and epochs are productive in their own way, not only is not one of them to be condemned in the light of history, but all are to be praised and venerated." [15] The alternative view, the view of progress, committed the crime of reducing ages and persons to fragments, ages and persons that had achieved only partial humanity, or in the philosophy of infinite progress, an infinitely small part, and so were "reduced to less than dust." On the contrary, "humanity in every epoch, in every human person, is always whole." [16] Every age was an age, every man a man, and none less than others.

Like Dilthey, Croce saw no reason why such an outlook should lead to despair. Far from "promoting fatalism" or "sanctifying the past," historicism enabled mankind to liberate itself from the thralldom of history. To understand the past was to be set free from its demands, and historicism was future-oriented, always looking ahead to the next victory, the next overcoming of opposites. In the spirit of Nietzsche, Croce scorned the notion that mankind would ever be at rest: "It would be safe to define progress, if we so wish, as an ever higher and more complex form of human suffering." [17]

But notice the phrase "ever higher." The difficulty in interpreting Croce's thoughts on progress arises from his unwillingness to relinquish some of the language and perhaps also the substance of neo-Hegelian idealism and of liberal progressivism. In the same chapter of *History: Its Theory and Practice* in which he denied the possibility of discriminating between "good" and "evil" facts in history he proposed an idea of progress "as the passage from the good to the better, in which the evil is the good itself seen in the light of the better." [18] To be consistent, he should have written "the passage from one good to another good, in which the evil is a former good seen in the light of a later." Even then, such a "passage" would be progress only in the sense of onward motion, not in the sense of qualitative improvement.

The confusion further deepens when we consider Croce as a practicing historian and as a political thinker responding to the challenge of Fascism. In his studies of the history of modern Europe and the history of historiography, Croce seems to place himself unambiguously among the believers in progress. He extolled the nineteenth century in Europe as the highest age in the history of human freedom, with special praise for the Italy of the Risorgimento. At the same time he regarded contemporary historiography as the highest stage in the progress of historical thought, and his own philosophy, by implication, as the highest point achieved in the history of philosophy, which was, as we have noticed, reducible to historical thought.[19]

In *History as the Story of Liberty,* written while Italy was living under Fascism, Croce made it clear that in later life he had also found a perspective from which to judge history *sub specie aeternitatis,* almost in the manner of Hegel. His discovery was the absolute, trans-historical value of liberty, conceived primarily as liberty of spirit. He now saw liberty as "the eternal creator of history and itself the subject of every history. As such it is on the one hand the explanatory principle of the course of history, and on the other the moral ideal of humanity." It had abided "purely and invincibly and consciously only in a few spirits; but these alone are those which count historically." [20] H. Stuart Hughes quite properly points out that Croce failed to explain whether he thought that his absolutization of liberty was "eternally valid, as Hegel had assumed" or "historically conditioned," like all other acts of valuation in history.[21] Despite his glorification of nineteenth-century liberalism, however, Croce did at least resist the ultimate temptation: in *History as the Story of Liberty,* he emphatically ruled out of court Hegel's doctrine of the progress of liberty. Although some ages were more liberal than others, liberty fought a constant battle against its enemies. In illiberal ages (such as the one in which Croce then lived) liberty continued its struggle just the same. Croce's book itself gave evidence that liberty in Italy had not become extinct. Like tyranny, liberty was a hardy perennial, inspiring select higher minds in every age.

But it is difficult to reconcile this picture of *libertas perennis* with Croce's eulogy of the nineteenth century and of his own philosophy, or to reconcile either one with the historicism that refused to distinguish between "good" and "evil" facts or "progressive" and "regressive" ages. The only possible conclusion is to assume that Croce was guilty of gross inconsistency, but also to suspect, with Hughes, that "ultimately, whatever Croce himself may say, the implications of his thought are relativist." [22] On the whole, he belongs among the opponents of the belief in progress, although he was never in any sense an apostle of anxiety or despair.

In Britain Crocean historicism was expounded by the Oxford philosopher R. G. Collingwood, not without original and critical contributions by Collingwood himself, although it is clear that he regarded Croce as the leading historical mind of the twentieth century. Collingwood echoed the Italian philosopher's objection to the idea of periods of greatness and decadence, attributing the notion of "dark" ages to scholarly ignorance. As for progress, he agreed that net improvement might be detected in history, but only in narrowly circumscribed areas of life, and only for limited periods of time. The general progress of humanity or even general progress from any one period of history to some later period could not be demonstrated by the historian, for lack of data and also because of the nature of certain categories of data, which did not lend themselves to objective judgments of value. The only sort of progress that the historian could uncover consisted of solutions to new problems that incorporated without loss the solutions to prior problems. Einstein, for example, included Newton, but went beyond Newton. In this way Collingwood substituted quantitative for qualitative judgments of value, positing that "more" is "better," and evading the historicist doctrine of the relativity of values.[23]

Relativism in historical thought can also spring from approaches that have little to do with historicism as such. The work of Spengler, for example, is, at least on the surface, ultra-relativistic, asserting the uniqueness and mutual incomprehensibility of the major historical cultures; but his methodology combined romantic philosophy with the techniques of the social or natural sciences, and owed almost nothing to the historicism of Ranke and Dilthey.* The same might be said of Pitirim A. Sorokin's *Social and Cultural Dynamics* and the earlier volumes of Arnold J. Toynbee's *A Study of History*. Their goal is to convert the study of history into a social science with predictive power, and nothing could be further from the world-view of *Historismus*.**

Opposition to efforts such as these from yet another perspective that turns out to be equally relativistic is mounted in the books of the British logician Karl R. Popper. Unfortunately Popper has chosen to confuse the issue by defining "historicism" as the approach to historical study of the social scientist. Historicism, he writes, is the view of the social sciences "which assumes that

* See above pp. 163–64.
** See below, pp. 199–202.

historical prediction is their principal aim, and which assumes that this aim is attainable by discovering the 'rhythms' or 'patterns,' the 'laws' or the 'trends' that underlie the evolution of history." [24] If one reads for "historicism" something like "historical determinism," Popper's indictment makes much more sense. Yet he manages to stumble into the same pitfall that captured Croce, by maintaining in one breath that all interpretations of history are strictly arbitrary and in the next that his own is morally and logically superior to that of the "historicists," or, in other words, of the determinists who find in history a pattern of inevitable progress or inevitable cyclical repetition. Like Croce, Popper is a bitter critic of totalitarianism and its theories of history; much of the emotional power of his argument obviously derives from his profound distaste for both.

As a philosopher, however, Popper has little in common with Croce; he belongs rather to the analytical movement in modern philosophy of science, which has discovered in its own way, independently of historicism, the relativity of values. On neopositivist assumptions he argues that history has no being or energy of its own, consisting exclusively as it does of the wills, actions, and desires of individual human beings. History as fact is meaningless, just as all facts are meaningless. "Facts as such have no meaning; they can gain it only through our decisions." Since the idea of progress is grounded in human normative decisions, progress may be said to occur only when individual human beings define their values and work to realize them in history. "If we think that history progresses, or that we are bound to progress, then we commit the same mistake as those who believe that history has a meaning that can be discovered in it and need not be given to it." [25]

But then Popper goes on to present his own view of progress as if it were universally binding, emulating Croce's absolutization of liberty. "For to progress is to move towards some kind of end, towards an end which exists for us as human beings. 'History' cannot do that; only we, the human individuals, can do it; we can do it by defending and strengthening those democratic institutions upon which freedom, and with it progress, depends." [26] In short, progress consists of human self-realization as facilitated by modern Western liberal democracy, by the life of the "open" as opposed to the "closed" society. Yet Popper freely admits that the decision to make history "mean" the struggle between open and closed societies, with the former preferable to the latter, is the result of an arbitrarily "preconceived selective point of view" that has interest and relevance to "us." One could just as easily make history "mean" the "history of class struggle, or of the struggle of races for supremacy, or . . . the history of religious ideas." [27]

Clearly, Popper can find no universally compelling reason to believe that progress consists of the growth of freedom rather than tyranny. The belief in progress is reduced, therefore, to a matter of private preference, with all theories of progress and non-progress equally valid in the eyes of science, historiography, and pure reason. In the end, all that Popper has left to complain about are the false claims of historical determinism, whether Hegelian, Comtian, Marxist, or Spenglerian, to objective knowledge of the allegedly necessary course of historical change, when, in fact, nothing is necessary, and the future lies open to every possibility. But his point of view is ultimately no less relativistic than that of Dilthey or Spengler.

The majority of professional historians in the twentieth century, it should be noted in conclusion, have returned to the program of Ranke (if, indeed, they had ever abandoned it), to the idea that the study of history is an end in itself, which has been assigned the task of reconstructing and explaining specific past events and ways of life in terms of their historical context. H. A. L. Fisher's confession in *A History of Europe* speaks for the academic historian everywhere: "Men wiser and more learned than I have discerned in history a plot, a rhythm, a predetermined pattern. These harmonies are concealed from me. I can see only one emergency following upon another as wave follows upon wave." [28] * Fisher's dictum is cited approvingly in G. R. Elton's recent neo-Rankean manifesto, *The Practice of History,* which goes on to caution historians against issuing theories of progress for two reasons. There are too many different lines of development at any one time, he writes, and our knowledge of the past is too incomplete to allow us to speak of wholesale progress. Moreover, general agreement on what constitutes progress is not to be expected, since "one man's better is

---

* But Fisher's Victorian upbringing—he was born in 1865—betrayed him only two sentences later, when he proclaimed that "the fact of progress is written plain and large on the page of history." Progress was not guaranteed, and "disaster and barbarism" might one day overwhelm the race; yet progress, if not a "plot" or a "pattern," remained nonetheless a "fact"!

usually another man's worse." All contrary claims notwithstanding, "progress and necessity are doctrines which cannot be derived from, can only be superimposed upon, the study of history." [29] *

Readers of J. B. Bury's *The Idea of Progress* will also not fail to recall his warning against assigning finality to the belief in progress. In his Epilogue, Bury pointed out that the study of history taught the endlessness of change: it was therefore reasonable to expect the day when a new doctrine would usurp the place of the idea of progress. "Another star, unnoticed now or invisible, will climb up the intellectual heaven, and human emotions will react to its influence, human plans respond to its guidance. It will be the criterion by which Progress and all other ideas will be judged. And it too will have its successor." Bury confusingly suggested later in the same Epilogue that the idea of progress logically required its own supersession for the sake of the continued improvement of the race, just as belief in progress at one time had superseded belief in providence. But this may have been something of a joke, "a disconcerting trick of dialectic," as Bury himself intimated. The argumentive force of his Epilogue stems entirely from its implicit avowal of the premises of historicism, its recognition that all values in history were relative to their times, and that nothing endured forever.[30]

Until about 1900 anthropology followed in the positivist tradition of Comte, Darwin, and Spencer, with some influence exerted, principally in the German-speaking world, by Hegelian idealism and the *Naturphilosophie* of Schelling. The master-idea was evolution, and the founders of anthropology as a scientific discipline were nearly all evolutionists.** Early in the twentieth century, however, dissenting voices began to be heard. By 1914, when the general public no doubt still thought of anthropology as the study of human progress from savagery to civilization with special emphasis on the

---

* Georg G. Iggers points out, however, that *Historismus* in all its forms has come under heavy fire in Germany since 1945 because of its presumed role in preparing the German mind for an acceptance of Nazism. Many historians are experimenting with "natural law" theory, others with the nomothetic method of the social sciences. See Iggers, "The Dissolution of German Historism," in Richard Herr and Harold T. Parker, eds., *Ideas in History* (Durham, N.C., 1965), pp. 310–20 and 323–24. This would seem to be the counterpart in historical studies of "Christian democracy" in politics.

** See above, pp. 39–40.

earliest stages thereof, professional anthropologists were already in process of dividing into a number of rival schools and tendencies, some of which expressed indifference or even hostility to the idea of progress. In the twentieth century proper, anthropology has been a fashionable major source of relativistic attitudes; and the study of so-called primitive peoples has done as much to deflate the pretensions of modern Western progressivism and ethnocentrism as the use of travelers' tales by the *philosophes* of the Enlightenment did to deflate the pretensions of Christianity and the *ancien régime* two hundred years ago.

Some of the new tendencies after 1900 clearly affect the belief in progress more than others. The attack on classical unilinear evolutionism by the exponents of multilinear evolution, diffusionism, functionalism, the psychological school, and the like, bears no necessary relevance to the problem of progress. But two prominent attitudes among the critics of nineteenth-century anthropology are quite relevant: an inclination to what can be described as anthropological historicism and a deep sympathy for the primitive world that springs from improved techniques of gaining insight into the primitive mind and also from modern man's disillusionment with himself. The two attitudes are intimately related. Historicism as it applies to anthropology insists on explaining specific past or contemporary cultures in their own terms, without reference to some larger hypothetical scheme. Sympathy for primitivism assails the idea that modern Western white culture is the standard against which all other cultures must be measured. In both attitudes, cultural and ethical relativism prevail over the notion of a general progress of mankind from "lower" to "higher" cultures or races. There may still be room for an idea of progress in certain aspects of life, but general progress becomes more or less inconceivable.

Evolutionism itself produced a few sceptical spirits in the post-Darwinian generation. No one, for example, respected Darwin and the principle of natural selection more than the Anglo-Finnish anthropologist and philosopher Edward Westermarck; he used the term "evolution" freely in his writings, and one of his most important books was *The Origin and Development of Moral Ideas* (1906), which can be compared in some ways with another large work that appeared in the same year, Hobhouse's *Morals in Evolution*. But Westermarck clung with tenacity throughout his career to the argument that moral ideas were, unlike the truths of science, a product of emotion, and therefore relative to whatever emotional state gave them validity for the believer. His studies of sexual customs and taboos, of ethnocentrism, and many other moral ideas both in primitive and historic societies, together with his own reflections as a philosopher, led him again and again to the same conclusions. Values were relative. Not much alteration had taken place in man's moral life through the centuries. In 1932 he wrote:

> The changes of moral ideas appear small when compared with the enormous progress in knowledge our race has made on its path from savagery to modern civilization. And the reason for this is that while intellectual evolution has been a perpetual succession of new discoveries, the changes of moral ideas have been no discoveries at all, but only been due to more or less varying reactions to the moral emotions.[31]

One could speak of diversity and, to a limited extent, of development or evolution, but not of improvement, since the supposed objectivity of moral judgments was a myth.

One of the first anthropologists to break more or less completely with evolutionism was Franz Boas; his career illustrates, much like that of Troeltsch, the conflict between relativism and liberal values in a mind that, just a few years earlier, would have had no difficulty in subscribing to a consistent theory of progress. Born and educated in Germany, Boas moved to the United States in 1887. From 1896 he taught anthropology at Columbia University, where he numbered among his students a sizeable percentage of the leading anthropologists of the twentieth century, including Ruth Benedict, Alfred L. Kroeber, Margaret Mead, and Robert Lowie. Boas' basic approach is well summarized by H. R. Hays: "Previously, the European investigator had cultivated a condescending viewpoint toward his humble informant's culture; the civilized way of life was the standard of reference. Boas tried to see the Northwest Coast Indian culture as simply another form of human achievement, valid in its own way."[32] The relativism of Boas' anthropological method was rooted, to a very significant degree, in historicism. He hoped that in the future it might be possible to work out at least a certain number of general laws pertaining to human development, but he discouraged anthropologists from system-building and simplistic formula-mongering on the grounds that theirs was a historical science concerned with the analysis of specific processes in actual societies. Generalizations could have no value unless based on careful empirical study. As in Dilthey's *Geisteswissenschaften,* or in such natural sciences as astronomy and geology, "the center of investigation must be the individual case."[33]

One of the few generalizations that Boas chose to risk, on the basis of his study of cultures, was the observation that history revealed no pattern of necessary uniform evolution throughout the world. "We rather see that each cultural group has its own unique history, dependent partly upon the peculiar inner development of the social group, and partly upon the foreign influences to which it has been subjected." The attempt to explain mankind as a whole, or even any single culture in its total historical development, would involve the scholar in an undertaking so complex "that all systems that can be devised will be subjective and unrevealing. . . . They will always be reflections of our own culture."[34] The scholar, therefore, had to practice humility, and Boas challenged both scholar and layman to look on cultures different from their own with sympathy and open-mindedness. We were forced to acknowledge

that there may be other civilizations, based perhaps on different traditions and on a different equilibrium of emotion and reason, which are of no less value than ours, although it may be impossible for us to appreciate their values without having grown up under their influence. The general theory of valuation of human activities, as developed by anthropological research, teaches us a higher tolerance than the one which we now profess.[35]

The "higher tolerance" allowed Boas to point out, for example, that the fierce solidarity, tribalism, and hatred of strangers noted in primitive societies had their exact counterpart in the nationalisms of modern Western man.[36] During and after the First World War he launched numerous attacks on modern nationalism and racism, coupled with eulogies of American freedom and democracy that appeared to contradict his own plea for tolerance of alien cultures. But he continued to argue that all cultures had "to solve their problems in their own ways. . . . The very standpoint that we are right and they are wrong is opposed to the fundamental idea that nations have distinctive individualities, which are expressed in their modes of life, thought and feeling." Such local differences were "of the greatest value to mankind as a whole, because they make for that variety in cultural life that is the necessary condition for a life worth living." The end of pluralism would be "fatal to human happiness and human progress."[37]

But what, in this case, constituted progress? Boas did not say, although his official position, as summarized in his long article on anthropology in the *Encyclopaedia of the Social Sciences* (1930), doubted the occurrence of any kind of progress other than scientific and technological.[38] He was caught on the horns of the old dilemma of whether the tolerant man should tolerate intolerance, and the newer dilemma of whether the relativist should be relativistic about the arguments of nonrelativists. It seems likely that he felt a strong desire to impose his liberal values on the pattern of world history, but he was too deeply enmeshed in anthropological-historical relativism to resolve the ambiguities of his position.

A somewhat more consistent relativism was espoused by Boas' student Ruth Benedict in her *Patterns of Culture* (1934), which added to historicism the insights of *Gestalt* psychology and Spengler's *Decline of the West*. Deploring Western ethnocentrism and showing the close interrelationship of cultural and psychological relativity, Benedict maintained that the "recognition of cultural relativity carries with it its own values, which need not be those of the absolutist philosophies." The traditionalist belief that any one set of cultural values or type of personality surpassed any other and the individualist's feeling of complete autonomy both turned out to be illusory, but as soon as men had embraced wholeheartedly the new approach to value, they would convert it into "another trusted bulwark of the good life." They would find a new basis for hope and tolerance in accepting "the coexisting and equally valid patterns of life which mankind has created for itself from the raw materials of existence."[39] Another Boas disciple, Melville J. Herskovits, directed similar criticisms against Western ethnocentrism, which he held responsible for the now outmoded concept of progress.[40]

The interest in cultural "patterns" or "configurations" evidenced by Benedict also appeared in the studies of Alfred L. Kroeber, who found food for thought not only in Spengler's theory of cultural integration but in his cyclical theory of history as well. Kroeber's *Configurations of Culture Growth* (1944) viewed with some sympathy Spengler's argument that cultures rose and fell according to a natural rhythm, and opposed the alternative assumption "that cultures inherently tend to progress"; but he preferred to steer a middle course between the two extremes, maintaining that "cultures do not necessarily either age or progress, but . . . do undergo variations in vigor, originality, and values produced." Moreover, there was a tendency for cultures, after having passed through several cycles of florescence and decline, to exhaust their possibilities, meet with invincible competition from younger and stronger cultures, and become extinct.[41] Kroeber's "middle course," on close scrutiny, seems much more Spenglerian than progressivist.

There still remains the question, dismissed by Spengler, of whether mankind as a whole progresses, despite the deaths of individual cultures. Kroeber confronted this question forthrightly in a comprehensive survey of anthropology published in 1948. The popular doctrine of progress, he acknowledged, was an untested conceit of modern Western man. Much of it rested on ethnocentric judgments of value that did not lend themselves to scientific confirmation. Along three lines, however, mankind had been able to achieve the sort of qualitative improvement that scientists could hope to define objectively as progress: "the atrophy of magic based on psychopathology; the decline of infantile obsession with the outstanding physiological events of human life; and the persistent tendency of technology and science to grow accumulatively." Here was "the residuum supported by fact when the emotional and aprioristic idea of a continuous and inherent progress of civilization is subjected to analysis." [42] That Kroeber's criteria for improvement were still entirely arbitrary and subjective does not seem to have occurred to him, nor is it clear that he wished to consider the type and quantity of progress actually achieved by mankind tantamount to general progress. But at least he cannot be ranked among the dogmatic unbelievers in progress in any form. His approach to the problem of progress is echoed approvingly in Julian H. Steward's *Theory of Culture Change,* and probably reflects the thinking of more anthropologists in recent years than does the approach of cultural relativism championed by Boas.[43] Some anthropologists go still further in the direction of a reaffirmation of progress.*

Nevertheless, it is quite apparent that twentieth-century cultural and social anthropology tends to be much less interested in the problem of progress, whatever conclusions various speculative minds may reach, than was the anthropology of the preceding century. The majority of scholars in the field restrict themselves to specialized

* See below, pp. 311–16.

empirical studies, and even theorists ordinarily avoid all-encompassing historical generalizations. Not a few contemporary anthropologists would concur with the Oxford functionalist E. E. Evans-Pritchard in feeling "a moral separation from the anthropologists of last century," who were not content to investigate and speculate, but also experienced a compulsion to justify their belief in progress. "We are less certain today about the values they accepted." Scepticism and a sounder view of the limits of scientific inquiry dictate that the interests of twentieth-century social anthropology "are more in what makes for integration and equilibrium in society than in plotting scales and stages of progress." [44]

Academic sociology since the turn of the century has followed a course parallel in most respects to that of scientific history and cultural anthropology, stressing value-free empirical research and preferring functional analysis to speculative evolutionism. There has also been an undercurrent, at least, of primitivism and considerable interest in the role of the irrational in social behavior.

Most nineteenth-century sociologists, as we have seen, expounded theories of evolutionary progress. Occasionally evolution did not result in progress, as in Ludwig Gumplowicz's conflict theory, and even led to retrogression, as some scholars have interpreted the concept of evolution from *Gemeinschaft* to *Gesellschaft,* from the "natural" communal order to the "rational" social order, in the sociology of Ferdinand Tönnies.[45] The decisive turn to a relativistic, ethically neutral sociology of function and process came with Max Weber in the first two decades of the twentieth century. In Weber, perhaps for the first time in the social sciences, a mind of the first magnitude extracted from Kant and the historicism of Rickert and Dilthey a methodology that unequivocally divorced the scholarly vocation from the prophetic. Science and reason had no means of spanning the chasm between value and fact, in Weber's judgment; therefore, the various scholarly disciplines, including sociology, stood obliged to abandon all claims of being able to arbitrate among competing value-systems. Values could be scientifically studied, and Weber did some of his best work in this field. But the normative correctness of any given value lay beyond the scholar's power to determine. "The professor," he wrote in 1917, "should not demand the right as a professor to carry the marshal's baton of the statesman or reformer in his knapsack." [46] Only the most ruthless professionalization of disciplines could ensure their integrity, their technical progress, and their social utility—for what use to society was a science that discovered only what its practitioners wanted it to discover?

From the point of view of sociology, then, a man's intrinsic values belonged to the realm of will, emotion, conscience, or fate; they were his to believe, as he preferred, but never to impose on his fellows in the name of science. Weber often referred to values and faiths as "gods," beyond proof or refutation, locked in endless combat in a polytheist world. The scholar as scholar dared not worship at their shrines, but as citizen and human being, he had as much right to his beliefs as the next man, provided that he kept them out of the classroom. "In the press, in public meetings, in associations, in essays, in every avenue which is open to every other citizen, he can and should do what his God or daemon demands." [47] Yet even in public forums, to say nothing of the private forum of the soul, the scholar knew only too well that from the perspective of science his beliefs constituted only one more value-system among multitudes.

Weber's methodology precluded any discovery of universal laws of historical evolution or progress. All sciences proceeded by means of heuristic devices, such as the "ideal types" proposed by Weber for sociology, which could never give more than partial explanations, since they all started from arbitrarily selected points of departure. A complete knowledge of human culture was impossible. In any event, no matter how much ground he covered the sociologist could by no means arrive at a scientific definition of "improvement" without deserting his vocation. "The use of the term 'progress' is legitimate in our disciplines when it refers to 'technical' problems, i.e., to the 'means' of attaining an unambiguously given end. It can never elevate itself into the sphere of 'ultimate' evaluations." Even in this limited sense its use was, Weber thought, "very unfortunate." [48]

But when all is considered, Weber did have a comprehensive conception of the development of Western culture, on which he passed a generally unfavorable judgment. Like Tönnies and many other German social thinkers of this century, he called attention to the sharp contrast between the traditional social order and the increasingly rationalized society of the modern West, with its efficiency, organization, and discipline, a society that had undergone *Entzauberung,* by which he meant conversion from supernaturalism to secularism. As an academician he approved, since *Entzauberung* had made posible the achievements of modern science and scholarship. But as a man, he feared that rationalization would be carried so far that it would destroy personal freedom and eliminate every trace of idealism from public life. "Weber could see no escape from this rationalization of human life," writes Raymond Aron, "except in a total and non-rational liberty . . . exercised in the political sphere against bureaucratic crystallization, and in the sphere of morality by decisions in cases of moral conflict and ultimately by a personal choice of supreme values." [49] Almost identical thoughts were given expression after Weber's death by Karl Jaspers.

All the same, Weber belongs to that select group of penetrating early twentieth-century minds who have made it difficult for the honest scholar even in his role as a private citizen to arrive at judgments of value. The mental habits of the study and the classroom are not easily shaken off in the parlor or the public lecture hall. It has also been unfashionable since Weber for most sociologists to believe in the possibility of objective knowledge of human progress.

The first two decades of the twentieth century saw, in addition to the best work of Max Weber, the appearance of a major system of sociology in Italy, devoted to the unmasking of the liberal "theology of progress." Its author was Vilfredo Pareto, a former businessman and civil engineer, who turned to academic life in the 1890s. In 1916 he published his magnum opus, *Treatise on General Sociology,* which was issued in a French translation the following year and in English in 1935. A liberal in earlier life, Pareto became disillusioned with the hopes of his youth and reached the conclusion that most human actions were not motivated by logic or reason; it was not a question in history of replacing the rule of unreason by the rule of reason, or tyranny by liberty, but only of watching, helplessly, the lateral movement of society from one state of equilibrium to another. The value-laden theories with which men justified their actions proved, under rigorous cross-examination, to have no scientific verifiability whatsoever, and were, instead, systems of self-delusion or rationalization, under cover of which men unconsciously pursued their interests. Pareto characterized the oscillatory rhythm of history as an alternation in politics between innovative and conservative governing elites, and in the sphere of culture between corresponding ages of rationalism and mysticism, or liberalism and faith. Both types of regime and both types of culture were guilty of the same basically irrational patterns of behavior, whether they appealed to "reason" or not.

On the surface, then, Pareto's system of general sociology was entirely relativistic and cynical. He felt a special contempt, however, for the social order of pre-1914 Italy, an order that appealed to the mythology of progress, freedom, democracy, and equality, but was in fact pluto-democratic, a conspiracy of capitalist profiteers and working masses to produce great wealth and share it unequally between them, at the expense of the rest of the population. The attacks he leveled at modern pluto-democracy, which far exceeded the requirements of his own sociological method, were motivated not by nostalgic yearnings for the feudal past, but rather by an almost Puritanical ethicalism, mixed with respect for logic and science, which compelled him to attach the highest value to honesty, courage, and self-restraint in public life.[50]

Nor did Pareto by any means fully abandon the liberalism of his younger years. He clearly preferred rational action to nonrational, as he preferred a logico-positivist system of sociology, such as his own, to any other. Although he became thoroughly disenchanted with nineteenth-century ideologies of progress, he was not even prepared to deny that reason had made some measurable progress since ancient times, chiefly in science and economic production, and to a much lesser extent in social and political life.

But the thrust of Pareto's thought was undeniably pessimistic. In the same section of his *Treatise* in which he conceded the possibility of the progress of reason, he also warned against "viewing the contingent observation of experience as something absolute." To believe in "the god Progress, the blessings of 'evolution,' and a golden age . . . located in the future" was no less irrational than to believe in retrogression and a golden age located in the past.[51] By precept and temperament, if not always by example, Pareto adhered to a cheerless scepticism; and no one would find it difficult to derive from his sociology a relentlessly relativistic *Weltbild*.

Among the sociologists whose principal work has been done during the last fifty years is the Polish scholar Florian Znaniecki. Cultural relativism was a fundamental premise of his thought. Reacting against the absolute claims of naturalism and idealism in his *Cultural Reality* (1919), he offered instead as the proper approach of the social scientist what he termed "culturalism." Man was a free, creative being, who made his own environment, and who was in turn conditioned by the products of his creative genius, the mores and institutions of culture. If modern thought "intends to avoid the emptiness of idealism and the self-contradictions of naturalism, it must accept the culturalistic thesis. It must maintain against idealism the universal historical relativity of all forms of reason and standards of valuation as being within, not above, the evolving empirical world. It must maintain against naturalism that man as he is now is not a product of the evolution of nature, but that, on the contrary, nature as it is now is, in a large measure at least, the product of human culture." Even the nature we see was seen "through the *prisma* of culture" and man could not "act upon nature otherwise than in culturally determined ways. . . . There is no way out of culture." [52]

In the spirit of Dilthey, and perhaps also of William James, Znaniecki rejoiced in the richness and variety of life. In his last book he looked forward to a unified but culturally pluralistic world of ceaseless diversification and experiment, safeguarded from catastrophe by the expert prognoses of sociologists.[53] As optimistic as Pareto was pessimistic, Znaniecki felt no better prepared than Pareto, however, to embrace an idea of general progress, since this would involve ranking the cultures of man in some sort of ascending order, whereas from the culturalist point of view, each had its own special and unique value.

Znaniecki's position is typical of those sociologists who have addressed themselves to the problem of progress in the last two generations. Karl Mannheim's doctrine of "relationism" and his other contributions to the development of the sociology of knowledge hover somewhere between pure relativism and pragmatic progressivism, and since his later thought by intention inclined to the latter, it will be studied in Part IV.* Nevertheless, Mannheim's logic was primarily the logic of *Historismus,* and he was never able fully to extricate himself from it.[54] Another closely related approach is the dichotomy between "civilization," which progresses, and "culture," which does not, in the work of Alfred Weber (Max Weber's younger brother) and of Robert M. MacIver.[55] Since what matters less (technique) advances and what matters more (values) does not, there is no possibility of speaking of the general progress of mankind. Even highly optimistic, liberal-radical sociologists such as William F. Ogburn, best known for his thesis of "cultural lag," have been

* See below, pp. 318–20.

prevented from expounding a theory of general progress because of their recognition of the subjectivity of any possible definition of progress.[56]

One more group of scholars cannot be passed by in a discussion of relativism in the *Geisteswissenschaften;* they are the specialists in the broad field of comparative civilizations, some of whom have had their training in sociology or anthropology, and others in history. The comparative study of civilizations might be undertaken by scholars with any number of methodological views, but in the main it seems to be confined in the half-century since Spengler's *Decline of the West* to thinkers with a bias toward positivism rather than historicism, toward cyclicalism rather than evolutionism, and—in the two very significant instances of Pitirim A. Sorokin and Arnold J. Toynbee—toward a transcendental rather than a relativist approach to ethical and metaphysical values. Cultural relativism and disbelief in progress remain, but for somewhat different reasons.

One example of this type of study has already been discussed, Kroeber's *Configurations of Culture Growth.** Although influenced by Spengler, it was written independently of the parallel investigations of Sorokin and Toynbee, which were published a few years earlier. The first three volumes of Toynbee's *A Study of History* reached their public in 1934, and three more appeared in 1939. Sorokin's *Social and Cultural Dynamics* was published in four volumes between 1937 and 1941. These attempts are at all odds the most substantial made since Spengler to study the major world civilizations comparatively and historically. Of the three writers, Kroeber was an anthropologist, Sorokin a sociologist, and Toynbee an historian.

Only a little will be said here about Toynbee's contribution, since he has removed himself to the camp of the believers in progress in his prolific postwar writings, but many literate people still know Toynbee only through the first six volumes of *A Study of History* and the competent abridgment of them made by D. C. Somervell in 1946; it seems useful, therefore, to examine Toynbee's earlier thinking at this point, at least briefly.** Even in these first volumes, however, his philosophical eclecticism and his characteristically wide range of vision prevented him from rejecting the idea of progress

* See above, pp. 192–93.
** See also below, pp. 291–94.

altogether. He assumed from the beginning a teleological view of history: mankind was here below for some inscrutable purpose not yet fulfilled. It had evolved, with great effort, from the apes, reaching a higher level of being than any ever attained by life in its upward thrust. The "primitive" societies were "seasoned athletes who have successfully scaled the 'pitch' below and are still taking a well-earned rest from their recent labours." The more than twenty "civilized" societies identified and studied by Toynbee represented as many attempts by mankind to climb still further, to reach a higher "ledge," to accomplish "the transformation of Sub-Man through Man into Super-Man," and so realize "the goal of human endeavours" and "see the Promised Land." There was no guarantee of success, and no way to forecast what lay ahead in any case, but Toynbee—linking the world-views of St. Augustine and Bergson (or Shaw)—had no difficulty in subscribing to the idea of a purposeful and progressive universe.[57]

But when he took up the less exalted task of studying the comparative history of civilizations, he became almost the model relativist, in the tradition of modern anti-ethnocentric anthropology. The notion of the unity of civilization, the "beanstalk" theory of unilinear evolution, the ranking of all civilized societies by the standards of the modern West, and the assumption that the West would automatically triumph and prove its world-historical uniqueness and superiority were all aggressively ridiculed. From the perspective of comparative history, the various civilizations, including the West, looked very much alike. They passed through more or less the same life cycles from genesis to growth to breakdown to disintegration. All the extinct civilizations and probably the seven still alive in the twentieth century could be seen as brave, but unsuccessful, efforts to reach a higher level of human achievement. Each tended to fall because of the same sort of human error: arrogance, the flagging of creative energy, and the resort to tyranny. Each was philosophically "equivalent."[58] The West, alone of all living representatives of the species, had not yet reached the final stage of its disintegration, the stage of the universal state or *pax oecumenica,* but most of the familiar symptoms of early disintegration could be read in its history since the sixteenth century.[59]

In one respect, however, Toynbee failed egregiously to maintain his posture of cultural relativism. As a believing Christian, he insisted that the figure of Christ was absolutely unique in history, and the

Christian doctrine of salvation alone offered a meaningful escape from the vicious circle of history. Other formulas for salvation, which proposed the rebirth of a civilization already in process of disintegration, the birth of another mundane civilization equally bound to the wheel of time, or philosophic flight, were to no avail. By accepting Christ man entered a new species of society altogether, a society not of this world, the *civitas Dei* proclaimed by St. Augustine. In so doing he did not escape the perils of earthly life, but rather seized "the initiative in order . . . to save the City of Destruction from its doom by converting it to the Peace of God." [60] How this promise of possible salvation not only for the person but also for his civilization might be fulfilled, or how it related to the larger scheme of world evolution sketched in the first volume, Toynbee did not explain until his postwar volumes, by which time he had revised his religious outlook drastically.

Much the same fusion (or confusion) of positivism and supernaturalism may be found in the study of comparative cultures made by the Harvard sociologist Pitirim A. Sorokin, although Sorokin was somewhat more consistent in his adherence to cultural relativism. He also differed from Toynbee in arguing that the largest unit of comparison was not the civilization but the sociocultural "supersystem," an integrated system of values that might become dominant for a period of time in a given geographical area among given populations, but was in no sense identical to the area, the populations, or their total sociocultural life. Sorokin distinguished three major types of sociocultural supersystems, the "ideational," the "idealistic," and the "sensate." Each was grounded in a different value-premise—the "ideational" in the assumption that the ultimate reality was supersensory and superrational, the "sensate" in the assumption that it was sensory, and the "idealistic" in the assumption that both the sensory and the supersensory were real and rationally-integrally related one to the other. These supersystems tended to take turns in dominating the life of civilized peoples, each exhausting its possibilities after a time by reason of the principles of "immanent change" and "limit." Theories of linear progress, or of any eternally unlimited linear processes, were dismissed as "fantastic," and even the notion of the linear growth of knowledge collapsed if one took the long view: such a trend was "likely to be a mere part of a long-time parabola or other nonlinear curve." [61]

Sorokin attempted to preserve an attitude of neutrality with respect to his three supersystems. Each in its years of greatest creativity produced works of high quality in philosophy, art, literature, and other departments of culture; each evolved characteristic forms of political and social relationships, economic life, and so forth. Each, however, suffered from an inevitable lopsidedness, since the separate truths of faith, reason, and sensory experience were only aspects of truth in the absolute.[62] Here, then, was a relativism based not on scepticism but on presumed knowledge of absolute reality, a presumption rare in twentieth-century scholars in any field, not excluding philosophy and theology.

In practice Sorokin showed a marked preference for "ideational" or "idealistic" supersystems, partly because of his personal belief in a supersensory realm of reality, and partly because he felt himself to be living in the last and most decadent phase of an era of "sensate" dominance. His many books on the twentieth-century world crisis subjected this "late sensate" culture to violent abuse in the uncompromising polemical style of the pamphlets of the Reformation or the Russian Revolution.* Sensate culture in its death agony was generating "poison gas" rather than "fresh air." It had done "its best in the way of degrading man to the level of a mere reflex mechanism, a mere organ motivated by sex, a mere semimechanical semiphysiological organism, devoid of any divine spark, of any absolute value, of anything noble and sacred." The time had arrived, therefore, for Western man to concentrate on the building of a new society dominated by the premises of ideational or idealistic culture. Hard times lay immediately before us, but still further ahead loomed "the magnificent peaks of the new Ideational or Idealistic culture. . . . The great sociocultural mystery will be ended by a new victory. *Et incarnatus est de Spiritu sancto . . . et homo factus est . . . Crucifixus . . . et Resurrexit . . . Amen."* [63]

It is clear that both Toynbee and Sorokin are anything but classical exponents of a value-free academic relativism. Other recent attempts to work out a theory of civilizational cycles, by scholars such as Philip Bagby and Carroll Quigley, follow a more strictly empirical approach, as did Kroeber, although their works provide

---

* The similarity of Sorokin's rhetoric to that of the Russian Revolution is not necessarily coincidental, since he was active in revolutionary politics as an agrarian socialist and assistant to Kerensky. After several years of ideological struggle with the Soviet regime from his position as founder of the department of sociology at the University of Petrograd, he was expelled from Russia in 1922. He resumed his academic career in the United States in 1924.

only bare outlines of the subject. Both Bagby and Quigley, it should be added, find modern Western civilization at the same approximate point reached by classical civilization in the fourth or third centuries B.C., and are therefore not inclined to prophesy its imminent death.[64]

A biologist, Charles Galton Darwin, also contributed to the literature of cyclical theory in his book *The Next Million Years,* reaching somewhat gloomier conclusions. He saw no solution to the problem of population growth in relation to food supply and predicted alternating eras of feast and famine until the end of history. The modern Western world lived in a golden age, achieved through rapid industrialization and the exploiting of virgin continents, but Darwin doubted that it could last much longer or could ever quite be equaled in the future. The return of starvation and slavery, the increasing ferocity of wars, and "a greater callousness about human life" were all virtually certain.[65]

I I

The Age of Anxiety

THE ALMOST BIBLICAL WRITINGS of Toynbee and Sorokin not only illustrate a phase of twentieth-century sociological relativism. They also introduce us to contemporary cultural anxiety and despair. Both scholars adopt a more or less relativist attitude toward the historic sociocultural systems. At the same time, the writings of both are permeated with the hostility to humanistic optimism that has become epidemic in the spiritual life of our century. The familiar term "age of anxiety" is not inappropriate, but perhaps "age of disappointment" would serve as well. Most of the leading thinkers of the first half of the twentieth century were full-grown men by the year 1914. They had heard repeatedly the promises of their fathers and grandfathers that the new century would be a golden age. When these promises could not be kept, filial disappointment was bitterly keen. In the analysis of Albert Camus, who belonged to a still younger generation than the young men of 1914, the problem of the twentieth century was, therefore, how to live without grace or justice, without the hope of a kingdom of God but also without the hope of a kingdom of humanity. All through the preceding century mankind put itself in God's place and pledged to create a just millennial order on earth. But the dream of utopia "has retreated into the distance, gigantic wars have ravaged the oldest countries of Europe, the blood of rebels has bespattered walls, and total justice has approached not a step nearer." [1]

Perhaps because the principal source of anxiety and disappointment in the twentieth century has been man himself, man's wars,

204

man's tyrannies, and (one might say) man's false religion of progress, twentieth-century despair tends, like Toynbee's or Sorokin's, to resemble the despair of romanticism rather than of naturalism. It recoils from man's works, especially his works of war, politics, and industry. It recoils from modern history, as the bearer of disasters, and from the belief in progress, as the bearer of prideful and illusory hopes. It urges leaps out of time, out of humanity, out of reason, into the divine, the eternal, or the existential. It is a new romanticism of despair, seasoned with elements of preromantic traditional value-systems arranged in various combinations and proportions, but always essentially romantic in its theatricality and distrust of reason.[2]

This twentieth-century neoromantic pessimism has flourished with special vigor in theology, existentialist philosophy, the *belles-lettres*, and political thought. Not a few thinkers have been able to give their despair expression in two or three fields at the same time, as in the Christian political philosophy of T. S. Eliot and Eric Voegelin, the neomysticism and anti-utopism of Aldous Huxley, the political and theological writings of Emil Brunner and Reinhold Niebuhr, and the plays, novels, and treatises of Jean-Paul Sartre. The sections into which the present chapter are divided for purposes of exposition may, therefore, seem somewhat arbitrary, and many of the thinkers studied in one section might just as easily be studied in another. The nature of the material makes untidiness inevitable.

One of the most unexpected spiritual events of the twentieth century has been a renaissance in theology, coupled with a powerful resurgence of interest in the problems of religion on the part of the intellectual avant-garde. Although "faith" in the traditional sense has continued to deteriorate, the first half of the twentieth century has witnessed the greatest outpouring of creative theological thought since the sixteenth-century Reformation. For the most part it has been a theology of reaction and disillusionment, provoked by the apocalyptic political events of the century. Just as the theology of the nineteenth century took comfort and inspiration from the apparent triumph of modern Western man, so the theology of the contemporary age reflects his apparent defeat. Several theologians have pointed to the belief in progress as the most pernicious of all the doctrines that sustained modern man in his idolatry and *hubris;* in recent decades the strongest attacks against progressivism from the perspective of cultural despair are to be found in the writings of theologians and philosophers of religion.

Pessimistic Christian thinkers have been all the more concerned to refute the belief in progress because they see it as a creed not merely blasphemous but also heretical. As noted above, the standard interpretation of the origins of the idea of progress in recent scholarship suggests that progress is a secularized version of the Johannine millennium or even of the Christian heaven.* It is therefore quite possible for a Christian thinker to feel a certain sense of responsibility for the belief in progress, which is only Christian faith "gone astray."

Whatever the merits of this argument, clearly no sincere Christian can fail to observe the peculiar affinities between his faith and the doctrine of progress. As an offshoot of Judaism and as an heir of Greek philosophy and Asian mysticism, Christianity is simultaneously sympathetic and hostile to the concept of earthly progress. On the one side, it challenges man to cooperate with the divine in the establishment of the "Kingdom of God," follows Judaism in attaching great importance to history, engages in missionary activity throughout the world, and preaches the "good tidings" of salvation. On the other side, it also anticipates the end of history, dwells on original sin and human finitude, stresses the transcendental and ahistorical character of ultimate being, and contains a seemingly ineradicable streak of otherworldly asceticism that condemns the pleasures and values of profane life. The founder of Christianity himself has been convincingly portrayed from biblical evidence as both a virile extravert and a mystical introvert. In the nineteenth century theologians emphasized the this-worldliness of Christian faith; in the twentieth century, they have emphasized its otherworldliness.

Although the principal reason for the abandonment of liberal, immanentist theology has been disillusionment with modern man, which for theologians began during the First World War, a departure from the conception of the earliest Christians as men of worldly hope and good cheer is observable in some theological circles well before 1914. The writings of Alfred Loisy and George Tyrrell have already been mentioned in this connection.** In German theology the crucial figures were Franz Overbeck, Johannes Weiss, and

* See above pp. 12–14.
** See above, pp. 86 and 90–91.

Albert Schweitzer, all of whom insisted on the eschatological nature of early Christian faith. In their reading of the New Testament, Jesus and his disciples were not attempting to inaugurate a terrestrial millennium, but on the contrary expected the imminent dissolution of historical time. The duty of God-fearing men was to prepare themselves for the approaching Judgment. Not all the theologians who accepted this revised view of primitive Christianity urged a return to it; most did not. But they laid some of the foundations for the conservative reaction of the post-1914 generation by discrediting the scriptural credentials of theological liberalism.[3]

The revolt against liberalism was led in the 1920s by the Swiss preacher and theologian Karl Barth, a student of Harnack and Herrmann, who became converted to "neo-orthodoxy" during the First World War. Barth first shocked the theological community in 1919 with his *Epistle to the Romans,* a radically eschatological interpretation of St. Paul. During a long and prolific career, he wrote more than thirty other books, most of them available in English translations, including his *Church Dogmatics* in twelve volumes, published between 1932 and 1959, but his influence was greatest in the 1920s and early 1930s, at first in the German-speaking world and soon afterwards in Great Britain and America. He died in 1968.

Barth's early theology developed two themes of special relevance to the belief in progress: the infinite gulf fixed between God and man, which rendered all merely human effort hopeless, and the timelessness of "last things." Moved by his reading of Kierkegaard and Overbeck, and by the failure of liberal Christian ethics in the crisis of the First World War, Barth concluded that the trouble with modern man was his incorrigible tendency to rely on himself instead of on God, to hear God speaking within earthly history, instead of hurling down his word from eternity. Human culture, human religion, human ethics were fatally flawed by original sin, and had no power to save or improve mankind. "Everything which emerges in men," Barth proclaimed, "and which owes its form and expansion to them is always and everywhere and as such, ungodly and unclean." [4] Nothing could be more pitiful than the "happy gentleman of culture who today drives up so briskly in his little car of progress and so cheerfully displays the pennants of his various ideals." The end to which history moved was not a literal day of judgment or the achievement of a millennial divine kingdom on earth, as in the Ritschlian theology, but an end above time. Man's relationship to God, then, could be best described as vertical, and the world came to an "end" whenever God enabled man to make contact with the life eternal. Meanwhile, in the world of time, the hope of progress was vain since there existed "nothing in the whole range of human possibilities . . . which is capable of realizing the moral objective, the goal of history. Our range of possibilities is certainly capable of being increased and broadened, but its relation to the final goal must continue to be as $1 : \infty$." [5]

Beginning in the mid-1920s Barth's thought gradually became less angular. In his *Church Dogmatics* and other writings, he qualified his proto-existentialist eschatology, reaffirming both the meaningfulness of history and the approach of a genuine end of time, as prophesied in the New Testament. The infinite distance between God and man remained true for Barth in one sense, but he came to see that in another it was annihilated by Christ, in whom God disclosed his "humanity." Man's culture, despite its many failures, had its own natural good, as illustrated by the music of Barth's favorite composer, Mozart; and time, despite its many disappointments, had its ordained tasks. Although the kingdom could not come within time, and the world as such could never become Christian, the proclamation of the kingdom by evangelism would bring the good news in due course to all men. As a child of God, moreover, the Christian lived in hope. From hope issued action, and Christian action was always "very different" from that of non-Christians. "For each new day and year the Christian hopes. He hopes that throughout the Christian world and the world at large there will always be relative restraints and restorations and reconstructions as indications of the ultimate new creation to which the whole of creation moves." [6]

In short, with the indispensable aid of grace the world might indeed see better days before the inevitable end, although such provisional betterment is difficult to equate with progress, if only because man also remained at all times capable of willful self- and world-destruction. Even in his later years Barth did not concern himself with the question of progress in its purely secular meaning. As he told an audience in 1946, "The Church will always and in all circumstances be interested primarily in human beings and not in some abstract cause or other, whether it be anonymous capital or the State . . . or the honour of the nation or the progress of civilisation or culture or the idea, however conceived, of the historical development of the human race." [7] In any case, the radical Barth of the 1920s, not the older Barth, wielded the greater influence on his contemporaries.[8]

Barth's countryman Emil Brunner shared many of his fundamental views, and expressed them to a wide reading public during a life very nearly as long and productive as Barth's. Like Barth, he denounced liberal theology because of its alleged contamination by man-centered values, its dependence on science, and its surrender to historical relativism. He represented man as "cut off from the tree of life" by the curse of original sin. "A contradiction beyond repair clings to our life. . . . All the ways of man are sinful ways." [9] Brunner displayed more interest than Barth, however, in ethical and social questions, as evidenced by some of his most important books, from *The Divine Imperative* (in German, *Das Gebot und die Ordnungen,* 1932) and his Gifford Lectures, *Christianity and Civilisation* (1948), to his study of the idea of progress in relation to Christian eschatology, *Eternal Hope* (1954). Most of his thoughts on the question of progress are fully reported in *Eternal Hope.*

The modern gospel of progress, according to Brunner, was nothing more than an illegitimate child of Christianity, feeding parasitically on the spiritual capital created by centuries of Christian faith and hope. In the middle of the twentieth century its demise could be definitively announced. "The two world wars and the rise of the totalitarian states have destroyed it. They have shattered the two main pillars on which it rested, belief in technics and belief in the state and organization as the means of guaranteeing man's progressive control of his future." Even in its heyday, the doctrine of progress was "essentially only a slightly concealed hopelessness." How could faith in posterity "irradiate my present situation, inspire my deeds, or satisfy my aspirations—the thought that at some distant date generations of mankind who are as alien to me as the ghostly inhabitants of the past will be sitting at that imaginary switchboard which will enable them to control their future?" [10]

Progress was undeniably detectable in certain limited areas of human endeavor, in technics and social-technical organization, as well as in knowledge. Freedom expanded. Nor could any Christian overlook the leavening effect of Christian truth itself—its effect, for example, on marriage, relations between masters and servants, and the treatment of women, children, the weak, and the sick. The

New Testament seriously proposed that the Kingdom of God could develop progressively in a concealed and imperfect form in time.

But the nature of man and history ensured that all such progress would be countered by the onward march of evil. The expansion of freedom made possible the greater misuse or curtailment of freedom, as illustrated by the history of Germany first under the Weimar Republic and then under Hitler. "The greater the resources which progress places in [modern man's] hands, the more dreadful must be his work of destruction." Even the Kingdom of God, which Christians sought to realize on earth, had its perennial antagonist in the person of the Antichrist of the *kerygma*. "Just as there is a progress in good there is also a progress in evil; on both sides there is growth and accumulation. . . . After God has sown His seed the enemy comes by night and sows *his* seed. And then both grow together until the harvest of judgment day." [11] Only in eternity could the Christian hope achieve its goal. Brunner challenged his readers to let the sober scientific prophecy of thermonuclear doom help reawaken in their imaginations the biblical prophecy of a literal sudden end of history. Such an end had become thinkable again, with the collapse of man's secular illusions; once again it had become possible to anticipate the second coming of Christ in glory and man's entrance into his eternal reward of heavenly bliss above and beyond time.

Neo-orthodox, or perhaps one should say neo-Augustinian, attacks on the belief in progress have also been mounted by American theologians, in particular by Reinhold Niebuhr, in such works as *The Nature and Destiny of Man,* based on his 1939 Gifford Lectures, and *Faith and History,* published in 1949. Niebuhr argued that the modern idea of progress constitutes a fusion of the classical faith in reason with the Christian assertion of the meaningfulness of history. For centuries the belief in progress flourished, rendering the Christian faith apparently irrelevant. "Then came the deluge. Since 1914 one tragic experience has followed another, as if history had been designed to refute the vain delusions of modern man." [12] The error of modernism had been to assume that growth and development necessarily lead to progress, ignoring (unlike Christianity) the ambiguities of history. The growth of knowledge, freedom, power, and organization had enlarged man's possibilities for falsehood as well as truth, for evil as well as good, for the witness of Antichrist as well as for the witness of Christ, without effecting any change at all in human nature. "The Antichrist stands at the end of history to indicate that history cumulates, rather than solves, the essential problems of human existence. . . . Both the *civitas Dei* and the *civitas terrena* grow in history, as Augustine observes." [13]

But Christian realism could not be allowed to lead to fatalism. Niebuhr severely criticized Swiss neo-orthodoxy for not taking a stronger stand on behalf of Christian action in the world. Man's fallibility, he wrote, did not relieve him of his duty to serve the *civitas Dei* in his temporal existence. "There is no individual or interior spiritual situation, no cultural or scientific task, and no social or political problem in which men do not face new possibilities of the good and the obligation to realize them." [14] For Niebuhr this meant, in effect, the obligation to work for democracy, social justice, and world unity. "The expansion of the perennial task of achieving a tolerable harmony of life with life under ever higher conditions of freedom and in ever wider frames of harmony represents the residual truth in modern progressive interpretations of history." The Christian had only to remember that the growth of his wheat would not prevent the continuing growth of the tares of Antichrist, and that history could not be purged of its corruptions and contradictions until after the end of time. On the contrary, "The antinomies of good and evil increase rather than diminish in the long course of history." [15]

Barth, Brunner, and Niebuhr, despite their differences, belong quite obviously to the same tradition in Christian thought, a tradition that runs back to the Protestant Reformation and ultimately to the Latin Fathers of the early Church. Theirs was a faith founded on disjunctions: the gulf between God and man, between man and nature, between here and hereafter. Another stream of thought in the post-1914 era, which tended to replace neo-Augustinianism among the theological avant-garde in the 1940s and 1950s, is Christian existentialism. Its points of departure are the eschatological position of Barth in his *Epistle to the Romans,* the works of Kierkegaard, Nietzsche, and Heidegger, and a radically romantic humanism that recalls the long struggle in nineteenth-century thought against the Enlightenment and against positivism. The most illustrious representatives of Christian existentialism are Nicolas Berdyaev, Rudolf Bultmann, and Paul Tillich, all members of the same generation as Barth, but *vieillards* (rather than *enfants*) *terribles.*

Christian existentialism does not present quite the common front against the belief in progress encountered in neo-orthodoxy. Each representative of the movement has approached the problem of progress in his own way. All-encompassing generalizations are impossible.

Berdyaev, for example, could legitimately be studied as a believer in progress, since he foresaw the possibility of man's improvement in historical time. His thought (like Tillich's) is remarkably eclectic, incorporating not only the existentialist influence but also a heavy admixture of Russian Orthodox and German mysticism and several varieties of socialism—Marxist, idealist, and syndicalist. Born in Russia in 1874, he passed through a Marxist phase in his twenties. In the decade after the 1905 Revolution he was a prominent lay leader of the Orthodox revival, in association with such thinkers as Sergius Bulgakov and Leon Shestov. At this time Berdyaev retained enough of the revolutionary idealism of his youth to adhere to a program for Christian action similar to those of Ritschl and Rauschenbusch in the West: he espoused an unambiguous doctrine of progress that envisioned the full realization of the kingdom of God on earth. World War I and the horrors of the period after 1917 in Russian history forced him to abandon much of his earlier optimism. For the last twenty-four years of his life he lived in exile in Paris, where he published a new book almost every year until his death in 1948, including nearly all the volumes that established him as one of the outstanding religious minds of the century.

Berdyaev may be accused of inconsistency in several important respects, but he remained a humanist throughout his career, and the most serious charge that he brought against his age was its tendency to dehumanize the life of man. Fallen man in his pride often chose evil in preference to good, but the Christ-event had offered all mankind the chance of redemption. Redeemed man partook of the supernatural qualities of personality, love, creativity, and freedom that belonged to God himself. It was the irremediable heresy of Barthian theology, according to Berdyaev, that it posited the total depravity of man and the unlimited autocracy of God, thereby dehumanizing Christianity.[16] Traditional Christianity had in fact repeatedly undervalued or opposed man's intellectual and aesthetic faculties and his quest for justice, personal love, and freedom. To this extent, Christianity had failed in its mission. The truly creative human person drew his powers not from nature or history, but from God: redeemed man and man's redeemer were existentially united, so that each lived in the other.

For Berdyaev, the traditional Christian view of man erred, above all, in its doctrine of divine determinism. If man were not absolutely free to accept or deny grace, to work with or against God, he became nothing more than an automaton. But man's will was free, unfettered either by "fate" or "predestination." Only in this way could the sorrows of history be explained without impugning the goodness of God, since they followed from fallen man's freely elected preference for evil. "Only for that reason is the world process a terrible tragedy, only for that reason is history bloody and at its center stands the crucifixion, the cross on which the Son of God himself was crucified." [17] Nevertheless God had never compelled man to sin, and history also bore witness to man's redemption from sin with God's help and redeemed man's free acceptance of the divine challenge to work creatively for world transformation. Good and evil struggled together dialectically in historical time, and would continue to do so, until time's end.

Berdyaev regarded the twentieth century as a period of transition between an exhausted, dehumanized modernity and a second Middle Age. The failure of man in the twentieth-century world crisis was a judgment not only of the modern epoch but of history itself, revealing the bankruptcy of every effort to create societies unredeemed by grace and love. Yet Berdyaev stressed that God's kingdom could be partially realized even in historical time, through the free choice and acts of God's children. He called for the replacement of totalitarianism and capitalist democracy alike by a syndicalist-socialist Christian world brotherhood. With every dehumanizing ideology and social order thoroughly discredited by history, "Christianity will again become the only and the final refuge of man" since Christianity alone "stands for man and for humanity, for the value and dignity of personality, for freedom, for social justice, for the brotherhood of men and of nations, for enlightenment, for the creation of a new life." [18]

Taken this far, Berdyaev's interpretation of history seems, on balance, progressivist. Neither good nor evil could triumph absolutely in historical time, but the first coming of Christ and the possibility of a future partial achievement of the divine kingdom on earth pointed to both past and coming progress. Several Christian

thinkers who adhere to similar views will be studied in Part IV.*
Berdyaev differed from them in the overwhelming importance he
attached to still another future transformation: the abolition of
historical time itself at the end of history. His hopes for terrestrial
improvement notwithstanding, his stronger hope was for eternity.

> The striving towards the future, characteristic of our world-aeon,
> brings about a quickening of time and makes it impossible to dwell
> in the present for contemplating the eternal. But heavenly life is in
> the eternal present and means victory over the torment of time. . . .
> This aeon moves towards a catastrophe; it cannot last for ever, it is
> self-destructive. There will come another aeon in which the bad quick-
> ening of time and the bad striving for the future will be replaced by
> a creative flight into infinity and eternity.[19]

Even the possibility of finite progress within history arose
through man's yearning for eternity and his faith in its eventual
conquest of time. Without the eschatological hope all secular effort
was pointless. Also, because the kingdom of God had its being in
eternity, the kingdom "comes not only at the end of time but at
every moment. . . . There are two ways to eternity—through the
depth of the moment and through the end of time and of the
world." [20] The mystic's direct experience of God was nothing less
than a piercing of the veil of time, and in this way, too, the world
(if only for an instant) ended, affording an ecstatic glimpse of the
heavenly life to come. It must therefore be concluded that Berdyaev
was essentially an apocalypticist, and only in a limited, ambiguous,
and provisional sense a believer in progress. Temporal hope had its
source and its end in the hope of eternity.

A more consistently existentialist view of history is offered in the
thought of the German New Testament scholar Rudolf Bultmann,
who is sometimes accused of merely translating Heidegger's philosophy into the language of theology. Bultmann studied under
many of the same liberal theologians as Barth did in the period just
before the First World War. In the 1920s he was regarded as a con-
federate of Barth in the struggle against liberalism, although his
own special field of interest lay in New Testament studies, rather
than theology as such. Following the method of form-criticism
originated by Johannes Weiss, he became convinced that most of
the dogmas of the traditional Christian faith were meaningless to

* See below, ch. 13, especially pp. 245–46.

modern man because they had been formulated in terms of a first-century mythological world-view, which had now been irrevocably superseded. If the Christian message were to be saved, it would have to undergo "demythologization." A new and more relevant language was required in which to convey its trans-historical truth. Bultmann discovered such a language in the thought of Heidegger, although he never became a convert in any sense to Heideggerian metaphysics. He first expounded his program for demythologizing the New Testament in an essay published in 1941. Since then he has written prolifically, one of the rare figures in *Geistesgeschichte* who have produced the greater part of their best work in old age.

Bultmann interprets the idea of progress as a remote result of the waning of the eschatological interest in medieval Christian thought. As Christians came gradually to disbelieve in a literal end of time, they transferred their attentions to the problem of the meaning of history. The historical process as a whole was invested with meaning by finding providence immanent and active in terrestrial time. From there it was but a short step to the complete secularization of providence, and the conversion of the old eschatological vision into a fallacious doctrine of ever-increasing welfare.

In the late nineteenth and twentieth centuries, however, philosophers of history and existence have reached a deeper understanding of the meaning of history, which brings us back to some of the nuclear insights of the New Testament, stripped of their mythic integuments. Bultmann admires Croce and Collingwood, who have seen the profound truth that "humanity is a whole within each epoch and in each human individual." But the reality behind the Christian myth of the day of judgment is grasped only in the existential concept of personal decision for Christ here and now. The eschatological event is Christ himself, his first coming and his countless second comings in the lives of men, whenever the responsible self encounters its redeemer in the presence of eternity. The unity of history "does not consist in a causal connection of events, nor in a progress developing by logical necessity," but rather in the fact that "every moment is the *now* of responsibility, of decision." Historicism shows us that we cannot stand outside history and comprehend its total meaning, as if we were detached observers: therefore

> the meaning in history lies always in the present, and when the present is conceived as the eschatological present by Christian faith the meaning in history is realised. Man who complains: "I cannot see meaning

in history, and therefore my life, interwoven in history, is meaningless," is to be admonished: do not look around yourself into universal
history, you must look into your own personal history. Always in your
present lies the meaning in history, and you cannot see it as a specta-
tor, but only in your responsible decisions. In every moment slumbers
the possibility of being the eschatological moment. You must awaken
it.[21]

As one might expect, Bultmann pretends to no knowledge of the
designs of divine providence. God's ways are hidden, and his plans
are inscrutable. "I cannot speak of God's actions in general statements; I can speak only of what He does here and now with me,
of what He speaks here and now to me." [22] It follows that all ideas
of general, universal progress are meaningless. Bultmann seems
closer in this regard to the world-view of historicism than to cultural
despair or even to Heidegger's existentialism.

Bultmann's thought invites comparison with that of his fellow
countryman Paul Tillich, who does not hesitate to supply an onto-
logical definition of God akin to Heidegger's definition of being, but
whose doctrine of *kairos* and the "New Being" steers the same middle course between progressivism and eschatological literalism pursued by Bultmann.[23] Both men agree that the meaning of history
lies in the divine-human encounter, which happens here and now,
and must happen over and over again, as the result of the penetration
of time by eternity. In such encounters, and in such momentary
triumphs, there is no progress; within history, man's struggle to
experience God and defeat evil is endlessly renewed. The resemblance between this position and Barth's original explanation of the
eschatological mystery in *The Epistle to the Romans,* which (as we
have seen) he later abandoned, should be obvious.*

Rejection of the idea of progress is not limited to Christian thinkers of the eminence of Barth or Tillich. Nearly all the contributors
to the symposium on the kingdom of God and history commissioned

---

* Shortly before his death in 1965, Tillich recorded his astonishment that the
"breakdown of progressivism" had been so sudden and so radical in the post-1914
years; he now felt "driven to defend the justified elements of this concept," which
earlier he had joined with many others in attacking. He concluded that it was
legitimate to speak of historical progress in technology, science and scholarship,
education, and "the increasing conquest of spatial divisions and separations within
and beyond mankind." All such progress was measurable quantitatively and might
continue "in an indefinite future." But he remained hostile to any notion of "qualitative" progress. *Systematic Theology,* III (Chicago, 1963), pp. 338–39 and 352–54.

in the mid-1930s by the Oxford Conference on Church, Community, and State agreed that general progress within historical time was either impossible or unlikely, a conclusion that almost certainly could not have been reached by any similar body of scholars thirty or forty years earlier.[24]

The disavowal of liberalism is equally apparent in Karl Löwith's *Meaning in History*, which discovers man "at the end of the modern rope," purged of the "modern illusion" of progressivism; and in Erich Frank's lament that "in this world, it is always Caesar who is bound to be victorious, while Christ will for ever be crucified." [25] Other attacks on the idea of progress have been made by Dean Inge in his Romanes Lecture of 1920, by Paul Althaus in his important treatise on eschatology (*Last Things*, 1922), in the "realized eschatology" of C. H. Dodd, in the bitter attacks on modern "gnosticism" of Eric Voegelin, and in Josef Pieper's prophecy of the coming of Antichrist in the form of a totalitarian world superstate.[26] For Mircea Eliade the belief in progress is a futile attempt to escape from "the terror of history," and one of his disciples, the American "death-of-God" theologian, Thomas J. J. Altizer, interprets sacrality as "eternal recurrence." In relation to the absolute, Altizer contends, "there is only one tense: the present. No longer can we dream that the path to the sacred is *backwards,* nor can we live in the vain hope that the true path is only *forwards:* the Center is everywhere, eternity begins in every Now." [27]

Christian existentialism both preceded and followed the appearance of existentialism as an acknowledged movement in recent philosophy. Every twentieth-century existentialist owes something to Kierkegaard and Dostoyevsky. The Christian existentialism of Berdyaev and the Jewish existentalism of Buber were both represented by substantial published works several years before any of the major writings of Heidegger, Jaspers, or Sartre had come to light. After about 1930 the flow of influence reversed itself, and theology fell increasingly into the debt of philosophy.

Whether theological or philosophical, theist or atheist, existentialism is a tendency rather than a distinct school of thought. Insofar as existentialists present a common front to the world, they share one conviction: the person is to be understood first and primarily as a subject rather than an object. The person is a field of being, who has his being in time. But the time of his being is existential time:

private as opposed to public time, although each person's private time intersects with the private time of others, so that it assumes a social dimension. Most existentialists oppose or ignore the idea of progress because conceptions of the universal meaning of history tend to convert the human subject into an object. A belief in progress ordinarily requires the believer to postulate a public time, in which the acts of men achieve value only by contributing to this or that objective world-historical process. The present is "sacrificed" to the future, the man is "sacrificed" to mankind. Ideas of progress, and above all ideas of determined or automatic progress, are therefore "inauthentic." [28]

Existentialism as a philosophy holds both a tragic and a joyful view of life. It despairs of man as existence because existence is finite and contingent. It despairs of social man because society, and above all modern society, seeks to subordinate the human subject to a rational, scientific, technological order that narrows freedom to the vanishing point. The romantic cultural despair of the nineteenth century lives again in existentalism, quite recognizably. Yet existentialism as a philosophy may also be seen as a philosophy of joy, in Nietzsche's sense. The person who insists upon his freedom is free to live, within the limits of mortality, "beyond good and evil," beyond the injunctions of any externally imposed law or morality, beyond history (as world process). In the same way the future of all men, in their existential social relations, becomes open, not in the sense that all things are possible, but in the sense that nothing forbids anything from being attempted. The three leading existentialist philosophers have found authenticity at one time or another in a remarkable variety of existential social "projects": Martin Heidegger, briefly, in German National Socialism; Karl Jaspers, in political and economic liberalism; Jean-Paul Sartre, in the French Resistance movement during the Second World War and in revolutionary Marxist communism today. On the question of man's relationship to cosmic being, also, the spectrum is very wide, from Heidegger's embryonic ontology and Jaspers' almost Tillichian theism to the absolute atheism of Sartre.

Existentialist philosophers have devoted very little space in their work to explicit refutations of the idea of progress, but its refutation is implicit in the premises of their thought. The fullest analysis of time and history from the existentialist perspective is provided by Heidegger in his *Being and Time* (1927), where a clear distinction is drawn between "clock-time," seen as a succession of inexorable measured "nows," and the temporality of personal existence (*Dasein*). Heidegger's discussion of his disagreement with Hegel and his debt to Dilthey further illuminates the contrast between the traditional and existential views of time.[29]

Most existentialist thinkers have followed Heidegger in his analysis, either independently or under his direct influence. This did not prevent Jaspers from elaborating a nondeterministic schema of world history that includes a conception of progress, based only partially on existentialist premises.* More typically, Maurice Merleau-Ponty developed the rudiments of a philosophy of history based on Heidegger and Max Weber that rejected "universal" history altogether and argued that historical study is always the study of a particular past.[30] Sartre has explained the collapse of modern man's belief in progress as a result not only of the cogency of existential analysis but also of the growth of atheism. As God has disappeared from Western thought, *a priori* values have also disappeared. The secular moralist who imagines that the Enlightenment ideals of humanity, honesty, progress, and so forth can survive the dethronement of God is very much mistaken. Existentialists strongly oppose the hope that God can be suppressed "at the least possible expense." [31]

It should be added that Sartre's radical political conscience has led him in recent years to proclaim that existentialism must incorporate into itself an explanation of the historical process, which he can find only in Marxism. He does not, however, view Marxism as a body of *a priori* dogmatic truth, nor does he embrace its deterministic idea of progress; rather, he suggests that Marx has provided modern man with heuristic principles that enable him to see plainly how men are transformed into objects by socio-economic exploitation. Sartre also believes that Marxism points out the only authentic way of social life in the modern age, which is the way of proletarian revolution. Because revolutionary groups have a fatal tendency, however, to harden into organizations and bureaucracies, which by their oppressiveness create, in turn, the need for further revolution, there is no real progress in history. In the analysis of Sartre's Marxism by Wilfrid Desan, "Man builds over the ages his tomb and his prison, resuscitates and liberates, only to destroy again and to rebuild anew in a never-ending cycle." [32]

The existentialist thinker who has approached the problem of progress with the greatest explicitness and fidelity to the existentialist

* See below, pp. 285–88.

idea of freedom is Albert Camus. Like Sartre, Camus subscribed to a radical atheism, which found the universe silent, opaque, and absurd. In the absence of God or any organic relationship between man's existence and cosmic existence, man was compelled to abandon the search for objective values and look for value within his own being. The purpose of life in an absurd cosmos could be nothing more than living; a man's only legitimate master was himself. Camus urged his readers not to imagine that salvation could be won by a leap into faith, in the manner of Kierkegaard, or by a leap into futurity, in the manner of the believers in progress. Such leaps always involved the sacrifice of the real to the pursuit of the unreal: life taught "indifference to the future and a desire to use up everything that is given." An honest world would be "peopled with men who think clearly and have ceased to hope." [33]

In his most ambitious work, *The Rebel,* first published in 1951, Camus examined the sources of modern man's failure. Where had he taken a wrong turning? The gist of Camus's answer is that modern man, finding himself bereft of God, committed the mortal error of divinizing history. Beginning with the doctrine of progress in the Enlightenment and with Hegel's immanentist philosophy of history, modern man came increasingly to believe that his task was to enslave present humanity for the sake of a future liberation or salvation, dictated by the quasi-divine "logic" or "laws" of progress. All means became legitimate to him, including mass murder. Hitler, for instance, "was history in its purest form." [34] Similarly, socialist rebellion had decayed into a revolutionary totalitarianism that decreed interminable servitude and suffering for the sake of a classless utopia to be realized in the remote future. Only the path of the free, uncollectivized rebel, who refuses to bow to fatality in any of its forms, protected man from falling into bourgeois conformity on the one hand or into a career of revolutionary violence and terror on the other.

The hero of Camus's novel *The Plague,* Dr. Rieux, is the rebel *par excellence* of his creator's mature thought. Rieux freely chooses to remain in the plague city to minister to its inhabitants, not so much to help his fellow men as to reaffirm his own stubborn resolve to resist the injustice and absurdity of the cosmos. He defines and preserves himself by rebellion. He has no illusions about the possibility of "winning" his fight. The war against terror, he says, can never end, but must be waged eternally by men who, without trying to become saints, refuse "to bow down to pestilences" and "strive their utmost to be healers." [35] Or as Camus wrote in *The Rebel,*

> Even by his greatest effort man can only propose to diminish arithmetically the sufferings of the world. But the injustice and the suffering of the world will remain and, no matter how limited they are, they will not cease to be an outrage. Dimitri Karamazov's cry of "Why?" will continue to resound; art and rebellion will die only with the last man.[36]

The work of Albert Camus calls to mind the powerful contribution of belletrists to the offensive against the belief in progress. "We later civilizations," as Paul Valéry remarked in 1919, ". . . we too now know that we are mortal." Many ancient empires had sunk into the earth without a trace. "We see now that the abyss of history is deep enough to hold us all. We are aware that a civilization has the same fragility as a life." [37] In the poet's vision, progress reduced to nothing more than greater technical precision, which man busily applied to his self-destruction.

Valéry is the classic example of the respectable twentieth-century man of letters: post-symbolist poet, philosopher, patriot, aesthetic aristocrat, leading figure in the Institute of Intellectual Co-operation of the League of Nations, honored by a state funeral in 1945 in the presence of General de Gaulle. But his intimations of disaster place him in the mainstream of contemporary thought, not in the far outposts manned in the nineteenth century by such "disreputable" writers as Baudelaire and Dostoyevsky.

To report fully the contribution of poets, novelists, playwrights, and artists to the assault on the belief in progress since 1914 would be a formidable undertaking indeed. The existentialists have made frequent use of literary forms to present their ideas: Gabriel Marcel has written extensively for the stage, Camus and Sartre are novelists and playwrights of the first rank. Some of the most vital political thought of the century has reached its public in the form of novels and romances of the dystopian future. Painting and sculpture since the early 1900s provide a visual record of the disintegration of the sense of form, value, and meaning in modern thought. The representative writer and artist (outside the Marxist countries) no longer takes his signals from the natural sciences, the warfare between reason and faith, or the struggle for social justice. His themes are the unconscious, alienation, absurdity, dehumanization, the malevolence of the machine, the nausea of time. The literature of historical engagement, in the tradition of Voltaire, Dickens, Tolstoy, Ibsen, and Zola, has flourished in our century only fitfully —perhaps most authentically in the 1930s, as a response to the immediate political challenge of Hitlerism and Stalinism, and the economic challenge of the Great Depression. The competing tradition of romantic cultural despair has otherwise clearly prevailed.

The most direct attack on the gospel of progress led by creative writers has emanated from a set of mind that one might define as "chronophobia." [38] The chronophobe experiences disgust or ennui or meaninglessness in his confrontation with time. He seeks to escape from the relentless pressure of history into an inner or greater world beyond time's reach. Existentialism, with its emphasis on the private "now" of decision and its counsels of resistance to the tyranny of "public time," is one of the chief philosophical sources of modern chronophobia. We have also discovered chronophobia in twentieth-century theological thought, which opposes a radically eschatological world-view to the immanentism and historical-mindedness of the theology of the nineteenth century. For the creative writer, an especially important additional influence has been the world-view of Jungian psychoanalysis. Jung's discovery of the psychotherapeutic value of the eternal "archetypes" in the "collective unconscious" led to a peculiarly modern form of gnosticism, as Thomas J. J. Altizer points out. "Like all forms of *gnosis*, it is grounded in a nondialectical negation of the world, finally dissolving reason, consciousness, and history in its search for a total consummation and liberation of the 'self'." [39]

Chronophobia is not something that existentialists, eschatologists, or psychoanalysts have forced modern writers to discover. It is very much in the air, and freely available to all. In the novels of Franz Kafka, for example, and most powerfully in *The Castle,* the world of everyday experienced time—each event in itself more or less plausible—becomes a nightmare of meaninglessness. In time, nothing that is real can happen. No progress can be made toward the truth. Reality is represented symbolically as the castle that Kafka's earnest and tormented hero cannot reach, although it seems but a short distance away. In Erich Heller's analysis, "it was the keenest wish of Kafka the artist . . . to write in such a way that life, in all its deceptively convincing reality, would be seen as a dream and a nothing before the Absolute." [40] Yet the Absolute does not speak to man; Kafka died a mystic *manqué,* a rejected suitor of eternity, condemned to the inauthentic life of time.

Much the same interpretative strategy may be applied to the plays and novels of Samuel Beckett. A disciple for many years of James Joyce, whose own work was obsessively chronophobic and antihistorical, Beckett devoted his first book to an evaluation of Marcel Proust and particularly to Proust's study of time. Beckett's best-known play, *Waiting for Godot,* follows closely the "plot" of *The Castle.* Vladimir, one of the two tramps waiting for the arrival of the mysterious Godot, is a figure not unlike Kafka's K. He hopes, and now and then his partner Estragon hopes with him, that Godot will come, that salvation from boredom and suffering will be theirs. But just as K. could not reach the castle, Godot never comes, and the cyclical absurdity of life continues without relief. In the mime play *Act without Words,* a desert castaway is tempted by a dangling pitcher of water, to which his attention is called by whistles from offstage, but no matter how he exerts himself, the water slips out of his reach. At last, he refuses to play the world's game. He sits motionless, staring at his hands, ignoring time's importunity, and approaching by world denial the peace that eluded Vladimir and Estragon. "Nothing," Beckett quotes Democritus in *Malone Dies,* "is more real than nothing." [41]

A rather different, and more subtle, flight from time occurs in the novels of Hermann Hesse. Throughout his artistic life, Hesse was a twentieth-century avatar of the introverted humanist of early German romanticism. Like some of the first German romantics, he tempered his mysticism with the classical ideal of cosmic harmony, and he also borrowed themes from psychoanalytical thought and the religion and philosophy of India. As an exile from Germany who lived throughout his productive later years in a remote corner of Switzerland, he foresaw the imminent collapse of modern Western civilization. Because of its soulless materialism, its lethal tribalism, and its idolization of science and technics, it had failed to make contact with the timeless inner world of ultimate reality. As Hermine observes in *Steppenwolf,* we could not live at all "if there were not another air to breathe outside the air of this world, if there were not eternity at the back of time; and this is the kingdom of truth." [42]

Steppenwolf, published in 1927, was set in contemporary Germany, but in an earlier novel, *Siddhartha,* Hesse explored the possibilities of mystical self-fulfillment in the India of Buddha's day; and in his last novel, *The Glass Bead Game* (1943), he depicted the neomedieval world of the distant future, where an esoteric order of artist-intellectuals had created a spiritual exercise uniting all departments of higher culture, "the sublime cult, the Unio Mystica of all the separate limbs of the Universitas Litterarum." Time seemed to have stopped at last. But even the Bead Game was not immune to the corrosive effects of time. The hero Knecht eventually resigned from his post as Magister Ludi, warning his colleagues that "even beauty . . . is transitory once it has become history and taken on an earthly form." [43]

Some of the most explicit attacks on the belief in progress from the perspective of chronophobia took place between the two world wars in English literature. The novelist and painter Wyndham Lewis, in this *Time and Western Man* (1927), quarreled vigorously, if not always discriminatingly, with what he called the "time-cult" in modern philosophy and letters. The "time-cult" included a considerable variety of minds, from Joyce and Gertrude Stein to Bergson, Spengler, and Alexander. All of them posited a universe of flux and restless action, which usually involved progress, against which Lewis upheld his own "spatial" and "visual" values of matter, line, and form, epitomized by Hellenic sculpture. Certain time-cultists sought to persuade us that "we are in the process of making a superior reality to ourselves," which at first might seem to furnish man with "new causes of self-congratulation," but was in reality a profoundly degrading and pessimistic philosophy. "You, in imagination, are already cancelled out by those who will 'perfect' you in the mechanical time-scale that stretches out, always ascending, before us. What you do and how you live has no worth in itself. You are an *inferior,* fatally, to all the future." [44]

English poetry found its most authentic voice of despair in the 1920s in T. S. Eliot, a transplanted American. *The Waste Land,* published in 1922, may be read at one level as a statement of the barrenness and futility of history. Scenes of postwar Europe alternate with glimpses of earlier ages, and all are equally "unreal," emptied of meaning and value; time is flattened out so that all eras appear to flourish simultaneously. Eliot followed *The Waste Land* with a still more despondent work, "The Hollow Men," in 1924–25. In 1927, after years of scepticism and spiritual torment, he found the security for which he had long been searching in the Church of England.

From this point his work became almost theological in its thrust, without abandoning its earlier chronophobia. "Burnt Norton," the first of his *Four Quartets,* contrasts the meaningless world of temporality mirrored in "the strained time-ridden faces" of modern men with the divine love, "unmoving . . . timeless and undesiring." The seemingly unbridgeable gap between the two was bridged by the Incarnation: in Christ eternity met time, linking creature to creator. Eliot also made use in "Burnt Norton," and in his play *The Rock,* of the image of the endlessly turning wheel, which represented time and history, but whose center was the "still point," our starting and ending point as servants of eternal being.[45]

Eliot's conversion to orthodox Christianity gave him an antidote to the vertiginous life of duration. At the same time, it challenged him to produce a philosophy of culture, somewhat analogous to the neomedievalism of the nineteenth-century romantics. *The Idea of a Christian Society* expounded his view of the relationship between religion and culture. In *Notes towards the Definition of Culture,* published in 1948, he proposed a distinction between "higher" and "lower" cultures and between ages of "advance" and of "retrogression." On this scale Christian culture ranked first, but only in those past centuries when Christianity had been faithfully lived. A people following "a religion of partial truth" might at times surpass "another people which had a truer light" by adhering with greater fidelity to their faith.

> We can assert with some confidence that our own period is one of decline; that the standards of culture are lower than they were fifty years ago; and that the evidences of this decline are visible in every department of human activity. I see no reason why the decay of culture should not proceed much further, and why we may not even anticipate a period, of some duration, of which it is possible to say that it will have *no* culture.

Even during times of "advance," Eliot added, there was no possibility of "greater sanctity or divine illumination becoming available to human beings through collective progress."[46] *

---

* Another poet who would certainly have joined Eliot in denouncing modern civilization and the whirl of time, had he not been killed in action in 1917 while still a young man, was T. E. Hulme. Indeed, he had already reached a position similar to Eliot's on the question of progress, which appeared in his posthumously published *Speculations,* a work of no little influence in the 1920s. Hulme deplored the "modern method of disguising . . . the futility of existence," which consisted

Aldous Huxley, a grandson of Thomas Henry Huxley, no doubt attracted the greatest number of readers of all the twentieth-century English chronophobes. His natural medium was satire, in which he operated with destructive facility in such novels of the 1920s as *Crome Yellow* and *Point Counter Point*. Both mysticism and progressivism came in for their share of ridicule. In 1936, however, he announced his conversion to the "perennial philosophy" of the mystics in his novel *Eyeless in Gaza,* which he followed in 1944 with another novel, *Time Must Have a Stop,* and in 1945 with a systematic exposition of his views, *The Perennial Philosophy.* He now maintained that man's final end was "the knowledge of the immanent and transcendent Ground of all being," a goal to which men in all periods of history had enjoyed access. "Only in regard to peripheral knowledge" had there been "a genuine historical development." The modern belief in inevitable progress amounted, in the last analysis, to "the hope and faith (in the teeth of all human experience) that one can get something for nothing." It was much more probable "that gains in one direction entail losses in other directions, and that we never get something except for something." Yet all progress in this world, real or illusory, made little difference to the perennial philosopher, who saw communion with eternity as the highest purpose of life.[47]

Huxley could not remain indifferent to the historical situation of modern man, however, and he devoted much of his later work to warnings of disaster and formulas for establishing world peace and world order. Even before *Eyeless in Gaza,* his *Brave New World* alerted a whole generation to the possibility of a technocratic counterutopia. In *The Perennial Philosophy* he suggested that the "reign of violence" in the world could be brought to an end by universal acceptance of a mystical world outlook and by universal rejection of "political pseudo-religions" and idolatrous "time-philosophies."[48] His last novel, *Island* (1962), depicted an East Indian utopian society uniting the highest values of Buddhism and Western culture. He also did much to promote the contemporary interest in world ecological planning. But all his efforts to improve civilization were grounded in the paradox that only a world outlook that saw the

of imagining that the wheel of time spiraled upwards, when in reality it only turned in a simple circle; the doctrine of progress was "a lasting and devastating stupidity." *Speculations* (London, 1924), pp. 34–35.

ultimate good residing in eternity rather than in the future could unify the human race and make a terrestrial utopia possible.

In one last example, chronophobia is fused with existentialism in the deeply pessimistic thought of the Rumanian expatriate writer E. M. Cioran. Born in 1911, Cioran came to Paris in 1937, and has lived in France ever since, publishing five volumes of essays somewhat in the style of Nietzsche, with resonances of Schopenhauer and Spengler. His titles make clear the orientation of his thought: *Survey of Decomposition, Syllogisms of Sorrow, The Temptation to Exist, History and Utopia, The Fall into Time.* He has also edited a collection of the works of Joseph de Maistre.

Cioran adheres to an almost Augustinian anthropology. The world's suffering is due to the incorrigibly bad will of man. By his very nature man strives to work his will in the temporal-historical process, and this striving itself brings about his ruination. Optimism, Cioran writes, is to be found "among botanists, specialists in the pure sciences, explorers, never among politicians, historians, or priests. . . . We turn sour in the vicinity of man. Those who devote their thoughts to him, examine him or try to help him, sooner or later come to despise him, hold him in horror." The best priests are in fact "voyeurs . . . of original sin," who experience a cynical pleasure in observing the follies of their flocks.[49] Man's will takes only two forms: either it is sickly and therefore ineffective, or it is aggressive, selfish, ambitious, and dynamic, in a word—evil. "We march toward hell with every step we take to transcend the vegetative life, whose passivity should serve as the key to all things, the supreme answer to all our questions; the horror such a life inspires in us has made us into this horde of civilized folk, of omniscient monsters who know everything but what is essential." Cioran suggests that man's search for knowledge is perhaps the darkest of all his passions, and the most dangerous. "We are born to exist," he writes, "not to know; to be, not to assert ourselves. Knowledge, having irritated and stimulated our appetite for power, will lead us inexorably to our doom. *Genesis* perceived our condition better than have all our dreams and our systems." In another passage, he condemns "that murderous curiosity which prevents us from conforming ourselves to the world," which by idealizing knowledge and action "has in the same blow ruined being, and, with being, all possibility of the golden age."[50]

227

Turning to man's "dreams and systems," Cioran follows the familiar strategy of defining theories of progress and visions of utopia as modern substitutes for Christian faith, although he also finds an important source of progressivism in the historical optimism of the Jews.

> Not content with having preached the idea of progress, they [the Jews] have even seized upon it with a sensual and almost shameless fervor. Did they expect, by adopting it without reservation, to benefit by the salvation it promises humanity in general, to profit from a universal Grace, a universal apotheosis? The truism that all our disasters date from the moment when we began to glimpse the possibility of "something better"—they will not admit.[51]

In any event, by transforming the terrible God of Judeo-Christian tradition into the historical world will, we have only made God all the more terrible. Cioran agrees with Nietzsche that ours is an age of decadence, above all in Europe, and with Spengler that the next step is the establishment of a new Roman Empire forged by the sword and ruled by a despot, of whom the herald and prototype is Adolf Hitler. "The dispersed human flock will be reunited in the keeping of a pitiless shepherd, a type of planetary monster before whom the nations will prostrate themselves, in a terror bordering on ecstasy. The universe on its knees, an important chapter of history will be closed." After a time, however, the new regime will collapse and man will return to barbarism and anarchy. "Dürer is my prophet. The more I contemplate the march of the centuries, the more I am persuaded that the only image capable of revealing its meaning is that of *The Horsemen of the Apocalypse.*"[52]

In such a world, the resolve to exist and to go on existing is quite irrational, an act of faith in defiance of truth. "Do you deign to breathe? You are approaching sainthood, you deserve canonization."[53] But the goal of existence is not the life of time. Cioran repeatedly confesses to a feeling of nostalgia for the prehistoric golden age when mankind lived in the eternal present. In saying "yes" to the temptation to exist, he also says "no" to the tasks of Prometheus. History

> is not the seat of being, it is the absence of being, the *no* of everything, the rupture of the living with itself. . . . It is within ourselves that we must search for the remedy for our evils, in the timeless principle of our nature. . . . Useless to climb back toward the ancient

paradise or to run toward the future: the one is inaccessible, the other unrealizable. . . . There is no paradise except in the deepest depths of our being, in the self of the self.

In the end, if we can believe him, Cioran is the complete mystic, clinging "to the world no better than a ring on a skeleton's finger."[54]

Nearly all normative political thinking since Rousseau has addressed itself to the formulation of ideological programs for human progress. The period between 1914 and 1945 was not exceptional in this respect. Democratic liberalism, fascism, social democracy, and communism struggled for ascendancy at the level of ideas and at the level of power. Visions of progress or utopia inherited from the nineteenth century provided the indispensable spiritual *élan*. Since 1945, ideologies have, by and large, fallen out of fashion in the Western world, a development that patently reflects the decline of political optimism in our time. But this decline was already quite obvious many years before 1945 among certain segments of the Western intelligentsia. It found expression in the revival, for example, of anti-progressivist theologies of politics, as mentioned earlier in this chapter. It is apparent in the value-free sociological approach to politics of such scholars as Pareto and Weber discussed in the preceding chapter. Yet another index to the decline of political optimism is the relative decline of interest in politics as a remedy for the human predicament: the twentieth century as a whole has given rise to a disproportionate number of thinkers who see no hope in the political process at all, and whose thought is fundamentally antipolitical.

The failure of political optimism is nowhere better documented than in the counter-utopism in twentieth-century imaginative literature. The counter-utopia as an established *genre* in fiction dates only from the years since the First World War, but noteworthy examples were published just before it, including some of the novels of H. G. Wells,* and Jack London's *The Iron Heel* (1907) and *The Scarlet Plague* (1912). Here novelists with socialist sympathies represented the menace to progress as proto-fascist plutocracy, the unholy alliance of totalitarian democracy and capitalism. More recently, counter-utopists have dwelled on the potentialities for dehumanization of socialism itself, and also of science and technol-

* See above, p. 169.

ogy conceived as ends in themselves by benevolent (or malevolent) technocrats. Evgeni Zamyatin's *We,* first published in 1922, pictured a counter-utopian Soviet society of the 26th century, anticipating George Orwell's *Nineteen Eighty-Four* (1949). One of the first counter-utopian indictments of technocracy was delivered by E. M. Forster in his short story "The Machine Stops" (1909), which envisaged a decadent humanity living underground in total dependence on its machine culture, and eventually eliminated, save for a few exiles on the surface, when the machines began to break down and could not be repaired. Dehumanization by totalitarian technocracy was the theme of Aldous Huxley's *Brave New World* (1932) and the Czech novelist Karel Čapek's *War with the Newts* (1936).

Since 1945, counter-utopian novels, plays, and films have been produced in such quantities that the counter-utopia is now almost a literary cliché. *Nineteen Eighty-Four* has enjoyed the greatest public success, in part because of its obvious political usefulness in the Cold War. Writers of "science fiction," which in the recent context is more often "anti-science fiction," have contributed such excellent studies in cultural despair as Ray Bradbury's *Fahrenheit 451,* Kurt Vonnegut's *Player Piano* and *Cat's Cradle,* and John Brunner's *Quicksand.* C. S. Lewis' *That Hideous Strength* and William Golding's *Lord of the Flies* are in the same tradition. On the screen, in addition to creditable productions of *Nineteen Eighty-Four, Fahrenheit 451,* and *Lord of the Flies,* a story by Robert Sheckley has been skillfully adapted by Elio Petri in his film *The Tenth Victim,* and Jean-Luc Godard has created *Alphaville* and *Weekend.*

Whereas most counter-utopias discover mankind enslaved by demagogic tyrants or the ubiquitous machine, *Weekend* is Godard's commentary on the increasing incivility of civilized life in the late 1960s, perhaps auguring a new wave of counter-utopias in which the causes of hell will be sought in man's instinct for anarchy and violence, rather than in his superabundant talents for organization. At one level, although other readings are possible, Eugène Ionesco conveys a similar message in his plays *The Killer* and *Rhinoceros.* In the former, a utopian housing development known as the "radiant city" is uninhabitable because it is stalked by a snickering homicidal madman. In the latter, men and women change one by one into angry pachyderms until only the uncertain hero remains in human form as the curtain falls.

On occasion, the counter-utopist interrupts his catalogue of horrors to fix some of the blame for modern man's decline on the belief in progress itself. One such interruption occurs in the middle of Aldous Huxley's *Ape and Essence* (1948), set in the ruins of California after the Third World War. The survivors of the war were a race of mutants who worship Belial instead of God. As their Arch-Vicar explained in a long conversation with a visiting New Zealander, the see-saw battle between God and Belial had been going on for "a hundred thousand years or so" until the Industrial Revolution and the sudden sharp rise in the world's population in the nineteenth century. "Almost overnight the tide starts to run uninterruptedly in one direction." Belial took charge of world history. The increase in population forced further increases in food production, which failed to keep pace with demand, until the coming of the "Higher Hunger . . . the hunger that is the cause of total wars and the total wars that are the cause of yet more hunger." All the while, mankind, prompted by Belial, lived under the illusion that its plunder of the earth's resources and its senseless proliferation constituted "progress." Progress was "the theory that Utopia lies just ahead and that, since ideal ends justify the most abominable means, it is your privilege and duty to rob, swindle, torture, enslave and murder all those who, in your opinion (which is, by definition, infallible) obstruct the onward march to the earthly paradise." Marx had said that force was the midwife of progress.

> He might have added—but of course Belial didn't want to let the cat out of the bag at that early stage of the proceedings—that Progress is the midwife of Force. Doubly the midwife, for the fact of technological progress provides people with the instruments of ever more indiscriminate destruction, while the myth of political and moral progress serves as the excuse for using those means to the very limit. . . . The longer you study modern history, the more evidence you find of Belial's Guiding Hand.[55]

Prophecies of man's self-dehumanization through the achievement of a technocratic superstate are not, of course, confined to imaginative literature. The progressive enslavement of man to his own machines and to his own baleful genius for rationalization and organization is a familiar theme in many diagnostic studies of the ills of modern civilization. Some see the continuation of this tendency to its ultimate, counter-utopian conclusion more or less inevitable; for others there is hope. Karl Jaspers was among the first thinkers of his generation to suggest, in *Man in the Modern Age* in 1931, that the most serious threat to the persistence of human life on earth was "the transformation of human beings into functions of a titanic apparatus," the apparatus of a technological mass-order, in which humanity, properly speaking, no longer existed. This mass-order could find no use for freedom or for the exceptional man; it compelled "a general levelling-down." One could legitimately fear "that the whole history of mankind is a vain endeavour to be free. Perhaps freedom has only existed for a real but passing moment between two immeasurably long periods of sleep, of which the first period was that of the life of nature, and the second period was that of the life of technique. If so, human existence must die out, must come to an end in a more radical sense than ever before."[56]

In recent years Jaspers has been echoed by many writers, both Christian and humanist, who share his anxiety. Erich Kahler, Friedrich Georg Juenger, C. S. Lewis, Alex Comfort, Jacques Ellul, Lewis Mumford, Joseph Wood Krutch, and Roderick Seidenberg come to mind at once as prophets of what Lewis terms "the abolition of man."[57]

Seidenberg's verdict is perhaps the most pessimistic. History, he writes, is no more than a period of transition from the organic life of instinct to the rational life of organization. Order breeds more order. "Organization, dissolving chaos and incoherence in ever wider arcs of life, moves toward universality." Once the final world order comes into being, it will no longer be subject, as were the orders of history, to change or "progress." The individual will be completely integrated into his society, reviving at a higher level the anonymity of prehistoric times. "The machine, as the most highly crystallized form of organization achieved by man, demands a degree of co-ordination in the societal relationships of man corresponding to its own high functional development." Unless a positive movement of the spirit can succeed in restoring to man his sense of communion with the divine and in preserving authentic freedom, which Seidenberg finds improbable, man's historical life is nearly over. "For the process of crystallization . . . is a converging, cumulative, essentially irreversible process that approaches a condition of stable equilibrium as its limit." The disappearance of man's faith, first in the spiritual world of animism, then in the geocentric cosmos, and finally in the dignity of the person and the existence of God "may be merely stages in his diminishing stature before he himself vanishes from the scene—lost in the icy fixity of his final state in a posthistoric age." [58]

Ellul adds that virtually all talk of "controlling" the pitiless advance of technology through a new "humanism," whether Christian or atheist, misses the mark, since technology progresses according to its own laws, which mere wishing or willing cannot repeal. A "monolithic technical world" is coming into existence, he warns, beyond man's power to stop or to regulate, "a dictatorship of test tubes rather than of hobnailed boots." [59]

Counter-utopian speculation also extends to fears that through biopsychological engineering even the human body and mind may be altered to conform to the demands of tyrants or the blueprints of social planners. Aldous Huxley offered some unpleasant forecasts along these lines in *Brave New World*. Gordon Rattray Taylor has surveyed the most recent possibilities in *The Biological Time Bomb,* including the counter-utopian aspects of "cloning" (production of exact duplicates of organisms by laboratory cultivation of body cells), the surgical bonding of machinery and living bodies to create "cyborgs," further refinements in the arts of brainwashing and thought control, "improvement" of racial stock through "gene surgery," and the military applications of bacteriology. The author concludes that the rate of scientific discovery in the twentieth century has been far too rapid, and that research and its applications must be severely curtailed if we are to avoid both societal and personal disintegration. But the odds do not work in our favor. Too much imagination and effort are needed. "Current indications are that the world is bent on going to hell in a handcart, and that is probably what it will do." [60]

Those political thinkers who are able to continue in one or another ideological, philosophical, or theological tradition often explicitly dissociate themselves from the idea of progress, even if they are not prepared to rule out all possibilities of future progress. As Judith N. Shklar has shown, one of the leading trends in twentieth-century political thought, especially powerful in the 1940s and 1950s, is "the liberalism of defeat," which brings together such thinkers as Wilhelm Roepke, Friedrich A. Hayek, Ludwig von Mises, Bertrand de Jouvenel, and Alfred Cobban in a common front against socialism and totalitarianism.[61] The conservative liberal despairs of the modern world and traces its decline to the French Revolution, Jacobinism, "totalitarian democracy," the St.-Simonian and Positivist doctrine of social engineering, scientism, and the belief in inevitable progress. The liberal virtues of freedom and reason have been ruthlessly subordinated, so the analysis runs, to the will of masses and the plans of experts. The predictable outcome is the fascist or socialist or democratic superstate, which is also a slave state. Thus, Albert Salomon, in a critique of St.-Simon and Comte, maintains that "the logic and tyranny of progress gave to the world the progress of total tyranny," and Walter Lippmann—an American counterpart in some respects of the European thinkers discussed by Shklar—warns that "the decline of Western society" cannot be arrested unless thinking men return to "the great tradition of the public philosophy" and reject "the Jacobin conception of the emancipated and sovereign people." [62] *

Even socialists in the Western democracies have been somewhat deficient in hopefulness in recent decades, and quick to find the nineteenth-century belief in automatic progress guilty of fostering dangerous illusions. George Orwell's later writings perfectly illustrated despondent socialism. In quasi-official statements of the socialist position, such as Richard Crossman's much quoted first article in the *New Fabian Essays* of 1952, the gospel of progress has had to accept severe criticism. Both the evolutionary and revolutionary philosophies of inevitable progress have been proved false, Crossman contends. "Judging by the facts, there is more to be said for the Christian doctrine of original sin than for Rousseau's fantasy of the noble savage, or Marx's vision of the classless society." He agrees with Reinhold Niebuhr that there has been no moral progress whatever in history. In the realm of social progress some periods surpass others, but "exploitation and slavery" are the "normal state of man," and the "brief epochs of liberty" scattered through the pages of history should be regarded as exceptional. The only continuous lines in which progress can be traced in history are "the social accumulation of knowledge" and "the enlargement, through this accumulation, of men's power to control both nature and one another," which can be used for good or evil. Future progress remains possible—but only through the constant application of human will and social conscience, and without any guarantee of success in the long run or the short.[63]

Or, as the Australian philosopher Eugene Kamenka writes in a recent symposium on Marxist humanism, "History is neither the

* On balance, however, Lippmann is a believer in progress. See below, p. 324.

story of the progressive unfolding of a spontaneously co-operative human essence nor is it the inevitable march toward a truly just and human society. History is the battleground of competing traditions, movements, and ways of life: it presents us with no total story and no final end." Like the exiled Trotsky, the socialist humanist must recognize that history and society "can confront us with one outrage after another; when they do, he will, like Trotsky, have to fight back with his fists." [64] *

It goes without saying that views such as Crossman's and Kamenka's are almost infinitely distant from the historical world outlook of nineteenth-century socialism. The belief in progress is not totally discarded, but at the very least it is seriously compromised. To quote Shklar once again: "Utopianism is dead, and without it no radical philosophy can exist." [65] Although signs of a resurgence of utopianism can be observed in political life since the 1960s, her judgment is quite sound for the years immediately following World War II.

We closed our chapter on nineteenth-century pessimism with a reference to Flaubert's *Bouvard and Pécuchet*. Another, much more serious, yet sometimes equally ironic debate on the values and prospects of civilization takes place in Thomas Mann's novel *The Magic Mountain* (1924) between Ludovico Settembrini and Leo Naphta. Settembrini is the grandson of a *carbonaro* and a valiant defender of the principles of the Enlightenment: reason, science, liberty, humanity, progress. His antagonist, Naphta, a Jew who has become a Jesuit, speaks for the world-view of the Middle Ages: faith, hierarchy, discipline, original sin, war. Naphta prophesies a coming hell of conflict and chaos that will make the revolutions of Jacobinism seem trivial by contrast. Mann's Jesuit is a prescient portrait of all the nay-sayers of the twentieth century. It is time now once more, in our own study, to listen to Settembrini.

---

* Trotsky did not conceal his disgust with the twentieth century. As he wrote near the end of his life, "This throw-back to the most cruel Machiavellism seems incomprehensible to one who until yesterday abided in the comforting confidence that human history moves along a rising line of material and cultural progress. . . . All of us, I think, can say now: No epoch of the past was so cruel, so ruthless, so cynical as our epoch. Politically, morality has not improved at all by comparison with the standards of the Renaissance and with other even more distant epochs." Leon Trotsky, *Stalin,* tr. Charles Malamuth (New York, 1941), p. xiii.

PART FOUR

The Survival of Hope,

1914-1970

Does man advance? Well, don't be dull!
How can Marx and civil rights,
Keats and Proust mass-produced,
Peasants lounging in the Louvre,
Yeomen at the pinnacles of power,
How can dams and Spanish beaches,
Sulfa drugs and ballot boxes
Help? Is every man alive saved
By your Western white man's progress?
Are sin and suffering done?
Have kingdoms come?
No. We might as well be pelvis deep
In paleolithic mud. Oh, it is dead.
The myth of progress perished years ago.
Only pinks and freaks
Can still be fooled.

—*W. Warren Wagar* (*1964*)

I 2

In Defense of Modern Man

T HE DUTCH PHILOSOPHER Bernard Delfgaauw begins his recent
study of *History as Progress* with three rhetorical questions: Is
it not improper to write of progress when so much tyranny and in-
humanity flourish in the modern world? Is the question of progress
not too subjective to lend itself to scientific inquiry? Is the whole
problem of progress not dead in any event, after the collapse of the
utopian illusions of past centuries? He answers each question with
a defiant "no!" For Delfgaauw, progress is real, definable, and
measurable, and although nothing guarantees future progress, he
finds the pessimists, who have dominated twentieth-century thought,
no less mistaken in their assumption of the necessity of man's
failure than the optimists of earlier eras, who insisted on the inevitability of man's success.[1] Delfgaauw is not alone among twentieth-century thinkers in his belief in general human progress, but
like most contemporary believers he feels constrained to picture
himself as a dissenter from orthodoxy, a mind working against the
grain of recent thought. Like Charles Frankel, he comes before
the court of history as a counsel for the defense, to plead "the case
for modern man." Like Erich Fromm, he has joined "the revolution of hope" against a resolutely cynical intellectual-literary establishment.[2]

In exploring the survival of progressivism in the period since 1914
in Western thought we set foot on what is, from the academic point
of view, mostly *terra incognita*. Few scholars have given thought
to the question of hopefulness as a theme in twentieth-century in-

tellectual history. Some have entertained the false impression that such a theme is not even present in Western thought since the First World War. In an age of despair, anxiety, and apocalyptic visions, they suggest that only inferior minds, or old men born and educated in the previous century, could possibly believe in progress. All the seminal thinkers, presumably, have abandoned hope.

But if the seminal thinkers of this century include Freud, Russell, Whitehead, J. Huxley, Schweitzer, Maritain, Teilhard de Chardin, Toynbee, Mumford, Mannheim, Bloch, and Marcuse; if the utterances of politicians and ideologues count for anything, whether liberal, fascist, or communist; and if the attitude of the "common man" can be taken in evidence, then the relative indifference of scholars to the theme of hopefulness in twentieth-century thought becomes difficult to defend. It grows all the more difficult if we add to our list such thinkers as Shaw, Bergson, Smuts, Dewey, Wells, and Lenin, whom we relegated in earlier chapters to *la belle époque,* but whose work and influence continued well beyond 1914. The twentieth century is, above all, an age of disintegration, of the collapse of the spiritual cohesion of Western culture; yet hope has its place in the confused spirit of our times as surely as despair. It has no doubt a smaller place, all things considered. But any study of the contemporary mind that ignores it distorts the truth badly. In a century where all faiths and anti-faiths are possible, where no option is automatically ruled invalid, the belief in progress has not failed to find passionate defenders.

From the perspective of economic and social history, for example, one may not unreasonably view the period since 1914 as an organic continuation of progressive trends under way long before 1914. Several recent social scientists and historians have done just this. As Jean Fourastié points out, progress in the technics of production has continued all through the twentieth century, and technological progress is the key to many of the good things in contemporary life, from improved standards of material well-being to better health, progress in learning, and the fuller realization of freedom and equality in social relations. Despite interludes of totalitarian tyranny, the mass of men enjoy a better life, perhaps, than was ever before possible. Despite the losses suffered during two great world wars and the economic depression of the 1930s, the productivity of labor and the output of consumer goods have risen steadily throughout the Western world. The richest nations of 1914 would be poor indeed by contrast to those of 1970. If material progress helps to explain the exuberant mood of *la belle époque,* it would be surprising not to find some of the same exuberance in twentieth-century thinkers, in an age of even more spectacular material progress, which has had even more favorable social repercussions from the democratic point of view.

Economic and social progress has been accompanied by continuing progress in science and scholarship. The two interact. By making possible the expansion of facilities for education and research, economic and social progress accelerates learning; and learning in its turn, by stimulating technical progress, accelerates progress in the economy and the social order. The prestige of the natural, behavioral, and social sciences has held firm among many sections of the intelligentsia and the general public, in spite of obvious counter-tendencies. The scientific mind, moreover, is a mind predisposed to believe in human progress, if only because it understands how rapidly science itself progresses.

The advance of material well-being, equality, and science has been especially marked in the two leading countries of the contemporary Western world, the United States and the Soviet Union, and it is here that the idea of progress has lived on most vigorously in recent decades. Both countries officially affirm their allegiance to the Enlightenment and to one of the principal ideologies of progress of the nineteenth century, the United States to republican liberalism, the Soviet Union to Marxism. Statesmen of both countries repeatedly invoke "progress" as a justification for their policies, and insist that hopeful effort will always triumph over fear and despair. American and Soviet spokesmen proclaim that their respective social experiments not only have contributed signally to the progress of mankind but will also serve as models for all future progress. "It is an article of the democratic faith," as Adlai Stevenson told his audiences in the 1952 American presidential campaign, "that progress is a basic law of life."[3] Governor Nelson A. Rockefeller of New York, in his inaugural address in Albany in 1963, asserted that all Americans "believe in stable, ordered change and human progress, in the perfectibility of the individual human being and of the human society."[4] Gustav A. Wetter provides similar quotations from Stalin, Molotov, and others in his survey of Soviet thought.[5] Wherever one turns, to business leaders, commissars, presidents, party chairmen, ministers of state, or technocrats of every type in American and

Soviet life, belief in human progress is *de rigueur*. Certainly the same cannot be said for the public men of Western Europe in our time, although even here the majority have no doubt paid their obeisance to the idea of progress in some form.

Such is the durability of the belief in progress that at least some thinkers in every Western country have found comfort or cause for renewed optimism in nearly every major historical event of the century, including its greatest catastrophes. The First World War gave rise to hopes for an end to all wars and the establishment of a new international order through which mankind could advance more rapidly than ever before. The futility of the war did not become fully apparent until the 1920s. By then, fresh hopes had been engendered by the Bolshevik experiment in Russia, the return of prewar prosperity, and the work of the League of Nations. Fascism and German National Socialism both represented themselves, with broad popular success, as movements of national or racial reawakening, and dreamed of millennial empires. In the depths of the Great Depression, liberals and socialists took courage from the success of the Soviet Five-Year Plans and the work of the New Deal in the United States. They found a new holy cause in the struggle against "totalitarianism" in the late 1930s and throughout the Second World War. The Allied victory in 1945 and the founding of the United Nations produced another surge of optimism. The years immediately after the Second World War saw the appearance of the world federalist and European union movements and the publication of a number of prophetic visions of the "coming world civilization," discussed in my book *The City of Man*.[6] The return of general, and indeed upward-spiraling, prosperity throughout the West in itself did much to counteract the depressing effects on the Western spirit of the cold war, with its explicit threat of thermonuclear world annihilation.

Following Sidney Pollard in his new book on the idea of progress, one might even contend that in the period since 1945 the world's mood has been reconverted to hope. The prevailing pessimism of the 1914–1945 era, he writes, was derived not from "any powerful new theories, but from the historical experience." Mankind had almost torn itself to pieces—and then

> after decades of destructiveness, when the centuries-old promise of science and rationalism appeared to have been broken . . . the cloud is suddenly lifted once more. Societies are making the most of their

resources, and wealth advances almost everywhere as fast as technology and resources allow. The major powers have avoided war, and are collaborating in areas in which their differing ideologies makes such collaboration hardly credible. . . . Objectively speaking, a Condorcet or a Godwin might not be dissatisfied, were he to visit the world to-day.[7]

If this seems a little farfetched, it is at any rate quite clear that the idea of progress has itself registered progress in recent years, more particularly since the early 1960s. Evidences of revival are everywhere: a new "secular" theology of hope, the emergence of new radical ideologies in the West and a more humanistic socialism in Eastern Europe, the neo-Marxist and neo-Freudian thought of Fromm and Marcuse, the posthumous fame of Teilhard de Chardin, the return to fashion of evolutionism and theories of change and progress in the social sciences, and the new charismatic politics of the Kennedy brothers, Martin Luther King, Malcolm X, de Gaulle, Castro, and Che Guevara. It is too early, in 1972, to say that the intellectual avant-garde has recovered its belief in progress, or even that hopefulness has clearly replaced despair throughout Western civilization, but no one can doubt that the mood of the West has changed significantly in very recent years. As late as 1966 I could write that "we live today, so it seems, in a glacial age of the spirit."[8] The "inevitable thaw" which I also predicted at that time may have already arrived.

Twentieth-century survivals and revivals of the faith in progress will be studied below in five chapters, the first focusing on the contributions of theology, the second on philosophy and theories of history, and the last three on the contributions of the sciences, with special attention to biology, the social sciences, and psychology. As in earlier parts of this book, some thinkers could be appropriately discussed in several places. Thus, Teilhard de Chardin appears as a "philosopher," although he also possessed the credentials of a "biologist" and a "theologian." Schweitzer, on the other hand, figures as a "theologian," rather than a "philosopher" or a "historian"; and Marcuse is left until the end, to speak for "social psychology," instead of "philosophy," "history," or "political science." No system of classifying minds is proof against the versatility of the human spirit.

13

Holy Worldliness

CHRISTIAN THEOLOGY in the twentieth century offers a view of all recent thought in microcosm. Early in the century it brimmed with hope and blended its faith with the positivism and anti-positivism of secular philosophy. After 1914 Christian theologians helped lead the assault on the belief in progress. No thinkers were more profoundly disenchanted by the march of history or more despairing of man's capacity for improvement. Some developed an interest in the relevance to Christian tradition of existentialism, when existentialism was the dominant school of European philosophy. Others turned to Jungian psychoanalysis and the "timeless" wisdom of Asian thought.

But Christianity has also shared in the conservation of the idea of progress in twentieth-century intellectual life. Once again, it serves as a model for other branches of thought. It has its older thinkers, its survivors of *la belle époque,* like Albert Schweitzer, who continued stubbornly to adhere to their pre-1914 hopefulness long after Sarajevo, and even after Hiroshima. It has its thinkers who fiercely denounce modern civilization, like Jacques Maritain, but who then proceed to disclose an idea of progress that promises to rescue mankind from the evils of modernity and carry the race onward and upward. It has also had its tidal movement of resurgent meliorism since 1960, a new "radical" or "secular" theology inspired by the life and thought of Dietrich Bonhoeffer, which reintroduces worldly hope as a fundamental category of Christian faith.

None of this need surprise the historian of Christian theology.

From earliest times Christianity has been inextricably associated with various movements in secular thought, sometimes with greater damage to its first-century *kerygma* than at others. From the beginning it has nourished both historical pessimism and historical optimism, through the changing interpretations given to the Christ-event, to the prophetic books of the Old Testament, to the role of the church, and to the myth of the millennium. Although we may be justified in arguing that the neo-orthodox theology of Barth and Brunner was more faithful to biblical belief than the liberal and modernist theologies of the post-Darwinian generation, any Christian doctrine of earthly progress can find a plausible *raison d'être* in the long history of Christian thought.

Catholic thinkers have encountered rather less difficulty than Protestants in resisting twentieth-century despair. Rarely as leftish as their Protestant contemporaries in the nineteenth century, they have often failed to swing all the way to the right in the twentieth, leaving a place in many instances for a qualified theology or philosophy of progress. The twentieth-century crisis in civilization can be seen quite consistently by Catholics both as a proof of man's helplessness without God and as an opportunity for continuing progress in which God and man freely collaborate.

What preserves the Catholic from the kind of despair common to Protestantism is, above all, the importance in Catholic thought since Aquinas of the classical doctrine of nature. Man has fallen but his essential nature is not vitiated by his fall: it retains its original goodness, although man's perverse inclinations may inspire him to oppose both nature and God's will. Moreover, history is protected from the meaninglessness that it tends to acquire in Protestant thought by the Catholic conception of the Church, which acts throughout the course of history to reveal new truth, to win new souls, and in all ways to serve in Christ's place in the world. Nor is such a church merely hypothetical. It exists visibly on earth, and its pope is allegedly Christ's apostolic successor. Catholic thought avoids, therefore, the extremes of this-worldliness and of otherworldliness. It cannot easily fade into pure humanism, nor can it easily succumb to world denial and a pure worldly pessimism.

Many twentieth-century Catholic thinkers have given little attention to the doctrine of progress, or, like Hans Urs von Balthasar in *A Theology of History*, have abandoned it, preferring to see the battle between Jerusalem and Babylon as perennial, with no possibility of the temporal triumph of either city.[1] Nothing in Catholic tradition prevents Balthasar from adhering to such a view, basically that of St. Augustine; the presence of the "good" in history does not require the emergence of the "better," with or without divine help. But other Catholics, particularly in France and Britain, have been able to fuse the optimism native to Catholic teaching with the modern secular gospel of progress. Of these, the most influential over the greatest number of years has been the French neo-Thomist philosopher Jacques Maritain.

Born in 1882, Maritain came to Catholicism and Aquinas by a circuitous path. He was raised a freethinking Protestant, fell briefly under the influence of Bergson, who persuaded him of the spiritual bankruptcy of positivism, and then discovered the mysteries of the Catholic faith with the help of Léon Bloy. The circumstances of his conversion and baptism in 1906 and his friendship with Bloy and Péguy seemed to prepare him more for the role of a poet or a mystic than for the role of a Thomist philosopher, but his appetite for intellectual discipline greatly exceeded theirs. By the end of *la belle époque* he had become a professed Thomist and a member of the faculty at the Institut Catholique in Paris.

Maritain turned his attentions to social thought in the mid-1930s. During the next quarter-century he produced in effect a Christian theory of progress, set in polemical opposition to "Protestant" pessimism, atheist humanism, bourgeois capitalism, and all the ideologies of totalitarianism, of both the Left and the Right. In their place he offered "integral humanism," a Christian social philosophy for the new world civilization that he saw emerging from the chaos of contemporary life. The new civilization would be secular in form, but Christian in inspiration, and the most advanced of all societies in history, both culturally and spiritually. That Maritain's views did not change after his retirement to a monastery in Toulouse in 1961 may be surmised from his most recent book, *The Peasant of the Garonne,* which warns against the excesses of "neo-modernism," but at the same time reaffirms the thinking of his more active years on the question of temporal progress.[2]

Uniting the Aristotelian idea of nature and the modern idea of progress, Maritain suggests that civilization is the natural process by which mankind fulfills itself materially, spiritually, and morally. "It answers to an essential impulse of human nature, but is in itself a work of our spirit and our freedom acting in co-operation with nature." [3] He also proposes several "laws of history" that explain in what ways and by what means progress occurs: the increase of good through man's exercise of his freedom, the growth of man's self- and world-awareness, his inexorable tendency to replace myth with reason, the progress of secularization and democracy, and most important, the expansion of moral conscience, by which "the explicit knowledge of the various norms of natural law grows with time." Moral progress has made possible, for example, the institution of monogamy, the prohibition of slavery, and the humanization of war and government.

But man, says Maritain, is also a rebel aginst the natural order and the divine will, resulting in the progress not only of good but also of evil. Man is perfectly capable of putting to sinful use all his evolved powers, and often does. If he were left to his own devices, his natural capacity for improvement would be more or less canceled out by his proclivities as a fallen creature for wrong-doing. Fortunately for man and history, however, man has not been left alone. As Christian faith has always taught, man enjoys the gift of divine grace, which acts in history to overcome the power of evil. The entire realm of nature, through grace, "is superelevated in its own order," or, in other words, endowed with strength and with purposes greater than would be possible for purely natural man. Thus, the leaven of the Christian gospel making its way within the depths of secular consciousness facilitated the progress of democracy in modern Western society, although its progress also accorded with the deep-seated demands of human nature, and is now beginning to occur in non-Christian countries, where it accords with identical demands. In the struggle between good and evil, the losses and waste resulting from the growth of evil are "not as great" as the gains resulting from the growth of good—"not as great, in the last analysis, but a pretty nuisance for all that." [4]

Progress within history is not, of course, the only or even the highest goal of Christian endeavor. Yet history has its own infravalent goals, virtuous in and of themselves quite apart from their place in the economy of salvation. "We must seek with all our power," Maritain affirms, "a genuine . . . realization in this world of the requirements of the Gospel." The temporal mission of the Christian is to construct "a Christian-inspired social and political order, where justice and brotherhood are better and better served." [5] In the Middle Ages the Church saw itself as both spiritual and temporal guardian of society, and attempted to build a sacral civilization, dedicated to the greater glory of God. The medieval experiment was well-intentioned, but it fell into somewhat the same errors as the theocratic societies of Asia. It too often ignored the distinction drawn in the Christian gospel itself between the sacred and the secular orders; it sought to subordinate the state to the ecclesiastical hierarchy; and it sacrificed freedom of conscience to unity of belief. Times changed, with the progress of the secular order, and the old Christendom of the Middle Ages slowly passed away. "After sixteen centuries which it would be shameful to slander or claim to repudiate, but which have completed their death agony and whose grave defects were incontestable, a new age begins." [6]

Rather, a new age is ready to begin. Maritain's "New Christendom" gleams on the historical horizon, still not here, but within man's reach, God willing. Modern secular society, in its revolt against theocracy, has severed its roots in the transcendental spiritual order, with consequences fatal to its moral health and even to its own faith in reason. In the coming new world society the legitimate growth of the secular order will continue, but the lost links with transcendence will be restored, especially through the work of the Christian laity, which must assume the responsibility in a secularized society for actually translating Christian principles into social action, although the spiritual guidance of the Church will still be needed. In the New Christendom, in contradistinction to the old, full freedom of conscience must be assured, church and state must be separated, all members of the body politic must be treated and held as equals, and an associative form of industrial proprietorship must replace both state and private capitalism to ensure economic democracy. The nations will establish a world political organization capable of guaranteeing peace on earth, and mankind will live in a single brotherly city, at once free and unified. By contrast with what lies open to man, with the social possibilities of the Christian gospel and the natural possibilities for man's progress in knowledge and culture, all the achievements of civilization down to the present time must be regarded as "extraordinarily primitive." [7]

Maritain's counsels of hope may have had more impact in North America than in his native France, if only because during most of the years of his greatest interest in social philosophy he was living and teaching in the United States, first at Columbia and later at Princeton. Many Frenchmen heard the gospel of progress from another Catholic layman, also a philosopher by training, the founder of the "personalist" movement and the intellectual review *Esprit,* Emmanuel Mounier. A full generation younger than Maritain, Mounier was active in the French Resistance during the Second World War and died in mid-career in 1950 at the age of forty-five. Two years before his death he issued a short book on the quasi-apocalyptic forebodings of twentieth-century man, which challenged the fashionable theological pessimism of the 1930s and 1940s. Prophets like Kierkegaard and Bloy had done well to assail the fatuous complacency of the nineteenth century, but in an age of man's despair with man it deeply troubled Mounier to hear Christians whispering "to one another with downcast faces and a sort of spiritual gourmandise, that the time of catacombs was drawing near, and that a show of interest in anything other than a subterranean life and martyrdom, was purely frivolous." Christianity had, for many, petrified into "some morose and bitter malady, as far removed from Christian inspiration as any of the more accommodating varieties [of Christian faith]." The need of the hour was for "builders of hopes and duties," for a "tragic optimism" that recognized the existence of evil but evinced unshakable faith in a "triumphant history." [8]

Mounier defended the belief in progress as a fundamentally Christian idea, and recognized even in the advance of technology an important contribution to the spiritual perfection of mankind. Man's very nature was to be "artificial," to use all the devices that his intelligence could invent to enlarge his knowledge and power. Far from being inhuman, technics led man "from a condition of immanent servitude to an inhuman nature, to a considered mastery over a humanised nature." Progress in social relations was also crucial to the doing of God's will on earth. Christians had much to learn from Marx and from modern communism, which strove for social justice and equality with biblical fervor, although Mounier could not endorse the anti-personalist tactics of Stalin. Above all, Christians had to recover their faith in man as a child of God, set free by God to make his own future. "Christianity gives man his full stature and more than his full stature. It summons him to be a god, and it summons him in freedom. This, for the Christian, is the final and supreme significance of progress in history." For why else had God decreed that man should live in time, instead of creating a perfect world instantaneously? "He wished man's liberation to be the fruit of the toil, the genius and the suffering of man, that he should savour one day the full fruit of this labour, these toils and this loving, and not receive it as an overpowering gift from Heaven. Humanity *farà da sè*, slowly, progressively." [9]

The French Jesuit theologian Jean Daniélou has provided a somewhat more orthodox view of progress than those of Maritain and Mounier in his book *The Lord of History*, centered—appropriately enough for a theologian—on sacred rather than secular history. Daniélou rejects the idea of continuous evolutionary progress, but he finds progress of a sort, and even continuity, in the actions of God in history. The sequence of biblical revelations constitutes a system of divine pedagogy. God revealed progressively fuller visions of himself and progressively higher values for human life throughout the books of the Old Testament, saving the highest truth and the supreme manifestation of his grace for the first century of the Christian era. Nonetheless, history did not end in the first century A.D. Much good work remained, and still remains, to be done. "What is now in progress is something invisible, yet supremely real, the building up in charity of that mystical body of Christ that shall be revealed in the last day." [10] In short, the progress of the Church of Christ is the meaning and purpose of history.

Daniélou borrows the phrase "initiated eschatology" from Donatien Mollat to characterize his view of the future progress of Christendom. Christ has come, and every moment bears within itself eschatological implications, as the existentialists insist, but "this work of Judgment which Christ has substantially completed has not yet produced its due consequences throughout mankind and throughout creation." [11] Daniélou distinguishes at least four areas in which progress has occurred and will continue to occur: in the conversion of individual souls to Christ throughout the world, in the Christianization of the non-Christian cultures, in the ever-improving clarification of dogma, and in the betterment of civilization itself, in accordance with natural law and Christian inspiration. The Church helps to humanize social conditions and bring peace, justice, and rightful order into human life. But it cannot, within time and after the Fall, create a terrestrial utopia. "What can be done is to make the existence of man more human, within the possible and practical limits of real life." This gives to history its due, neither more nor less. "Indeed, the historical approach to reality was originally a Christian discovery; and even today, it is only in virtue of Christian presuppositions that history can attain to its proper standing, because Christianity alone represents a positive and irreversible increment." [12]

Daniélou adds, in passing, that "the conceptions of scientific development and of biological evolution have no relevance for the life of the spirit," [13] a rap on the knuckles of his fellow countrymen and fellow Jesuit, Pierre Teilhard de Chardin, whose philosophy of evolutionary progress has done more perhaps to rehabilitate the progressivist faith in the postwar era than the work of any other thinker. As expounded in *The Phenomenon of Man*, however, Teilhard's thought seems more metaphysical and Bergsonian (or Alexandrian) than Christian, and consideration of it will be deferred until the next chapter.*

Several prominent British Catholic writers have contributed thoughts on progress rather similar to those of Maritain and Daniélou. Among Jesuit scholars, Martin C. D'Arcy endorses both the idea of the gradual maturation of civilization and the idea of sacred history as a process of progressive, if also finite, spiritual perfection, prefiguring the greater perfection to come at the end of time.[14] Maritain's counterpart in the English laity is Christopher Dawson. He joined the Catholic Church (like Maritain) in his mid-twenties, after a traditional education at Winchester and Oxford, and has devoted most of his life to a study of the relationship between religion and culture. His *Progress and Religion* (1929) was among the first books to argue an organic relationship between the Christian idea of history and the modern belief in progress, which he found to be a "religion" in its own right, although after 1870 it had rapidly disintegrated and by the 1920s could be proclaimed virtually dead.

For Dawson, the task of the Church in the twentieth century is to recall Western man to his spiritual roots. Only the West, and specifically Western science and the Christian faith from which European culture has descended, can solve humanity's material and spiritual problems, but before it can save the world, Western civilization must first rediscover its unity of purpose and its sense of the transcendent meaning in history by returning to its ancestral faith.

* See below, pp. 279–83.

Certain Protestant pessimists notwithstanding, Christianity "does not deny the existence of progress. . . . On the contrary it teaches that throughout the ages the life of humanity is being leavened and permeated by a transcendent principle, and every culture or human way of life is capable of being influenced and remoulded by this divine influence." Seen from the highest perspective, the final goal of human progress on earth is nothing less than world spiritual integration, the achievement of a Christian cosmopolis, of which the imperial societies of ancient Asia and medieval Christendom were forerunners. To the Church of Rome, and to the Hellenic tradition of science and scholarship, which has always collaborated with Christianity in the making of Western culture, "we must look for the creation of a new world civilization, which will unite the nations and the continents in an all-embracing spiritual community." [15]

In avant-garde Protestant theological circles in the years between the two world wars, belief in progress was all but anathema, and it remained rare for some years after 1945. But this applies only to the most "advanced" thinkers, with an emphasis on the situation in the German-speaking world. In the United States, for example, writers such as Harry Emerson Fosdick, E. E. Kresge, Sherwood Eddy, and Shirley Jackson Case kept the tradition of the social gospel alive for many years.[16]

The promulgation of a Christian idea of progress, coupled with what might be termed Christian imperialism, the demand for a Christianized world order, is also central to the thought of Kenneth Scott Latourette and, in Britain, of John Macmurray and John Baillie, contemporaries whose major works belong to the interwar and early postwar years. All of them are impressed by the progress of Christianity through the centuries, its influence on Western culture, its penetration of the non-Western world, its social conscience, and its historical destiny as the sole religion of mankind and the buttress of the coming world order.

Both Macmurray and Baillie have written extensively on the subject of the belief in progress, which they agree is an entirely Jewish and Christian idea, implicit in the structure of Christian faith. Macmurray refers all modern efforts to improve human society, including Soviet communism, back to the teaching of Jesus, "centred from the beginning in [the] idea of the Kingdom of Heaven." Discovering the real meaning of the Kingdom and the conditions for its establishment on earth leads to "the realization of the meaning of the religious impulse in man" and "a programme for the conscious development of human society. . . . Real religion is the first condition of social development, and the establishment of Christianity in the minds of men is the first condition of conscious progress." By contrast, the other "world" religions are pseudo-religions, glorifying force like Islam, or drifting off into dreams and illusions that deny the world, like Buddhism. The will of God is that the meek shall inherit the earth and build the heavenly kingdom of Jesus in the form of a socialist world commonwealth.[17]

Baillie, one of the foremost Scots theologians of the twentieth century, shared Macmurray's view of the Christian sources of modern progressivism and indeed all modern culture in his book *The Belief in Progress,* but was more emphatic in rejecting modern secular versions of the Christian hope. Their temporary success was due in part to the "apparent defeatism that characterized so much of traditional Christian thought." But the authentic faith of Jesus Christ

> does offer us a very confident hope for the future course of terrestrial history. . . . We must recover that sense of standing on the threshold of a new historical economy (or dispensation), that sense of a noble prospect opening out before us, that sense of the power of the Spirit and of the inexhaustible resources now available to us, that adventurous zeal for the renewal of humanity and that confidence in ultimate victory of which the New Testament is so full.

The souls of men, of both the West and the East, could be won by Christian missions; society, both Western and Eastern, could be Christianized; and whatever good might come to mankind from pagan culture, "the further progress for which Christians may hope can only be that which radiates from the Christian centre of history, and can be nothing else than the progressive embodiment in the life of humanity of the mind that was in Christ and 'a growing up in all things unto Him who is the Head.' "[18] *

* In much the same spirit, the Swedish-American theologian Nels F. S. Ferré writes in *Christianity and Society* (New York, 1950) that history is God's chosen way of perfecting man and all creation. "Progress can mean only the effecting of Christian fellowship. . . . The Christian faith, above all, needs to be spread from pole to pole, and to become the steady background of a universal ethos. An effective missionary outreach is a necessity for world well-being. . . . We must extend Christ's reign, God's love, over all life, inner and outer." Ferré, pp. 144, 211, and

Among Continental thinkers, some of the optimism of German social Christianity survived in the thought of one of its chief prewar critics, the Alsatian theologian, philosopher, and missionary Albert Schweitzer. We have already met Schweitzer as the biblical scholar whose book *The Quest of the Historical Jesus* (1906) gave the *coup de grâce* to the widely held theory that the "historical" Jesus, as opposed to the Pauline Christ, had been essentially a moralist and social reformer seeking to establish God's kingdom on earth.* But Schweitzer did not go on from this point to despair of the world or to share Jesus' anticipation of the imminent end of time. In his well-publicized career after 1913 as a medical missionary in French Equatorial Africa and in his later writings, Schweitzer was one of the century's most resolute believers in progress.

His philosophy of progress and civilization and his diagnosis of the ills of modern man were most fully developed in *The Decay and the Restoration of Civilization* and *Civilization and Ethics,* both of which appeared in 1923. In the first book he focused on the question of what had gone wrong with Western civilization. The World War, he wrote, was not the cause of the collapse of the West, but merely a symptom of a deeper illness that had stricken Western man in the nineteenth century. The philosophy of the Enlightenment, which had long provided Europe with values and goals for life and a characteristic world-view, had come under severe attack ever since the middle decades of the nineteenth century. This attack had directed itself not only against the content of Enlightenment thought but against all varieties of ethical philosophy as well. Analysis and criticism replaced constructive thinking in philosophy; the historical study of origins and development replaced efforts to set goals for civilization. All that man left to himself was the possibility of a scientific knowledge of "reality." He became obsessed with the notion that all things were determined by the operation of positive laws of matter and history. The realm of freedom was abolished. At the same time, science and technology advanced spectacularly, giving weight to the notion that material progress was the highest and only good. The individual began to be thought of as a mere

275. Without abandoning his vision of a unified world order as the next stage of human progress, Ferré has recently altered his views, in a manner reminiscent of Toynbee, with regard to the other world religions. See his *The Universal Word: A Theology for a Universal Faith* (Philadelphia, 1969), especially pp. 170–71.

* See above, p. 207.

factor of production, something to be organized and exploited by the economic machine. "Man's ethical energy died away, while the conquests achieved by his spirit in the material sphere increased by leaps and bounds." He allowed himself to be swept along by the trends that he himself had set in motion and had labeled inexorable. He clutched at the straw of nationalism, a vain and perverse substitute for a public philosophy, and the end result was a world war in which millions died uselessly and mankind sank "helpless in the stream of events." [19]

The Western world, then, had fallen into decay between 1850 and 1920, but Schweitzer did not see what had happened as the inexorable grinding of the mills of fate, or anything more than a temporary interruption in the general progress of civilization. The Renaissance and the Enlightenment represented, for him, ages of vigorous progressive development. In particular, the rationalism of the Enlightenment had been "the greatest and most valuable manifestation of the spiritual life of man that the world has yet seen." [20] Its one fatal shortcoming lay in its adherence, like Confucianism and Stoicism, to the cosmic monism that insists on an alleged harmony between nature and ethics, and forces the latter into a relationship of dependence on the former. It was this false assumption that had brought the optimistic ethical world-view of the Enlightenment crashing down under critical scrutiny in the nineteenth century, after the noble failure of Kant and Hegel to resolve its paradoxes.

Schweitzer's formula for the restoration of civilization followed logically from his diagnosis. Mankind could save itself from the disastrous consequences of ethical paralysis only by furnishing itself with a deeper and stronger ethical consciousness, invulnerable to the sort of criticism that had demolished the monistic ethics of the Enlightenment. Schweitzer recommended a stern dualism. What man could know of the world, as scientist and scholar, had to be sharply distinguished from his will to live, in which his sense of the good and also his attitude of world negation or world affirmation had their origins. The will to live was suprarational: even more, it was from God, "the mysterious divine personality which I do not know as such in the world, but only experience as mysterious Will within myself." The ethical thought that could be derived from such an encounter was "not justified by any corresponding knowledge of the nature of the world" but rather stood

255

on its own, having the authority of mystical experience and not of science. "A new Renaissance must come," Schweitzer proclaimed, "and a much greater one than that in which we stepped out of the Middle Ages; a great Renaissance in which mankind discovers that the ethical is the highest truth and the highest practicality, and experiences at the same time its liberation from that miserable obsession by what it calls reality, in which it has hitherto dragged itself along." [21]

For the ethical philosophy that he sought, Schweitzer drew heavily on the teachings of Jesus. Although the historical Jesus had failed to imagine that the kingdom of God could be realized on earth except through supernatural means, in every other respect his ethical teachings remained valid. He had taught unconditional love of self and fellow man; he had denied the world as he found it, in order to liberate himself from its errors and become free to act. Negation had led to affirmation, the gospel of man's perfectibility through love. With the passage of centuries the Christian world abandoned its early prophecies of the end of time, and with the help of the modern doctrine of progress that originated in the discoveries and inventions of the Renaissance, Christianity today "accepts the idea that the kingdom of God must be established by a process of development which transforms the natural world." The ethic of active self-devotion taught by Jesus "makes it possible for Christianity, inspired by the spirit of the modern age, to modulate from the pessimistic to the optimistic world-view." [22] Schweitzer's revised version of the gospel combined the activism of Jesus (but not his historical pessimism) with the modern belief in progress (but not its cosmic naturalism). The Christian ethic was redefined in non-theological language as "reverence for life" grounded in "world-and life-affirmation."

Schweitzer made perfectly clear his allegiance to the idea of progress. The improvement of mankind was to take place on earth and in time. "Civilization, put quite simply, consists in our giving ourselves, as human beings, to the effort to attain the perfecting of the human race and the actualization of progress of every sort in the circumstances of humanity and of the objective world. . . . Civilization originates when men become inspired by a strong and clear determination to attain progress, and consecrate themselves, as a result of this determination, to the service of life and of the world." [23] Three kinds of progress could be discerned, each contributing in a different way to the lessening of the strain of the struggle for existence. The first was the development of rational human control over the forces of nature, the second the improvement of social organization, and the third, progress in spirituality, the highest and ultimate goal of civilization. Since the first two were, in the last analysis, only means to the third, they had no ethical value or purpose unless they were subordinated to the needs of the spirit. "All progress in discovery and invention evolves at last to a fatal result, if we do not maintain control over it through a corresponding progress in our spirituality." [24]

At the same time, spiritual growth was impossible without the restoration to its central position in life of ethical thought and the spontaneous adoption by individuals throughout the world of a deep and adequate ethical philosophy—founded on the Christian principle of active reverence for life. Schweitzer did his best to provide a personal example of his own teachings by devoting his last fifty years to missionary work in Africa. He looked forward to a world spiritual renaissance, the evolution of the modern nation-state into a truly "civilized" state, and the abolition of war. In his old age he gave the full weight of his support to the world propaganda for nuclear disarmament, winning the Nobel Peace Prize in 1952. No other thinker of our time more closely approached sainthood.

Of course Schweitzer had little influence on theology in the years immediately after 1914. He could be seen, not without injustice, as a venerable relic of the era of prewar theological optimism. His mystical, immanentist view of God and his attachment to the belief in historical progress appeared rather old-fashioned even in the Anglo-Saxon countries. Yet before Schweitzer's death in 1965, new currents of thought that resurrected his optimism were already flowing in Protestant theology. The new school, which is distinctly post-Barthian and post-Bultmannian alike, despite a certain indebtedness to both Barth and Bultmann, has come to be known, not always approvingly, as "secular theology." [25]

The basic premise of the new secular theology is the need for Christianity to abandon its privileged position as the "official" religion of Western civilization, in fact to transcend its function as a religion altogether, and engage itself fully in the world, accepting the world as being fully secularized and no longer dependent on the rites and creeds of the Christian churches. Secular theology is ultra-Protestant in that it carries a step further the objections of the Reformation to the well-defined Catholic distinction between the "sacred" and the "profane" orders, with its corollary of the hierarchical superiority of the former to the latter. The world as a whole is made the theater of Christian activism, and a new optimism arises, which at times resurrects the belief in terrestrial progress.

In the English-speaking world secular theology is most familiarly expounded in Bishop Robinson's *Honest to God* (1963) and in Harvey Cox's *The Secular City* (1965), but its roots are German and Swiss. As Eric Mascall points out in *The Secularization of Christianity,* it may well be seen as a movement of reaction against the "extreme revelationism and supernaturalism of the school of Barth, Brunner and Heim." [26] Yet it also owes something to Barth, and in particular to his celebrated attack on conventional "religion" as something man-made, pagan, and escapist. Like Barth, the new theology will have nothing to do with the notion that Christ's church offers man a place of pleasant refuge from the world's trials. Further and more obvious inspiration has come from Tillich's assault on traditional Christian supernaturalism and from the existentialist and demythologizing theology of Bultmann. Another influence has been the later thought of Friedrich Gogarten, notably his study of secularization as a theological problem, which he published in 1953, and in which he defined secularization as the historicization (*Vergeschichtlichung*) of human existence, its emancipation from both mythology and metaphysics.[27]

The first clearly secular theologian was Dietrich Bonhoeffer. Born in 1906, Bonhoeffer studied under Adolf von Harnack at the University of Berlin, and later became a friend and disciple of Barth, so that he had the closest possible contact with both Wilhelmian social Christianity and post-1914 neo-orthodoxy. Not yet thirty when Hitler seized power in 1933, he soon became a leader in the anti-Nazi movement in German Protestantism, and later a highly placed member of the German political underground, for which he was arrested in 1943 and executed in 1945. If Schweitzer serves as the Christian exemplar of sainthood in our century, Bonhoeffer is clearly its Christian exemplar of the martyr. His posthumously published *Letters and Papers from Prison* stands as one of the great documents in modern spiritual history.

Bonhoeffer's life and thought are inextricably associated. He had witnessed with his own eyes the conversion of a vast, civilized Christian nation into a tribe of pagans, who were quite capable of turning a deaf ear to the Christian gospel of love and peace—as, for that matter, had all the nations of Christian Europe during Bonhoeffer's boyhood, in the years of the First World War. The times plainly called for the most direct and unambiguous action by Christians: not for withdrawal into the little world of the holy church, but for action in the secular sphere, which Bonhoeffer saw unmistakably to be the center of life in the modern age, and the place, above all, for Christian witness. The end of time might come some day; God might be "wholly other" and all of man's works a tower of Babel, as Barth had written in the 1920s; but the business at hand was to institute Christian action in the world here and now. Bonhoeffer's principal theological works, *The Cost of Discipleship* (1937) and *Ethics* (1949), were calls to worldly engagement in an age that found the rites, myths, miracles, and mysteries of the Christian churches irrelevant to its purposes and alien to its concern.

Bonhoeffer did not condemn the secularization of human life, as such. On the contrary, it represented a step forward in history, the coming of age of mankind, its liberation from dependence on mythology. The medieval notion of the two spheres, the sacred and the secular, against which Luther had struggled, had no authority in the New Testament. The world was not divided into two separate "realities," one godly, the other earthly or even satanic: there existed only one world, and all of it was Christ's, the part that denied him fully as much as the part that accepted him. If Christianity was employed "as a polemical weapon against the secular, this must be done in the name of a better secularity and above all it must not lead back to a static predominance of the spiritual sphere as an end in itself." The great work of the Christian church, therefore, was not to provide solace or refuge for the afflicted. It was charged, rather, with "a mission of correction, improvement, etc., a mission to work towards a new wordly order," to set the secular life "free for true worldliness," to enable secular institutions to "attain to their own true worldliness and law which has its foundations in Christ." [28]

Although he rejected the dichotomy of the sacred and the secular, Bonhoeffer found it necessary to introduce two other dichotomies into his thinking. The events of history constituted the realm of the "penultimate," those beyond history the realm of the "ultimate." Radical neo-orthodoxy stripped the former of value; old-fashioned theological liberalism neglected the latter. The correct Christian attitude, in Bonhoeffer's theology, was to see the penultimate as valid and full of promise, but given its validity by the ultimate. Christian life was the "dawning of the ultimate" in the sphere of the penultimate, or the worldly. Bonhoeffer quite consciously borrowed a second dichotomy from Catholic thought—the distinction between the "natural" and the "unnatural," which further refined the meaning of the "penultimate." "The natural," he wrote, "is that which, after the Fall, is directed towards the coming of Christ. The unnatural is that which, after the Fall, closes its doors against the coming of Christ." Or again: "The natural is the form of life preserved by God for the fallen world and directed towards justification, redemption and renewal through Christ. . . . So long as life continues, the natural will always reassert itself. In this context there is a solid basis for that optimistic view of human history which confines itself within the limits of the fallen world." [29]

Bonhoeffer's reference to an "optimistic view of human history" is typical of his whole approach to life. He lived in desperate times, and he saw "secularization" at its diabolical worst, in the Third Reich. His plight somehow recalls that of one of the first great prophets of progress, the Marquis de Condorcet, who died in the prisons of the French Revolution. Bonhoeffer was perhaps not quite a prophet of progress, but he embraced the modern world with love and enthusiasm. Mankind had come of age, and ours was the strenuous task "of living every day as if it were our last, and yet living in faith and responsibility as though there were to be a great future." He admonished Christians to claim the future for themselves instead of abandoning it to their opponents. There could be no impiety in thinking a better world possible; only on the day of judgment itself, and not a day sooner, might Christians "gladly stop working for a better future." [30]

A great many theologians active in the Western countries over the past twenty years might wish to be known as followers of Bonhoeffer in some significant sense, and as secular theologians. Some have pushed his neo-Barthian idea of a "religionless Christianity" to the point of proclaiming "the death of God," a phrase that can mean everything or nothing. Some emphasize the need for clerical and lay activism in current affairs. Others preach "the new morality." All tend to be men of good hope, in marked contrast to the theological temper of the 1920s and 1930s. Few have more than a little to say about the belief in progress as such, and there is an obvious reluctance to return to the language of liberal theology, which is still accused of having betrayed the Christian gospel to nineteenth-century positivism and naturalism.* But especially among those who find it necessary to develop a theology of history, ideas arise that bear a strong spiritual affinity to the visions of the older liberalism. Three examples may be given, from books by Harvey Cox, Arend van Leeuwen, and Jürgen Moltmann.

Cox, a Harvard theologian of the younger generation, published the work by which he is best known, *The Secular City,* in 1965. Like Robinson's *Honest to God,* which it much resembles, *The Secular City* is an interpretation for the English-speaking world of ideas that are mostly not original with the author, but derived from German sources. The influence of Barth and Gogarten is clearly in evidence, also that of Max Weber, but Cox is preeminently a student of Dietrich Bonhoeffer. With Bonhoeffer, he dwells on the phenomenon of secularization, and finds it a progressive movement running all through history, from "tribe" to "town" to "technopolis," each higher form of social organization a step in the liberation of man from myth, or, to use Weber's term, in his *Entzauberung.* The rise from a magical world to a secular world constitutes the course of human progress. "Secularization implies a historical process, almost certainly irreversible, in which society and culture are delivered from tutelage to religious control and closed metaphysical world views. . . . It is basically a liberating movement." Man ceases to lie at the mercy of a mysterious cosmos and becomes his own master.[31]

Secularization occurred in the Christian West sooner and more rapidly than in Asia for the simple reason that the West was Christian. Cox repeats Bonhoeffer's equation of secularization with the impulse given to history by the biblical faith. The ancient Hebrews were the first among the world's peoples to abandon the myths of cosmic naturalism, idealism, and otherworldliness and adhere, instead, to a rigorously disenchanted and desacralized world-view, in which all things belonged to the same single realm of being, and

---

* Some secular theologians remain too close to the ahistorical eschatology of Bultmann to qualify as believers in progress at all. See, for example, Ronald Gregor Smith, *Secular Christianity* (New York, 1966), pp. 126–28.

history became charged with meaning and possibility. The biblical outlook, fulfilled in the modern world, is therefore anti-religious and anti-mystical. Nothing is sacred because everything is sacred. The Christian delight in secularity, in Bonhoeffer's "holy worldliness," must not, however, be confused—following Gogarten's distinction—with "secularism." Secularism is a closed world outlook, marked by fanatic refusal to remain open to transcendence, by an attitude of intolerance and exclusiveness that in fact duplicates the dogmatism of the religions of the mythological age.

The high point of Cox's iconoclasm is reached in his discussion of the superiority of the "secular city," the desacralized urban complex of the twentieth century, to the tribal and town life of past ages. He fires salvo after salvo at the intellectual "snobs" and "aristocrats" who despair of the modern world and pine for the past. The much deplored depersonalization of man, the collapse of I-thou relationships, is reinterpreted as the achievement of a liberating anonymity, which frees the individual in large urban areas from the exhausting task of relating existentially to everyone he meets, so that he can enjoy selective I-thou relationships with those whom he has the time to know intimately. Anonymity also provides privacy, emancipation from the surveillance of the meddlesome neighbor. The much deplored "rootlessness" of modern life is reinterpreted as urban man's freedom from the changeless, immobile life of tradition. Urban man is liberated by affluence to travel, to seek his fortune wherever he may find it, to become responsible for himself instead of accepting the bondage of provincial custom, the prescribed life style of peasant, monk, knight, or lord. The biblical faith, which originated among a freebooting nomadic people, once again supports in every way these progressive tendencies in contemporary life. Modern secular society offers more freedom, more responsibility, more opportunity, more choice than the societies of the past: it therefore offers more hope, and hope is the heartbeat of the Christian gospel.

What, then, of the Church? Cox does not hesitate to identify the coming of the secular city with the coming of the kingdom of God. He claims to differ from the social gospellers of Rauschenbusch's generation in affirming that the kingdom is indeed God's kingdom, and not the result of man's effort on God's behalf, or the work of a purely immanent God. Cox's God is both transcendent in Barth's sense and hidden in Bultmann's, but his hand is visible in history, and from his word as revealed in the Bible the secular city arises. The mission of the Church is to build and carry forward the secular city throughout the world, to defend it against its enemies, and to rebuild it whenever it crumbles or falls. The Church needs for its task a theology of revolution and social change, and it must become "God's avant-garde," always in the foreground of the struggle for liberation, justice, and self-determination. Cox has the Bergsonian and pragmatic vision of the future as open-ended, as brimming with hope, but the secular city at its best already fulfills that hope, since it is a city by definition that sets man free to fashion his own future. In the secular city, as far as it has progressed in our own time, we can "discern certain provisional elements of the promised Kingdom."[32]

Obviously, for Cox, the modern liberal-democratic, urbanized, and affluent society, with its almost ritualistic loyalty to the post-1789 revolutionary tradition, performs certain functions analogous to those performed for Hegel in his century by Hohenzollern Prussia, or for Marx by the ideal of the classless society. Like Hegel's Prussia, but unlike Marx's utopia, the secular city is not scheduled for future delivery by the logic of history: it is already, albeit imperfectly, here; and no higher order of commonwealth is apparently destined to replace it within terrestrial time.

Another important document of the new theology, which Cox hails as "the best corroboration of the main thesis of *The Secular City* I know about,"[33] is the Dutch theologian Arend van Leeuwen's *Christianity in World History*. Van Leeuwen draws on the same theological sources as Cox, and also on his countryman Hendrik Kraemer.[34] He develops at length the familiar contrast between the historically-minded civilization of the modern West, rooted in biblical faith, and the "cosmic" or "ontocratic" societies of the ancient Western world and of Asia, rooted in a metaphysical apprehension of the totality of being. Again, secularization is seen as an irreversible process springing from Judaism and the New Testament. It not only transforms Europe and America but also in our time permeates the non-Western world. It is "the spirit of 'Christianity incognito' . . . at work. . . . The technological revolution is the evident and inescapable form in which the whole world is now confronted with the most recent phase of Christian history. In and through this form Christian history becomes world history." A world civilization is in the making, for which the sec-

263

ularized, progressive Western society of the modern age serves as prototype. There is no turning back for the society that has once "eaten of the tree of Western civilization": it will henceforth find itself caught up in history, and "thrust forward along a devious way, with gorges and precipices on either side, yet leading upwards towards broader prospects and more copious expectations," realized at the cost of ever-mounting effort and risk.[35]

Van Leeuwen challenges the Christian churches not to resist secularization but to strive for the fullest possible realization of the secular society. No return is possible to the old age of "religion." Proclaiming the gospel in the modern age means, therefore, to guide the revolutions of technology and secularity in a Christian direction, to ensure that progress is for man, and not man for progress. As in Cox's thought, the foremost goal of secularization appears to be freedom. "Where the voice of Christ is understood, the technological revolution is conceived as history and not as process: as a series of continual human decisions and not as happening inevitably 'in the course of nature': . . . under God's judgment and grace and not in bondage to biological and sociological laws." [36] The entrenched ecclesiastical orders must be "dynamited" from their comfortable positions, and Christians must unite in a holy war not only against those who would turn back the clock of history but also against "the technocracy," the men and institutions who see technology as an end in itself, rather than as a servant of human freedom. There must be a never-ending struggle for continued secularization in the Christian sense, for democracy, enlightenment, world law and peace, the conquest of poverty, and social justice.

If Christianity does not seize the leadership in this struggle, secularized offshoots of the Christian faith, such as Marxism, will act, far less adequately, in its place, for every movement "which aims to proselytize the world nowadays works to an explicit or implicit philosophy of history, with itself as the fulfilment of the historical world-process, conducting mankind to the final goal of global unity and peace." [37] Even the religions of Asia seek to historicize their teachings, after the Christian example. But in fact there is no substitute for the gospel of Christ, and van Leeuwen, like Kraemer, will have no part of schemes for collaboration on terms of equality among the world's "religions," since Christianity is not a religion, like other religions, but the word of God. As the faith that initiated the whole process of secularization, Christianity is alone fully qualified to lead mankind to the coming secularized planetary civilization.

In Germany itself Bonhoeffer's thought has been continued in the work of Jürgen Moltmann.[38] Born in 1926, Moltmann teaches theology at Tübingen; his *Theology of Hope* (1964) has provoked considerable controversy in Germany and also in the United States, where a translation appeared in 1967. Its most striking feature is its forthright assault on the eschatological thought of Barth, Althaus, and Bultmann, which Moltmann accuses of being eternity-centered and basically Hellenic, rather than directed toward the future in the hopeful spirit of the biblical faith. This kind of thinking, he warns, has always been useful in subduing the church of Christ to the purposes of the established order, but it must be strenuously resisted. Hope of progress through Christian action, hope for the kingdom of God and the resurrection of the dead stand at the center of the gospel of Christ. "There is therefore only one real problem in Christian theology, which its own object forces upon it and which it in turn forces on mankind and on human thought: the problem of the future." [39]

Traditionally, Christians have tended to think of eschatology as concerning only the end of time. The historical future was neglected by the orthodox and became a lively interest for fanatical sects and revolutionary groups, inside or outside the pale of the faith. "Hope emigrated as it were from the Church and turned in one distorted form or another against the Church." [40] It was the outstripping of Christianity by an intoxicated millenarianism growing from Christian roots that brought about the secularization of life in modern times. The secularized world order of today owes its success to its emancipation from the priest-ridden sacral society of the Middle Ages and the Reformation, a society that put the Christian church in much the same position occupied by the *cultus* of Jupiter in pagan Rome.

But Moltmann does not wish to appear a defender of even the contemporary status quo. The secular society supplies affluence and freedom, but it also threatens the individual with depersonalization, especially in his working life, and whenever he encounters the institutions of the mass order. To provide relief from the nonhumanity of the rationalized social order, the powers that be encourage the Christian churches to offer a romantic haven of subjective and interpersonal experience, in which the individual can lower the pressures built up during the working day. In place of organization and routine, he finds intimacy and fellowship. Since the churches are mass institutions in their own right, they also become subtle instruments for the more complete institutionalization of modern life.

These are roles that the church of Christ must have the courage to refuse. It must become an "exodus church," dedicated to reforming the social order, not conserving it. Christian hope "leaves the existing situation behind and seeks for opportunities of bringing history into ever better correspondence to the promised future." It demands not only the salvation of individual souls "but also the realization of the eschatological hope of justice, the humanizing of man, the socializing of humanity, peace for all creation." [41]

Wherever one turns in the theological literature and discussion of the 1960s, the talk is of "hope," "promise," "history," "secularity," and the "kingdom of God." Not to be outdone by their Protestant colleagues, Catholics come forward with secular theologies of their own, such as that of Edward Schillebeeckx, and Catholic scholars earnestly consult St. Thomas to discover evidences of a "theology of hope" in the *Summa Theologica*.[42] Rabbi Heschel speaks for many contemporary Jews in representing the climax of Jewish hopes as "the establishment of the kingship of God" and "universal redemption." This is not, he writes, "an event that will take place all at once at 'the end of days' but a process that goes on all the time. . . . The vision of a world free of hatred and war, of a world filled with understanding for God as the ocean is filled with water, the certainty of ultimate redemption, must continue to inspire our thought and action." [43] More recently, stirred by the Israeli victory in the Six-Day War of June 1967, Heschel has argued that the state of Israel itself must be seen as "a profound indication of the possibility of redemption for all men." He likens the war to the six days of creation, which were followed, as the war of 1967 may also be followed, by a day of rest, an era of peace for all mankind.[44] *

* See also the gentler vision of Jewish/Israeli communalism as the prototype for a world community in Martin Buber, *Israel and the World* (New York, 1948); and *Paths in Utopia* (Boston, 1949). Nor have the Unitarians, who were among the staunchest believers in progress in the nineteenth century, abandoned their faith. A recent poll disclosed that 95.2% of North American Unitarians and Universalists believe "there has been progress in the history of human civilization." The evidences of progress considered most decisive by the 36,000 church members in the survey were the growth of science and knowledge, an increase in moral sensitivity, and the emergence of a world community. Unitarian Universalist Association, *Report of the Committee on Goals* (Boston, 1967), p. 26.

14

Philosophies of Hope

THE MOST PROMINENT MOVEMENTS in twentieth-century academic philosophy—logical positivism, linguistic analysis, phenomenology, and existentialism—are all clearly hostile or indifferent to the belief in progress. This contrasts so sharply with the situation before 1914 that one is tempted to give up the search for ideas of progress in recent philosophy without even beginning to look. Such a temptation should be strenuously resisted. The thought of all those who belong to one of these characteristic twentieth-century schools is often not exhausted by the dogmas of the schools. In any event, there are other schools; and all philosophy—*grâce à Dieu!*—is not scholastic. Philosophers in our century show fully as much enthusiasm for the idea of progress as theologians, if we are only prepared to make our definition of "philosopher" reasonably flexible. The general public has also displayed a lively interest in reading new philosophies of progress, especially in very recent years.

Philosophies of progress in the twentieth century may be divided into three distinct categories: humanist, cosmological, and historical.* The humanist movement, chiefly in Anglo-American thought, has enlisted a number of able natural and social scientists, men of letters, and professional philosophers, including (in a "lay" capacity) several leading logical positivists, from Bertrand Russell to A. J. Ayer. As a movement of thought, humanism combines the atheism

---

* A fourth category, Marxist philosophy, is treated in later chapters. See especially the discussions of Ernst Bloch, pp. 331–33, and of Herbert Marcuse, pp. 343–48.

of the nineteenth-century rationalist tradition with a celebration of the idea of progress and many other stock liberal values. Evolutionary cosmology, carrying on in the spirit of German idealism and French vitalism, has continued to attract first-rate minds in our century, such as Alfred North Whitehead and Pierre Teilhard de Chardin. In addition to important technical achievements in the field of historical methodology, philosophy of history has produced a great wealth of speculative systems embodying ideas of progress. The speculative philosophies of history of Karl Jaspers, Lewis Mumford, Erich Kahler, and (since the 1940s) Arnold J. Toynbee are all philosophies of hope.

Of course one should not expect to find, in so many disparate minds from so many different intellectual traditions, the unity of mood, style, and premise that characterizes the philosophy of *la belle époque*. But whatever the sources of their insights, these twentieth-century philosophers of progress reach remarkably similar conclusions. From the broad cosmopolitan perspectives of their thought, most of them contend that the human race has advanced and will probably continue to advance toward an organic planetary civilization. The world civilization will be an open society, freely formed by free men: not a utopia, but a society engaged in limitless progress toward still higher goals. Apprehensions of the relativity of positive cultures tend to be overcome in the end by the vision of a transcendental and unitary human destiny. In some instances, as in the thought of Kahler and Teilhard, traces even of nineteenth-century historical determinism survive; for them progress becomes something almost irresistible. Many academic philosophers, doubtless the majority, show no interest. But nothing could be more mistaken than to imagine that in the twentieth century the wells of inspiration of progressivist meta-historical and meta-biological thought have run dry.

The greatest protagonist of rationalist humanism in twentieth-century thought has been Bertrand Russell. Born in 1872, and educated at Cambridge, Russell had already published several substantial volumes of philosophy by the outbreak of the First World War. Most of his thoughts on history and human progress appeared later, in more than thirty books written as contributions to public enlightenment and improvement rather than as philosophy. Russell limited philosophy to the tasks of logical analysis. At the same time he felt compelled to follow the *philosophes* of the Enlightenment, whom he much resembled, in playing the role of social prophet, educator, and propagandist for a rational world order. The spirit of Voltaire permeated all his writings on matters that he considered outside the range of formal philosophical inquiry, from ethics and politics to education and history.

Like Voltaire, Russell could not contemplate the past and present conditions of man's existence without feeling profound disappointment. Man's passions time and again failed him, plunging him into war, tempting him to exploit his fellows, driving him to seek power and fame as ends in themselves. Russell lacked even Voltaire's deist faith in the rationality of the cosmic order. In the vein of late nineteenth-century naturalist pessimism, he pictured man struggling to survive in a completely indifferent universe, defiant of "the irresistible forces that tolerate, for a moment, his knowledge and his condemnation, to sustain alone, a weary but unyielding Atlas, the world that his own ideals have fashioned despite the trampling march of unconscious power." [1]

Yet it was not the universe that most dismayed Russell. The greatest enemy of man was man himself, and the root of all earthly evil was man's Hobbesian lust for power, whenever this lust became purely selfish, directed to no good but one's own. "The study of history from the building of the pyramids to the present day," Russell wrote in 1954, "is not encouraging for any humane person." As a political animal, man had improved not at all. On the contrary, "I do not think that the sum of human misery has ever in the past been as great as during the last twenty-five years." At mid-century mankind might stand "on the brink of final and utter catastrophe." [2]

Also like Voltaire, however, Russell discovered grounds for hope in mankind's higher qualities, and even grounds for a belief in general progress, despite the cruelty, greed, irrationality, and violence that dominate political history. He rejected Spengler's doctrine of historical cycles and all deterministic schemes for explaining the past and predicting the future. No inexorable destiny hung over man. Human "nature," largely the product of long cultural conditioning, could be changed in desirable ways by education, and man's so-called aggressive instincts could find other means of self-expression than war. His capacity for love and sympathy, properly harnessed to his intellect, could convert him into a sociable, free, and happy creature. Already the conquest of external nature by science and technology had brought the good life within mankind's reach. Step by step light had banished darkness, and civilized strength had replaced primitive weakness. "If bad times lie ahead of us we should remember while they last the slow march of man, checkered in the past by devastations and retrogressions, but always resuming the movement toward progress." The belief in progress was elevated by Russell to something very much like a psychological necessity. "The human race," he wrote, "has emerged slowly from the condition of a rare and miserable hunted animal, but if we suppose that it has no further journey to make, that there are no greater perfections to be achieved in the future, and that we are approaching a dead-end, something deeply instinctive and immeasurably important will wither and die." To hand on the treasures of human achievement, "not diminished, but increased, is our supreme duty to posterity." [3]

True to his rejection of determinism, Russell could imagine less desirable futures. A year before the publication of *Brave New World,* he sketched, by way of warning, the portrait of a scientifically managed world society much like Huxley's that would annihilate freedom and extinguish man's joy in life. In Russell's social philosophy science and technics were ethically neutral, available equally to bad men and good: only when they served the cause of human freedom and happiness could they offer hope for the future.[4] Until his death in 1970 Russell was a leading figure in the campaign for British unilateral nuclear disarmament, as well as an advocate of general world disarmament and the creation of an effective world government, failing which he feared that "scientific man is a doomed species." [5]

But in the main, throughout his incredibly long career, he was a bringer of good tidings. Most of his books on social problems end with a vision of the happy world that man could have for the asking. In one of the last, written at the age of eighty-nine, he anticipated a united world civilization capable of art, literature, and science that would soar to

> hitherto undreamt of heights. . . . The liberation of the human spirit may be expected to lead to new splendours, new beauties and new sublimities impossible in the cramped and fierce world of the past. If our present troubles can be conquered, Man can look forward to a future immeasurably longer than his past, inspired by a new breadth of vision, a continuing hope perpetually fed by a continuing achievement. . . . No limit can be set to what he may achieve in the future.

I see, in my mind's eye, a world of glory and joy, a world where minds expand, where hope remains undimmed. . . . All this can happen if we will let it happen.[6]

The various national humanist associations, particularly in North America and Britain, have for years enjoyed the support and leadership of many professional philosophers, including A. J. Ayer, Sir Karl Popper, Herbert Feigl, Sidney Hook, Corliss Lamont, and Charles Frankel. On the whole, the thrust of their thought has been to defend and reaffirm the idea of progress, though seldom as a law or necessity of history. As one prominent humanist spokesman (H. J. Blackham) writes, "The heady idea of progress which made our fathers and grandfathers patronize the past is of course as clean gone as an exploded shell." Humanists no longer think in terms of a utopian future. But "the difference between better and worse is so enormous" that no man, least of all no humanist, should succumb to despair. In particular, Blackham points to science and democracy as achievements of the modern age, "which make human progress and self-direction on a world scale practicable." [7]

The most incisive and hopeful manifesto of humanist beliefs from the pen of an American philosopher is Corliss Lamont's *The Philosophy of Humanism*. After surveying the historic varieties of humanism, he defines humanism in the twentieth century as "a philosophy of joyous service for the greater good of all humanity in this natural world and according to the methods of reason and democracy." The highest ethical goal of man is "the this-worldly happiness, freedom and progress—economic, cultural and ethical— of all mankind." Personal happiness must be sought primarily in working for the happiness of the species. From our collective efforts on behalf of humanity a world-wide humanist civilization may emerge, higher than any civilization in history. The universal humanist society will be founded on the principles of political, economic, social, racial, and sexual democracy. An expansion of education and culture, moreover, will result "in a cultural flowering comparable in achievement to the outstanding epochs of the past and going far beyond them in breadth of impact," since sums will be invested in cultural activities "proportionate to what present-day goverments allocate to armaments and war." Using their own intelligence and cooperating freely with one another, men "can build an enduring citadel of peace and beauty upon this earth." Even before the close of the present century, it is possible that "the human race will emerge onto the lofty plateau of a world-wide Humanist civilization." [8]

Another American philosopher, Charles Frankel, has carried the attack to some of the most fashionable "enemies" of modernity in his book *The Case for Modern Man*. Too many thinkers, he argues, have allowed the imagination of disaster to destroy their faith in the essential intelligence and goodness of mankind. Against the relativism, otherworldliness, and pessimism of men such as Jacques Maritain, Reinhold Niebuhr, Karl Mannheim, and Arnold J. Toynbee, Frankel upholds the basically eighteenth-century "faith of reason," which he studied in an earlier book and defends unblushingly here.[9] Although most of his intellectual opponents actually subscribe to a belief in progress themselves, he adheres for the most part to the classical doctrine of progress taught by the *philosophes* of the Enlightenment. Human progress, thus, can be measured in secular terms; mankind is indefinitely perfectible, which does not mean capable of effecting its own "salvation" or "redemption" in the theological sense; and social progress proceeds by deliberate, rational, piecemeal reconstruction and reform of human institutions. It would be "a grisly impertinence" to speak in the twentieth century of automatic progress, and inadvisable to assume that man's struggle for the good life enjoys "the special benediction of nature." But no modern liberal can deny that the ideals of reason and progress "have at least a fighting chance." [10]

The Anglo-American humanist movement harks back to the rationalism of the Enlightenment, with generous expropriations, nearer to the present day, of nineteenth-century British positivism and American pragmatism. One important Continental humanist philosopher, who belonged to no school, and who had quite different philosophical forebears but exerted a broad influence throughout Europe, North America, and Latin America, was José Ortega y Gasset. Ortega studied extensively in Germany before the First World War, where he fell under the influence of Goethe, Nietzsche, and the neo-Kantians. He was also exposed, with every other philosopher of his generation, to the thought of Henri Bergson, and in the 1930s he belatedly encountered Wilhelm Dilthey's writings. His role in Spanish intellectual life closely paralleled that of Croce in Italy. Literary critic, editor, philosopher, liberal republican, Good European, and the best-known Spanish thinker of the century, he is often regarded as an existentialist and historicist. Yet he was also, like Nietzsche, a great yea-sayer to life, a vitalist in somewhat the Bergsonian vein, and at the very least ambivalent on the subject of progress.

Ortega saw his function in Spain very much as if he had been an eighteenth-century *philosophe*. His was the self-appointed mission of bringing light and new energy to the Spanish people, of waking them from their lethargy and restoring them to the family of progressive Western nations. Many of the strongest influences on his philosophy tended to lead him away from the nineteenth-century gospel of progress—especially his contacts with historicism. Yet he could not have allowed any philosophical movement to negate his affirmation of life and his enthusiasm for European culture. For all his apprehensions about the drift of European politics and society in the twentieth century, he always remained a philosopher of hope.[11]

The central theme of Ortega's philosophy might be described as "synthesis." Surveying the conflicting claims of "rationalism" and "relativism," he arrived at the halfway position of "perspectivism." The struggle between "culture" and "life" he sought to pacify with his doctrine of "vital reason." In approaching the belief in progress, he looked for a formula that would save the optimistic spirit of Western culture without draining present life of its own immanent and self-validating importance. He proposed, on the one hand, that life was "valuable, no less than justice, beauty or beatitude, for its own sake." If, by progress, one meant the doctrine that relegated "real existence" to the "subordinate level of a mere transition towards an utopian future," then Ortega had to reject the idea of progress. He who would affirm any part of life must affirm it all. On the other hand, real existence was historical through and through. "Man, in a word, has no nature; what he has is . . . history." Man was "a substantial emigrant on a pilgrimage of being, and it is accordingly meaningless to set limits to what he is capable of being. . . . To progress is to accumulate being, to store up reality." Any given individual, at any time, "is never the first man but begins his life on a certain level of accumulated past. That is his single treasure, his mark and privilege. And the important part of this treasure is not what seems to us correct and worth preserving, but the memory of mistakes, allowing us not to repeat the same ones forever." [12]

Yet who could tell if mankind would progress in the future?

273

The future might hold in store any number of possibilities for man, from the best to the worst. "The error of the old doctrine of progress lay in affirming *a priori* that man progresses toward the better. That is something that can only be determined *a posteriori* by concrete historical reason." [13] In brief, progress was neither predictable nor inevitable; yet Ortega held out the hope that any given moment man could retrospectively measure his past progress, judging not by some abstract theory of progress, but by a careful historical survey of what had actually happened. As for the future, progress lay within man's power to achieve, although history guaranteed nothing.

The most comprehensive, and academically the most unfashionable, philosophies of progress in the twentieth century are those which contend that man's history can be understood only in terms of a greater cosmic process. In academic circles, needless to say, the validity of all metaphysical inquiry has fallen under the most acute critical scrutiny. But major cosmologies of progress are still being expounded, in the tradition of Spencer, Fiske, Bergson, and Alexander. Some of the more recent examples, such as those of J. C. Smuts, J. E. Boodin, A. N. Whitehead, Pierre Lecomte du Noüy, and Pierre Teilhard de Chardin, have captured the public imagination to an extent rarely equaled by even the most distinguished of academic philosophers of other persuasions. As in *la belle époque,* twentieth-century cosmologists tend to betray the continuing influence of scientific, and especially of biological, thought. They also recall the German philosophers of the romantic era in their representation of God as the creative principle immanent in both nature and history.

We have already examined the thought of one twentieth-century cosmologist, J. C. Smuts.* In the 1920s and 1930s, much imagination and eloquence were expended in defense of metaphysical views of progress by the British philosopher Alfred North Whitehead, and —somewhat more ingenuously—by his Swedish-American contemporary, J. E. Boodin. Their writings convey an almost Wordsworthian sense of the fundamental harmony and unity of nature; everything works out for the best, protected by the wisdom of providence.

Whitehead's cosmology remains the more interesting of the two. He first established himself, in collaboration with Bertrand Russell,

* See above, pp. 73–74.

as one of the century's leading philosophers of mathematics. But after coming to Harvard in 1924, at the age of sixty-three, he published a series of remarkable books in which he expounded a system of ideas far removed in method and spirit from Russell's—although both men were prophets of progress. Whitehead's philosophical vocabulary harks back to Hegel and Plato, despite his strong interest in the implications for metaphysics of modern physics and biology. His "philosophy of organism" often seems to belong more to the ancient thought-world than to the modern.

Whitehead's universe consisted of a hierarchy of organisms, whose ground and "lure" to creative advance was God, who, in one aspect of his divine nature, was also a finite entity in process of evolution, as the world itself evolved. God, the atoms, living creatures, and men were all "actual entities" struggling in common to fulfill their potentialities in a cosmic order suffused with timeless values. In any one cosmic epoch, progress and decay took place continuously, so that the universe did not move *en bloc* toward one final goal. Yet the system of natural laws of a given epoch, as well as its systems of organisms, evolved to a point of maximum development, before becoming infertile and passing away.[14] In the present epoch, the highest known level of organization had been attained by mankind.

For Whitehead, the doctrine of evolution taught by modern biology helped to refute, rather than support, a purely materialistic metaphysics. The motions of matter could not in themselves, he proposed, have led to the emergence of higher structures of activity; matter did not evolve. What had occurred was the purposeful and progressive evolution of more complex from less complex organisms, a process conceivable only in an organismic cosmos given the possibility of upward growth by a ceaselessly urging and adventuring divine principle. Nor could the evolution of the lower animals and of mankind have happened through the competitive struggle of individuals alone, as some evolutionists tended to think. The "neglected side" of evolution, Whitehead explained, "is expressed by the word *creativeness*. The organisms can create their own environment. For this purpose, the single organism is almost helpless. The adequate forces require societies of coöperating organisms." Mutual aid, therefore, constituted an essential factor in the advance of life from lower to higher levels.[15]

Even in the sphere of history proper, Whitehead could see a

steady growth of the arts and faculties of persuasion, over against the arts and faculties of force. Religion and morality had shown even more persistent progress than science. The disappearance of slavery in the world, including the disappearance of serious defenses of slavery as an institution, illustrated the way in which ideas had changed through the centuries to bring about "the gradual purification of conduct." The thousand years since the Dark Ages, in particular, had witnessed—thanks in part to the leaven of Christian doctrine and of pagan philosophy—an "increased sense of the dignity of man, as man. There has been a growth, slow and wavering, of respect for the preciousness of human life." Reverence for ideal ends and respect for life had secured "that liberty of thought and action, required for the upward adventure of life on this Earth." If one, therefore, mapped the movement of civilization on a large enough scale, it might well wear the aspect of "a uniform drift toward better things," despite many temporary retreats and disappointments visible at closer range.[16]

In the event that man learned to use humanely the vast new powers extended to him by modern science and technology, "there lies in front a golden age of beneficent creativeness."[17] Science and religion were obliged to work hand in hand, enlightening, instituting social reform, and striving for that union of adventuresomeness and love of harmony that alone could ensure true progress. God urged mankind forward, but he did not compel: the future was man's to shape as he saw best. The only course he could not elect was to rest comfortably on past achievement. "Advance or Decadence are the only choices offered to mankind. The pure conservative is fighting against the essence of the universe."[18]

Both Boodin and Whitehead reflected the impact of evolutionary biology on cosmological speculation, but evolutionary and historical themes received only intermittent attention in their work. Indeed, most of the extensive critical literature on Whitehead ignores his idea of progress and devotes little space to his thoughts on evolution. Other twentieth-century cosmologists, however, have built their systems more directly on biological foundations—especially those who started out in life as professional biologists. A few, like Sir Julian Huxley, remained primarily men of science throughout their careers, and will be discussed in the next chapter.* But several biologists in our century have produced speculative works of suffi-

* See below, pp. 297–304.

cient depth and subtlety to warrant treatment here as contributions to metaphysics.

One of these biologist-philosophers was Edgar Dacqué, a man little known outside Germany, but a controversial and significant thinker in his time. For many years he served as director of the museum of paleontology in Munich. His first book, a history of evolutionary theory since antiquity, appeared in 1903; most of his metaphysical works were published in the 1920s and 1930s. He died in 1945, his museum demolished in an air raid, and his name still unfamiliar abroad. But, as Ernst Benz points out, Dacqué anticipated much of the thought of Teilhard de Chardin, especially his synthesis of Christian faith and evolutionary theory, although both men were ultimately only following in the tracks of the German *Naturphilosophen* of the romantic age.[19]

For Dacqué, man's appearance on earth represented no accident but the provisional fulfillment of the divine will for nature. Almost like Christ in orthodox Christian doctrine, man had in fact existed from the beginning. His form was implicit in the lowest orders of life, and the whole evolutionary process manifestly advanced toward him step by step through the eons of geological time. Yet even man was not the end of the process, but only a rough draft, destined to be replaced in time by "perfect man who, in a new way, shall be the true image and facsimile of God. Present earthly man will become 'extinct.' . . . In his innermost being, he really 'aims' toward this new form of man, he carries it within himself, he wants to give expression to it, as subhuman nature once aimed for and was pregnant with him."[20]

A similar teleological view of evolution found expression in the last books of the French biologist Pierre Lecomte du Noüy, who, like Dacqué, was a convinced Christian. Lecomte du Noüy launched his philosophical career in 1939 with a book later translated into English as *The Road to Reason* (*L'Homme devant la science*), followed by *The Future of the Spirit,* which was awarded a prize by the Académie Française. In 1942 he escaped from occupied France and came to the United States. A final volume appeared in 1947, *Human Destiny,* which brought him the same sort of public acclaim in America that he had already won in France. He died the same year.

Speaking as a Christian and as a biologist, Lecomte du Noüy argued that the evolutionary process could not be explained in terms of physical laws of chance. The progressive thrust of evolution, culminating thus far in mankind, made sense only if one hypothesized not Bergson's *élan vital,* but the intervention of a supreme intelligence by whose design creation moved toward its own total spiritualization. The great stages in natural history were those of the subatomic particle, the atom, life, and man, governed by the laws of quantum mechanics, thermodynamics, biology, and morality, respectively. "Each stage represents an enrichment and a liberation. Everything comes to pass as if the Spirit had been unable to fulfill itself except progressively, abandoning along the way the scaffoldings which had become useless as a result of the emergence of more perfect forms evolving slowly toward an ultimate, and still far distant, perfection." [21]

Throughout the evolution of being, freedom of action steadily grew. Contrary to popular misconceptions, the goal of evolution was not successful adaptation and the consequent stabilization of the species—a true *cul-de-sac*—but open-ended change leading to ever higher levels of organic freedom. Only one species remained that had failed to achieve stasis: *Homo sapiens;* and it was clear to Lecomte that the emergence of this single species constituted the whole purpose and meaning of the evolutionary process down to the present time. In man, evolution had entered a new phase, no longer biological, but moral and spiritual, in which the supreme task was to defy nature, to take the strenuous upward path from animality to freedom, to abandon the crude struggle for existence enjoined by nature in favor of a higher struggle for spiritual fulfillment. Man's dignity consisted of his power of choice and his role "as a worker for evolution, as a collaborator with God. . . . Not only his own fate, but the fate of evolution is in his hands. At any moment, he can choose between progression and regression." [22]

But Lecomte was too good a Christian to suppose that morality itself evolved, beyond the teachings and the example of Jesus Christ. "The perfect man is not a myth; he has existed, in the person of Jesus. Others have very nearly approached perfection—some of the prophets and martyrs—but their number is infinitely small as compared to humanity, and it is humanity that must be ameliorated." In the lives of the saints one might see "the forerunners of the superior race which is to come." Through centuries of spiritual effort "man will gradually learn to appreciate the higher joys derived from his purely human faculties, until the distant day when the others will repel him. Our attachment to sensual pleasures which recall our origin affords the proof that we are still at the beginning of human evolution." The goal of the divine will was

the realization of a morally perfect being, completely liberated from human passions—egotism, greed, lust for power—hereditary chains, and physiological bondage. . . . Man, with his present brain, does not represent the end of evolution, but only an intermediary stage between the past, heavily weighed down with memories of the beast, and the future, rich in higher promise. Such is human destiny.[23]

The gospel of the spiritual superman proclaimed by Dacqué and Lecomte du Noüy appears also in the writings of Pierre Teilhard de Chardin. Much of the great enthusiasm shown in the past fifteen years for Teilhard can be explained only by pointing to the man himself, and the intrinsic value of his vision. But there are other considerations. Nearly all his books and essays were written in the same years (from the early 1920s to the late 1940s) as those of the other cosmologists studied above, but because he belonged to the Society of Jesus, which refused to grant his principal works an *imprimatur,* none could be published until his death in 1955. The intransigence of the ecclesiastical authorities, which did not end in 1955, seemed so outrageous to liberal Catholic opinion that Teilhard became, after his death, something of a standard-bearer for many of the forces within the Church that found her sadly behind the times. Even more important, he was one of the first major thinkers whose writings became available in quantity at just the period—the late 1950s and early 1960s—when most Europeans began to look hopefully toward the future again, after several decades of almost unrelieved despair and self-condemnation. They had, says Ernst Benz, "got tired of existentialism and theological dialectics." Teilhard "opened again the dimension of hope for our time." [24]

Another appealing dimension of Teilhard's thought has been its obviously sincere eclecticism, which embraces Catholic and Protestant, scientist and mystic, Christian and atheist, liberal democrat and Marxist revolutionary, bringing them together in a system that seeks to dissolve all creedal differences. Although he was poorly versed in German philosophy, Teilhard's cosmology of progress calls to mind not only the *Naturphilosophen* but also the theodicies of Leibniz, Kant, and Hegel; it is an apologia for the cosmos, a "hymn of the universe." It has won enthusiastic disciples in every Western country, and inspired innumerable critical studies.[25]

Pierre Teilhard de Chardin was born in 1881 in central France, and ordained in 1911. Like many other Jesuits he devoted the greater part of his life to science. He was a geologist and paleo-anthropologist of some note, who participated in many expeditions in Asia and Africa and published extensively in scientific journals. He drew inspiration from the thought of Bergson, whose *Creative Evolution* was the most widely discussed book in the France of his late twenties, but the strongest influence on his thought, apart from Christianity, was at all odds evolutionary biology. Even Christianity, in Teilhard's hands, seems at times to take second place to biology; it was not the science of life that had to be interpreted in harmony with the science of God, but just the reverse. Theology had to be recast to meet the demands of biology, although Teilhard found no ultimate contradiction whatever between the two, and every-thing in his thought flows together with a gentle but compellingly messianic fervor.

Most broadly, Teilhard's cosmology affirmed the origin of all being in God and its destiny to return to its source at the end of time. The course of evolution was a process of divinization, of the rising of creation, attracted by its creator, to godhood. Through in-creasing complexity of structure reality achieved steadily higher levels of consciousness, which, as it folded in upon itself, multiplied and concentrated spiritual energy at an ever-growing cosmic tempo.

Mankind represented the furthest forward advance of evolution, and the highest order of complexity and consciousness in the uni-verse. Spreading across the whole surface of the planet, mankind had created an interdependent world society, adding a "noosphere" or envelope of mind to the already existing spheres of earth, water, air, and life. As the result of progress in technology and science, ad-vances in transport and communications, inexorable demographic growth, and the increasing rate and volume of thoughts exchanged, the human species was undergoing "hyper-centration," a process of compression that could not be resisted or reversed.

Confined within the geometrically restricted surface of the globe, which is steadily reduced as their own radius of activity increases, the human particles do not merely multiply in numbers at an increasing rate, but through contact with one another automatically develop around themselves an ever denser tangle of economic and social relationships. . . . Nothing, absolutely nothing—we may as well make up our minds to it—can arrest the progress of social Man towards ever greater interdependence and cohesion.[26]

But the "totalization," "collectivization," or "socialization" of mankind accorded with the purposes of the universe, and was intrinsically progressive. Through the folding in upon itself of the human mass, physically and spiritually, reality moved upward toward its apotheosis.

Teilhard was able to take comfort in even the most calamitous events of the century. Both world wars, which had started as movements of fragmentation, contributed in spite of the bad will of the warriors to the further unification of the race and the further coordination of all the cells in the social organism. Each had been a salutary "turn of the screw." "The more we seek to thrust each other away, the more do we interpenetrate." [27] The totalitarian experiments in Russia, Germany, and elsewhere reflected the same inner impulse toward unity, although they were marred by crude resorts to coercion. Only persuasion could create true and lasting bonds in the human body politic; yet the dictators appealed to modern man's irresistible desire for social order. Even the atomic and hydrogen bombs gave Teilhard a sense of joy; at one and the same time, they had rendered warfare obsolete and confirmed the titanic creative power of human minds working in concert. Nor, obviously, could he deplore what he was among the first to call the "population explosion," since this, too, accelerated the pace of collectivization.

What concerned Teilhard most was the concentration of the noosphere, the progress of the collective mind and spirit of mankind, especially in recent centuries. The simple fact that modern men apprehended the immensity of time, had some inkling of the nature of cosmic evolution, and conceived of the possibility of progress, gave them an incalculable advantage over ancient men. As he wrote in 1920, "between the behaviour of men in the first century AD and our own, the difference is as great, or greater, than that between the behaviour of a fifteen-year-old boy and a man of forty." [28] Little by little the human race had developed a collective memory, a generalized nervous system, and a faculty of common vision: in short, a world mind freely advancing toward unanimity, drawing impartially on the will for progress of Christians and Marxists, liberals and totalitarians, and all those dedicated to the future, the first members of a new species, *Homo progres-sivus*.* This new type of man, possible only since the appearance of the theory of progress and evolution, was "the man to whom the terrestrial future matters more than the present . . . scientists, thinkers, airmen and so on—all those possessed by the demon (or the angel) of Research . . . scattered more or less all over the thinking face of the globe." The comfortable people who wished to conserve the past were destined to pass away; those who could conceive of the world only "as a machine for progress—or better, an organism that is progressing . . . will tomorrow constitute the human race." [29]

Belief in progress, then, was more than an attitude to be held or a pose to be struck. Teilhard knew well the despair of many of his contemporaries. So-called enlightened people, he wrote, nowadays scorned the old faith in man. "The nineteenth century had lived in sight of a promised land. . . . Instead of that, we find ourselves slipped back into a world of spreading and ever more tragic dis-sension." But whatever temporary reasons we might have for dis-couragement, to abandon the belief in progress would be disastrous, both for progress itself and for the mental health of the species. It was not mere well-being (*bien-être*) that mankind should hunger for, but also more-being (*plus-être*): the expectation of the enlarge-ment of being "can alone preserve the thinking earth from the *taedium vitae*." Faith in the future was "not dead in our hearts. Indeed, it is this faith, deepened and purified, which must save us." Such faith would help preserve "in Man his will to act," and guide him to the conquest of the world and communion with God. "It is upon the idea of progress, and faith in progress, that Mankind, today so divided, must rely and can reshape itself." [30]

Teilhard's vision of the immediate future, for all the many essays and chapters he devoted to the theme, lacked precision. In *The Phenomenon of Man,* he made clear that the translation of modern man to higher levels of fulfillment was not inevitable, except per-haps for a small portion of the race. But if all went as he hoped, future mankind would live in peace, evil would steadily diminish,

---

* The idea of a "world brain" or "racial mind" was also advanced, independently of Teilhard, by H. G. Wells, who may have had some influence on the younger man, and by the American philosopher Oliver L. Reiser. Both Wells and Reiser advanced the idea in interpretations of human history that were permeated by progressivism. See Wells, *World Brain* (London, 1938); and Reiser, *The World Sensorium* (New York, 1946).

disease and hunger would be conquered by science, man would acquire full control of his planet and his further evolution, science and religion would be harmonized, and the racial mind would bring all the separate centers of human consciousness into freely accepted unanimity.[31] Yet Teilhard failed to flesh out his picture of the unified world order in convincing detail; the idea of a "hyper-centrated" planetary society, totally collectivized and organized, raises more questions than it answers, particularly with respect to the place of the individual, and also with respect to law and government, matters for which Teilhard had little concern.

In the still more distant future, the earth would become uninhabitable, but by then mankind would have transformed itself into a single organism. "As the Earth grows older, so does its living skin contract, and even more rapidly. The last day of Man will coincide for Mankind with the maximum of its tightening and in-folding upon itself." Fully evolved, the human organism would become what Christ had been on earth, the godhead incarnate, a hyper-person within time extending itself toward the hyper-person outside of time. Mankind and mankind's Father would fuse ecstatically.

Is it not conceivable that Mankind, at the end of its totalisation, its folding-in upon itself, may reach a critical level of maturity where, leaving Earth and stars to lapse slowly back into the dwindling mass of primordial energy, it will detach itself from this planet and join the one true, irreversible essence of things, the Omega point? A phenomenon perhaps outwardly akin to death: but in reality a simple metamorphosis and arrival at the supreme synthesis.[32]

Etienne Gilson, seconded by Jacques Maritain, construes Teilhard's system as "one more Christian gnosis, and like gnoses from Marcion to the present, it is a *theology-fiction*." [33] In any event, it interprets the Incarnation and the Parousia in a new way: mankind by imitating Christ literally becomes Christ, and the Second Coming is the Christification of the race, not a second visit to earth of God's only begotten Son. Well might the orthodox shake their heads!

Teilhard's influence is reflected in his many critics and admirers, and also in the appearance of neo-Teilhardian systems of philosophy and theology, which make creative use of some of his fundamental insights. The most ambitious is expounded in the three volumes of the Dutch philosopher Bernard Delfgaauw's *History as Progress,* first published in 1961–64 and recently made available in a German translation. Delfgaauw offers a three-layered concept of progress. "Progress is, from the cosmic point of view, growth in the complexity of corpuscular structure; from the vital point of view, growth of vital consciousness; from the human point of view, growth of freedom. Growth in complexity is progress, since in this way man originates. Growth in freedom is progress, since in this way man becomes himself." [34]

The tasks for the future for mankind as a whole are self-evident. The world, Delfgaauw writes, needs political and economic integration to promote peace and prosperity for all its people; a world federal government is the only solution. Freedom of choice and of self-development must be enlarged; chiefly this will happen in the Marxist countries through increasing political democratization and in the liberal countries through increasing socialization, or economic democratization, until the systems on both sides closely resemble one another. Social justice, as it becomes more firmly established, will serve in turn to enhance the inner freedom of man to pursue truth, a still higher dimension of freedom. "It is to be expected that man will feel the need for spiritual deepening ever more and more, that art and philosophy and the search for scientific understanding of these insights will occupy ever more time." Christians, atheists, and pantheists will, and should cooperate in building the future human commonwealth. As Teilhard de Chardin taught, they are all engaged in the same progressive work, the perfection of the world, which for Christians is a divinely ordained task. For all men there is only one choice: either to accept defeat and abandon civilization to inexorable decay, or to seek "progress in freedom, in a realization of the unity of mankind in justice and truth.[35]

History continues nature, and philosophies of historical progress are only philosophies of cosmic progress limited to the part played by man in evolution. Delfgaauw's philosophy is perhaps more concerned with history than with biology, despite its debt to Teilhard de Chardin. Earlier we examined several other recent speculative theories of history that reached conclusions hostile to the belief in progress.* The present chapter closes with a survey of twentieth-century ideas of history that favor progress, from both philosophers and practicing historians. Although contemporary professional philosophers often treat theories of history with the same contempt

* See above, pp. 192–93, 199–203.

that they accord cosmological theories, this is a field of speculative thought in which our century has been more productive than even *la belle époque*. The preoccupation with political history characteristic of the late nineteenth century has given way to a broader concept of the responsibilities of historians, and a broader concept of the nature of history itself.

One of the most remarkable converts to progressivism was Karl Jaspers, who qualifies in many respects as a typical exponent of existential philosophy, but who unfolded in later life—in his efforts to philosophize freely beyond the limits of any "school" ancient or modern—a conception of history powerfully seasoned with hopes that could have sprung only from the traditional Western belief in human progress.

Jaspers was born in Germany in 1883. He did not turn to philosophy professionally until the 1920s, when he accepted a chair at Heidelberg. His greatest contribution to the existentialist movement, a work in three volumes entitled quite simply *Philosophy*, appeared in 1931. In his old age he became interested in both the history of philosophy and the philosophy of history, concerns that for our purposes culminated in the publication of his book *The Origin and Goal of History* in 1949.

As becomes a philosopher with intellectual roots in the thought of Kant, Nietzsche, and the critical historicism of *la belle époque*, Jaspers could not embrace a positivist or determinist theory of progress. He included in *The Origin and Goal of History* a refutation of all totalizing doctrines of the meaning and the unity of history. The idea of progress, at least the idea of progress of ordinary parlance, was dismissed as a "popular" theory of the rise of science and technology, with no wider applications. No single image of history could exhaust its meaning or see it whole, since man's history had only begun, and its course could not be construed by analogy with natural events, but belonged to the realm, ever unpredictable, of human freedom. The very fact that man's existence was historical could be explained only by his finitude.

Man is finite, imperfect and unperfectible. . . . Man's imperfection and his historicity are the same thing. . . . There can be no ideal state on earth. There is no right organisation of the world. There is no perfect man. Permanent end-states are possible only as a regression to mere natural happening. In consequence of perpetual imperfection in history, things must perpetually become different.

Any effort to understand history as a whole, as for example any effort to propose a law of general progress, would denigrate existential life, condemn whole peoples and epochs to inauthenticity, and claim falsely for the proposer the wisdom of God.[36]

On the contrary, wrote Jaspers, every struggle to perceive the meaning of history captured only a fragment of infinity; the future remained open; and history could be grasped in its totality only at the end of all time. "Man follows his great highway of history, but never terminates it by realising its final goal. The unity of mankind is, rather, the bourne of history. That is to say: Achieved, consummate unity would be the end of history." [37]

But the ghost of the Absolute haunted Jaspers, as it had haunted Dilthey, Troeltsch, and Meinecke before him. Although it could not be known or defined, the Absolute existed. The same was true of the meaning of history. "My outline is based on an article of faith: that mankind has one single origin and one goal. Origin and goal are unknown to us, utterly unknown by any kind of knowledge. They can only be felt in the glimmer of ambiguous symbols. Our actual existence moves between these two poles; in philosophical reflection we may endeavour to draw closer to both origin and goal." [38] In brief, Jaspers thought it worth his while to essay the impossible, and better to "draw closer" than to remain indifferent to the problem of the meaning of man's existence as a historical being.

Much of *The Origin and Goal of History* reduces, then, to a speculative effort predestined to failure and yet somehow seen by its author as a more adequate view of world history than no view at all. Part One ventures a schema of the history of mankind from earliest times, and Part Two a discussion in depth of man's present condition and his future prospects in light of observed trends. Jaspers could hardly claim finality for his vision, but he did claim to base his interpretation of history on a knowledge of "its factual, perceptibly given, unique shape, which is not a law, but the historical arcanum itself," an interpretation of the structure of history grounded in empirical evidence.[39] Studied concretely by the historian as observer, the past unveiled itself in two vast "breaths," the first extending from the great technical inventions of prehistory, such as speech and tool-making, through the age of the early civilizations to the "Axial Period," 800–200 B.C., when independently in China,

India, the Near East, and Greece, ancient man first became "conscious of Being as a whole, of himself and his limitations," and took "the step into universality," radically spiritualizing his cultures. Technological discovery preceded social organization, which in turn preceded spiritual discovery. The second "breath" had begun in the Renaissance with the Scientific Revolution, now spreading throughout the world. The new science and technics of Western man might well lead, "through constructions that will be analogous to the organisation and planning of the ancient civilisations, into a new, second Axial Period, to the final process of becoming-human, which is still remote and invisible to us." [40]

As Jaspers went on to define his schema, it assumed the familiar form of an outline of human progress, at least along certain highly significant lines. Mankind had first evolved in various isolated centers of civilization, and the highest point it reached in its age of cultural plurality was the so-called Axial Period, the age of Confucianism, Buddhism, the Old Testament, and Greek philosophy. In the modern era advances in knowledge had brought about the globalization of mankind. The second "breath," therefore, if it repeated on its higher, planetary level the experiences of the past, would see the development from its higher technology of a new planetary civilization, "the one world of mankind on the earth," and finally its own higher Axial Period. History might "lead in its entirety, even in the guise of boundless disaster and to the accompaniment of danger and ever-renewed failure, to Being becoming manifest through man and to man himself, in an upward sweep whose limits we cannot foresee, laying hold of potentialities of which we can have no foreknowledge." [41]

In homelier language, Jaspers attempted to conceive of the future as a period of greater fulfillment of human freedom, made possible by democratic planning, the voluntary federation of nations, and an ethic of love and confidence in man's infinite possibilities. Even in a unified world mankind would continue to pursue different varieties of truth, through historically distinct philosophies and creeds. A universal world order, he wrote, would be feasible "only when the multiple contents of faith remain free in their historical communication, without the unity of an objective, universally valid doctrinal content." [42]

Jaspers at times expressed the fear that perhaps none of this would come to pass, and that mankind would be transformed into a giant ant heap by the pressures of technology.[43] * But not forever, even if worst came to worst. "Man," he added, "cannot get lost entirely, because he is created in the 'image of the Deity'; he is not God, but he is bound to Him with oft-forgotten and always imperceptible, but fundamentally unsunderable, ties. Man cannot altogether cease to be man." He might slumber for a time, but his anxiety for his humanity would always awaken him. The very fears of modern man that his humanity could be lost were in themselves "a demonstration of our humanity, that wants to rid itself of evil dreams." Dread of a dehumanized future "may perhaps prevent it." [44] On balance, then, Jaspers emerges as a believer in progress, without intending to be; above all, as a philosopher of hope, rather than despair or cynicism; and as an heir of the German Christian metaphysical tradition of Kant and Hegel, despite his involvement in historicism and existentialism.

Jaspers' concept of axial ages and his view of the planetization of civilization receives support, much of it independent, in the writings of the American humanist scholar Lewis Mumford. Like Jaspers, Mumford devoted far-ranging attention in his earlier thought to the dilemmas of technological society, urging that modern Western civilization stood in peril of imminent collapse because of its self-enslavement to a dehumanizing technical culture that had no place for the person. In his books *The Condition of Man* (1944) and *The Conduct of Life* (1951), he saw wisdom in the cyclical philosophy of history; he preferred to speak of the need for "renewal" rather than for "progress."

But in a later book, *The Transformations of Man* (1956), written after the appearance in translation of *The Origin and Goal of History,* Mumford sets forth a rectilinear vision of history. He divides history into four eras: the archaic, the civilized, the axial (following Jaspers), and the modern. Although modern culture threatens to substitute mechanical values for human, it has also unified the human race for the first time, terminated the cyclical rhythm of "Old World" history, awakened in mankind a new confidence in human possibilities, and promised release from the ancient misery of mass poverty. In the main, the course of world history has been progressive, from the narrow life of savagery to the universalism of axial thought and modern civilization. Now

* Jaspers' pessimism is discussed above on pp. 231–32.

Mumford sees mankind standing at a fork in its historical road. Barring nuclear disaster, it has only two directions in which to travel. If it takes one turning, it will create a post-historic, post-human insect society, a society of exalted mechanism, in which progress cannot occur. If it takes the other, it will advance toward an organic world culture, in which the integrity of the person is fully respected, the nations live under one law, the economy meets human rather than market needs, and life is valued and understood in its totality.

Mumford clearly views the organic world culture, the second turning, as a higher stage in human development than any hitherto achieved by man. If man chooses an organic rather than a mechanical destiny,

> work and leisure and learning and love will unite to produce a fresh form for every stage of life, and a higher trajectory for life as a whole. . . . World culture will weld the nations and tribes together in a more meaningful network of relations and purposes. But unified man himself is no terminal point. For who can set bounds to man's emergence or to his power of surpassing his provisional achievements? . . . Every goal man reaches provides a new starting point, and the sum of all man's days is just a beginning.[45]

Mumford's sense of an undetermined future is challenged to some extent in the otherwise similar philosophy of history of Erich Kahler. A German scholar who came to the United States in 1938 as a refugee from National Socialism, Kahler gave a series of lectures on world history at the New School for Social Research in New York in 1941–42, later published as *Man the Measure,* which represents one of the most impressive efforts by a recent historian to see the human experience as an organic whole. Condemning the relativism of the modern historical outlook, Weber's conception of value-free scholarship, and the "devaluation of values" in contemporary thought generally, he contended that history is neither meaningless nor open-ended. It presented itself to the scholar looking from holistic perspectives as an evolutionary process moving in a definite direction, to which mankind must conform.

> This direction is not merely the work of man's will, nor can he alter it by an act of will any more than the phases of youth, maturity and age can be altered within an individual. Man can determine himself, but only in so far as he recognizes the processes of his nature, as he foresees

the tendency of development, smooths its path and exploits its forces. . . . The only guide for our plans and action is to recognize the direction of human development.[46]

More precisely, history was the evolution of the quality that distinguishes man from animal, his spiritual power to detach the self from the nonself and to transcend the self. The growth of the human spirit had not always or necessarily resulted in progress, i.e., moral betterment and the increase of happiness, but it was the precondition of specifically human progress, and Kahler saw spiritual evolution as good in and of itself, despite man's not infrequent misuse of his higher faculties. The course of history was upward, even if progress was spiraliform rather than linear.

Kahler shared Mumford's diagnosis of the ills of modern civilization; we live, he wrote, in an age of massive disintegration, characterized by the ascendancy of mechanism and the abandonment of spiritual values. Seeking knowledge of the desirable future from the observable past, we must plot a course for the years ahead based on "the forgotten human idea that brought about all that we comprehend under the term civilization." This principle was not static, but "a living, moving, expanding impulse and quality of man," which could be conserved only by striving for its realization on a "higher level." There could be no return to the mythology and the authoritarianism of the past: rather the future belonged to those who understood that "man has come of age" and that a social-democratic "Kingdom of Man" must be built on earth, fulfilling and crowning the spiritual evolution of mankind down to the present era.[47]

The exact process by which such a civilization would emerge lay within the province of human freedom to determine. Yet "the trend of the whole process, the general direction in which events are moving, and the alternatives they are pointing to for choice and decision, can definitely be seen quite a distance ahead—in that respect the course of history is predictable." [48] Either modern Western civilization would blow itself to bits, or the West would elect to join with the other peoples of the globe in creating a supranational world order, toward which all of man's history had strived for millennia. But even the failure of the West, which now leads the world, would not deter history from continuing its upward course. The "basic human substance" of the Slavs, Asians, and Africans "is not so spent as ours," and if the West destroyed itself, "the long dormant vital power of their masses, awakened and trained by Western methods, may—perhaps after a new 'dark age'—take up the torch and carry it further." [49] In any case, history would persist: its meaning was no mystery, and its patience was well-nigh inexhaustible.

It may fairly be argued that each of the philosophies of history studied above, whether Jaspers', Mumford's, or Kahler's, represents the seasoned reflections of later life; the same is true of the recent thought of Arnold J. Toynbee, whom we have met as a major exponent of the cyclical theory of history.* Despite certain fundamental concessions to progressivism, the Toynbee of the first six volumes of *A Study of History* (1934–39) counts in most respects as an opponent of the belief in progress. But after the Second World War, in the essays collected in his book *Civilization on Trial* (1948), and in the concluding six volumes of *A Study of History* (1954–61), Toynbee shifted his ground and emphases enough to qualify, on balance, as a prophet of human progress in the modern tradition. He agrees with Christopher Dawson that his new outlook should be viewed as "a change from a cyclical system to a progressive system." [50]

The later thought of Toynbee is progressivist on two scores: it develops at greater length, and with more specific formulas for future action, the vision first sketched in Volume One of *A Study of History* of the historical civilizations as an intermediate stage between primitive man and a higher man yet unborn; and it now identifies at least one major, and to Toynbee overwhelmingly important, line of unceasing progress in the history of man from the earliest times down to the present—the progress of religion.

Let us examine the second point first, since it has absorbed far more of Toynbee's own attention, as well as the attention of his critics. In the earlier volumes of his *Study,* he showed that the "higher religions" of mankind, such as Christianity and Buddhism, served as chrysalises in which the disintegrating civilizations of one generation had undergone metamorphosis and from which new civilizations of a younger generation had grown. But his focus as a practicing historian had been on the civilizations themselves, as the proper units of historical study; in Volume Seven in 1954, he raises the question of what happens when the historian focuses on the

* See above, pp. 199–200.

religions, and their history. Why not regard the civilizations as wombs of religion? He has also finally managed to disengage himself from what he now condemns as the provincial Western belief that only Christianity possesses the word of God and access to salvation. There are several "higher religions," Toynbee now affirms, appealing to different types of personality but striving toward the same transcendent goal. They must all "resign themselves to playing limited parts, and must school themselves to playing these parts in harmony, in order, between them, to fulfil their common purpose of enabling every human being of every psychological type to enter into communion with God the Ultimate Reality." [51]

In terms of a view of world history, Toynbee's new-found theological relativism makes it quite a simple matter for him to propose that history may be reconceived as a process by which mankind has progressively achieved a fuller understanding of God and has acquired steadily higher powers of self-determination and access to the means of divine grace. Christian faith has accumulated and transmitted to successive generations "a growing fund of illumination and grace," and the other higher religions have made comparable contributions "to the growing spiritual heritage of Mankind on Earth. In this matter of increasing spiritual opportunity for souls in their passages through this earthly life, there [is] assuredly an inexhaustible possibility of progress in This World." It should also be noted that "spiritual progress . . . will incidentally bring mundane progress in its train . . . far greater than the utmost that could be attained by aiming direct at a mundane goal." [52] World history, then, becomes not only the record of the rise and fall of civilizations but, more significantly, the record of the gradual ascent of man's religious life, from the most primitive religious cults to the highest world religions—Christianity, Buddhism, Hinduism, Islam, and (in a footnote added in 1961) Zoroastrianism and Judaism.[53]

Toynbee does not contend that religious progress equals progress "in the absolute." Although spiritual life is the highest activity of man, and the ultimate reason for his earthly pilgrimage, there are other activities, in some of which rectilinear progress cannot be seen—in art, for example, and in forms of political organization (if one takes, as Toynbee does, the universal state as the most "advanced" type of political organization). In art even the criteria for progress cannot be stated.[54] Nor has man become any less vulnerable to the lure of original sin. "So far as one can guess, human beings are no better, and saints are no more frequent, in the present-day world than they will have been in, let us say, the Lower Palaeolithic Age." [55]

Nonetheless, progress in religion is clearly significant enough to Toynbee to constitute, when all other considerations are weighed in, net gain for mankind as a species. He also does not hesitate to suggest the possibility of further progress in the future that involves far more than a simple rectilinear continuation of man's spiritual advance. The cyclical rhythm of civilizational rise and fall is a "natural law," but the higher law of freedom, made available to man by God, may enable him to transcend the age of the civilizations entirely and reach a new and higher ledge in his climb toward the heights. This "next ledge" is the organic world civilization prophesied by Mumford and Kahler, an ecumenical society without war, the exploitation of man by man, or the fierce rivalry of segmental religions and ideologies; an ecumenical society not subject, as all past civilizations have been subject, to the law of cyclical decay.

In the final reckoning, Toynbee is appealing for saints, for a new race of men surpassing present or past men precisely as Nietzsche's Superman rose above the common herd. With Jaspers he fears that human unity may come at the disastrous price of reducing men to anthood by one or another form of "conditioning." The alternative is to call forth a race of saints, in numbers such as the world has never known. The saint will not be, and has never been, a perfect man, but simply "a human being who has raised himself above the average level of human goodness. . . . Sainthood, thus described, is a well-attested historical phenomenon, and the human beings who have risen to this higher spiritual altitude have done so in different degrees. What some human beings have achieved in some degree must be a practicable objective for others; if grace has been offered to some souls, it will have been offered to all." With the help of these latter-day saints, mankind will be able to live as one family for the first time in history. Such a *union sacrée* will be

Man's finest achievement and most thrilling experience up to date. From the new position of charity and hope which Man will thereby have won for himself, all the past histories of the previous divisions of the human race will be seen, in retrospect, to be so many parts of one common historic heritage. They will be seen as leading up to unity, and as opening out, for a united human race, future prospects of which no human being could have dreamed in the age of unfettered parochialism.[56]

Even the conventional practicing historian has sometimes committed himself in recent decades to a belief in general human progress, although not always to the same belief professed by the philosophers of history. In British academic circles, historians such as E. H. Carr and J. H. Plumb have spoken eloquently in defense of the progressivist gospel—Carr in his 1961 Trevelyan Lectures at Cambridge, Plumb still more recently in his essay, "The Historian's Dilemma." Both not only affirm their own faith in progress but also identify disbelief in progress as a major threat to the social health of modern civilization. Plumb likens historians and philosophers who reject the idea of progress to "death-watch beetles, sapping the strength and confidence that history should give to leaders of society." A belief in progress is "the one certain judgment of value that can be made about history. . . . If this great human truth were once more to be frankly accepted . . . history would not only be an infinitely richer education but also play a much more effective part in the culture of western society." Carr quotes with approval his fellow historian A. J. P. Taylor's quip that all the fashionable talk nowadays of the decline of civilization "means only that university professors used to have domestic servants and now do their own washing-up." For Carr the idea of secular progress —in the absence of a sense of divine purpose—has become a necessity for the very survival of civilization.[57]

Acceptance of the doctrine of progress among historians has been still more widespread in the United States. The tenets of the "New History" of James Harvey Robinson * strongly pervaded American historiographical thought at least until World War II. Charles Beard, Harry Elmer Barnes, and even Carl Becker (after a brief despairing flirtation in middle life with relativism) all embraced progressivism in the Robinsonian spirit.[58] Sidney B. Fay closed his presidential address to the American Historical Association in 1946 with the judgment that "progress is not constant, automatic, and inevitable in accordance with cosmic laws, but is possible and even probable as a result of man's conscious and purposeful efforts." [59] Even if many self-styled intellectuals have suc-

* See above, p. 122.

cumbed to gloom, as the Harvard historian Crane Brinton pointed out in his *Ideas and Men* in 1950, the survival of "the basic optimism of the eighteenth century is evident from any newspaper, any periodical, any lecture platform, especially in the United States." Twentieth-century men are still "children of the Enlightenment." [60]

The popularizers of history in our century outside the ranks of the professional academic historians have also done their share to keep the belief in progress alive. H. G. Wells proclaimed the truth of progress in *The Outline of History* in 1920. Surveying history from earliest times, the English novelist and biographer Vincent Brome, whose books have included a life of Wells, discovers overwhelming evidence for the material, moral, social, and intellectual progress of mankind.[61] The greatest of all modern popularizers of history, Will and Ariel Durant, end their book *The Lessons of History* with the declaration that although individual men today are no better or happier than their predecessors in primitive times, the human race has progressed.

> We are born to a richer heritage, born on a higher level of that pedestal which the accumulation of knowledge and art raises as the ground and support of our being. The heritage rises, and man rises in proportion as he receives it. History is, above all else, the creation and recording of that heritage; progress is its increasing abundance, preservation, transmission, and use.[62] *

---

* A curious *aperçu* into the optimism of historians is provided by President S. I. Hayakawa of San Francisco State College, who, during the disturbances on his campus in 1969, told the columnist Drew Pearson: "Students of history don't revolt. They know that every generation has its problems and its setbacks. They know that we make progress slowly, but we do make it. They know history. It's the students who study English or art who have been unreasonable. They don't understand history." Quoted in Pearson's syndicated column in the American press, April 1, 1969.

15

Science and the Human Prospect

HE SCIENCES—natural, social, and behavioral—continue to enlist a high proportion of the best minds in every Western country and throughout the world. The living far outnumber the dead. More than 200,000 entries appear in the latest editions of *American Men of Science* and *Directory of British Scientists.** Heavily subsidized by government and industry, scientific research affects every part of daily life. Scientists nowadays are more likely to be compared to a swarm of ingenious and industrious bees than (as in the last century) to a heavenly host of angels bringing light and grace from the Almighty.[1] But hope for salvation by science has not altogether dissipated in the twentieth century, nor have all scientists lost interest in the question of human progress. Too much energy and intelligence have been invested in scientific work in our century, and the possibilities of science for human betterment are clearly too great to allow the complete collapse of optimism in the scientific community. Perhaps no segment of the thinking population shows itself less prone to despair.

Nor has intensifying specialization overwhelmed the scientific imagination. The majority of professional scientists predictably have few intellectual interests apart from their work, but an articu-

* See also *World Who's Who in Science* (Chicago, 1968), a compilation of thirty thousand sketches of prominent scientists from antiquity to the present time. All continents and eras are represented, and only the most distinguished figures have been selected for inclusion; yet fewer than half, in the editor's words, are "historical," which apparently means "dead."

late minority speak out on every conceivable topic. They popularize scientific research, they write science fiction, they philosophize and theologize, they enter the political arena, they pontificate on war and peace. The spirit and methods of the natural sciences are also applied increasingly to the study of man and society; the twentieth century is the golden age of psychology. Themes commonplace in the more general writings of twentieth-century scientists include the belief that only science can solve the social problems that science itself has in great measure created, that civilization will be saved when the cultural gap between the exact sciences and the other sciences has been closed, that human welfare can be "engineered," and that the evolutionary process in some sense promises or points the way to infinite human progress. Not one of the hopes kindled by writers studied in Part Two above fails to make its appearance somewhere in the books of twentieth-century scientists, especially those written in the English-speaking world, with its long tradition of politically active scientist-radicals, and also in the Soviet Union, where science receives unprecedented official encouragement, and where even the ruling ideology is advertised as a science.

In these last three chapters, natural scientists will command our attention first, followed by thinkers in the social sciences (anthropologists, economists, sociologists, and political philosophers and ideologists), and finally by several behavioral scientists, whose metapsychological theories of progress draw on most of the departments of twentieth-century thought: philosophy, religion, politics, history, and the sciences of man.

In the natural sciences biology no longer enjoys the prestige it earned in the nineteenth century, but biologists have continued to show a special interest in the problem of progress. In the preceding chapter we explored the cosmological thinking of Dacqué, Lecomte du Noüy, and Teilhard de Chardin. Less elaborately, but not less hopefully, many other biologists have directed their attention to the question of the relevance of genetics and evolutionary theory to human needs. The results of their inquiries would, on occasion, satisfy the shade even of Herbert Spencer. Of greatest interest is what comes very near being a school of philosophically inclined biologists in Great Britain and the United States, who rose to prominence both as research scholars and as prophets in the years between the two world wars.

Some of the members of the group, such as the Princeton biologist Edwin Grant Conklin and T. H. Huxley's grandson Sir Julian Huxley, are best described as liberal humanists, who discover the law of human progress in the upward trends of natural evolution and call upon biology to supply mankind with the elements of a new cosmic religion. Others, such as H. J. Muller in the United States and J. B. S. Haldane and Joseph Needham in Britain, turned to Marxism in the 1930s, which led to a sometimes incongruous fusion of biological and sociological themes in their writings on progress.

Muller's *Out of the Night* (1935) is the most striking book produced by the Marxist contingent. Published when he was a professor of zoology at the University of Texas and at the same time the organizer of a genetics institute in Moscow, *Out of the Night* argued that mankind's use of its intellectual powers and its faculties for social cooperation had produced considerable progress in the course of history, but progress chiefly in the biological sense: survival, proliferation, enhanced control of the environment. In man's way of reckoning, however, progress could mean only the increase of the sum total of human happiness, and here far less improvement could be measured.

Today, Muller declared, mankind was summoned to choose between chaos and decay on the one hand and the establishment of a scientific world commonwealth on the other. Man's knowledge and power had grown more rapidly than his techniques of social cooperation; at the same time, he faced the prospect of radical genetic degeneration as the result of his failure to take control of his own future biological evolution. Natural selection no longer operated as it had done in the past, so that all sorts of undesirable mutations—always more numerous than beneficial ones—found their way into the human gene pool. Muller's solution was twofold: speaking as a convert to Marxism, he advocated the revolutionary transformation of the socio-economic order to eliminate social injustice, warfare, and the egoistic temper of capitalist culture; and speaking as a geneticist, he sketched a program of positive eugenics for the improvement of mankind's genetic material.* In a truly co-

---

* Muller placed most of his hopes in artificial insemination; the sperm of the "best" individuals would be selected for the program, and women would choose to conceive one or more of their children in this way. "How many women," he exclaimed, "in an enlightened community devoid of superstitious taboos and of sex

operative, socially- and scientifically-minded world commonwealth, all mankind would at last enjoy the well-being and happiness from which it had been excluded, save for a few privileged members of the ruling classes, by the imperfect development of its social consciousness in past millennia. In such a new world, man would shape himself "into an increasingly sublime creation—a being beside which the mythical divinities of the past will seem more and more ridiculous, and which, setting its own marvelous inner powers against the brute Goliath of the suns and planets, challenges them to contest." All the progress still possible through manipulation of the external environment, great as this might be, dwindled to insignificance by contrast with what could be done to improve man himself. "Some hundreds of millions of years of happy endeavor" were in store for "the god-like beings whose meager foreshadowings we present ailing creatures are." [2]

The problem of progress has also been a social and continuing concern of Sir Julian Huxley. From the first chapter of his *Essays of a Biologist* (1923) to the last chapter of his *Essays of a Humanist* (1964),[3] he has returned repeatedly to the problem, always maintaining a stout faith in the continuity of the evolutionary process, from inorganic matter to mankind, and in mankind's cosmic obligation to carry evolution to its furthest possible limits. In his philosophical position, Sir Julian succeeds in seeming more Victorian than his grandfather. Whereas Thomas Henry Huxley rejected the doctrine of ethical continuity between biological evolution and human progress in his 1893 Romanes Lecture, his grandson, who delivered his own Romanes Lecture exactly fifty years later, strongly defended it—a position that late nineteenth-century thinkers would have found rather more congenial than they found Thomas Henry's.[4]

Sir Julian's deepest conviction is of the unity of being. From the subatomic world to the mind and spirit of man, the cosmos consists of a single "world-stuff" in process of evolution. Most broadly, progress may be defined as improvement that leaves open the path to further improvement. In the history of life, this improvement has comprised increase in control over the environment, increase

slavery, would be eager and proud to bear and rear a child of Lenin or of Darwin! Is it not obvious that restraint, rather than compulsion, would be called for?" H. J. Muller, *Out of the Night: A Biologist's View of the Future* (New York, 1935), p. 122.

in independence of the environment, and the intensification of mental capacity, resulting in the expansion and higher organization of knowledge. One by one the species in their upward development along particular lines have turned into blind alleys beyond which additional progress is no longer possible. *Homo sapiens* is "the only organism capable of further major advance or progress." "He finds himself in the unexpected position of business manager for the cosmic process of evolution." What the world-stuff can achieve henceforth rests—at least on this planet—on the shoulders of man, and man alone.[5]

But man's awesome responsibility is not one that he must bear without hope of cosmic guidance. "The evolutionary point of view makes it clear," Huxley insists

> that progress is neither a myth nor a will-of-the-wisp, still less a dangerous delusion: it is the desirable direction of change in the world, and desirable ethically as much as materially or intellectually. It also establishes the reassuring fact that our human ethics have their roots deep in the non-human universe, that our moral principles are not just a whistling in the dark, not the *ipse dixit* of an isolated humanity, but are by the nature of things related to the rest of reality.

Although in man "psychosocial" progress replaces, in large part, the older genetic progress, the criteria of improvement remain the same. "We can justifiably extrapolate some of the main trends of progress into the future, and conclude that man should aim at a continued increase of those qualities which have spelt progress in the biological past—efficiency and control of environment, self-regulation and independence of outer changes, individuation and level of organisation, wholeness or harmony of working, extent of awareness and knowledge, storage of experience, degree of mental organisation." [6]

Huxley sees the present era as a peculiarly crucial point in human evolution. Now that man is aware of the process by which he has advanced from subhuman savagery to civilization, and the connections between that process and the plan of the cosmos, he is in a position to take his future into his own hands, with full consciousness and responsibility for all the ages to come. He stands in pressing need of a new belief system, "a religion without revelation" that can supply for him the spiritual force with which to confront the spiraling challenges and opportunities of the future. Huxley has himself sketched the outlines of such a new faith in several books and essays. It will be a religion both naturalistic and scientific, valid for all men, true to reality as modern science divines reality, inspired by a vision of evolution that "shows us our destiny and our duty. . . . The central belief of Evolutionary Humanism is that existence can be improved."[7]

In the immediately foreseeable future, obedience to evolutionary imperatives demands the implementation of full equality of individual opportunity, economic planning, world government, population control, and a program of negative and positive eugenics. All this will serve to maximize the racial evolutionary potential, enabling evolution to "become a single joint enterprise of the human species as a whole." The coming world polity will have no one uniform culture founded on existing Western or Eastern models, but will be formed from "a single unified pool of tradition" and a "world orchestration of cultures."[8] Nevertheless, it is difficult to imagine Huxley's world order as anything but a fundamentally Western society, with a universal religion derived from science (mainly biology) and with the institutions and values of the European democratic-liberal welfare state.

Since the Second World War, except for Huxley, the most significant contributions to the biological philosophy of progress have been made by scholars not appreciably younger than those already discussed, but less doctrinaire and—one might say—less metaphysically inclined. The foremost contemporary American expositor and synthesist of evolutionary theory, George Gaylord Simpson, takes Huxley to task for assuming an automatic correspondence between the tendencies of prehuman evolution and the evolutionary duties of man. While agreeing with Huxley on many other points, he maintains that man can determine what constitutes progress for him only on the basis of his own choices and desires: so far as we know, the evolutionary process is purposeless, and the criteria of progress at the human level cannot be extracted from the general trends of natural history. Simpson's outlook on progress and evolutionary ethics is shared by another outstanding American evolutionist who has written extensively in the postwar period on the same topics, Theodosius Dobzhansky. Except for his profound admiration of the thought of Teilhard, Dobzhansky follows Simpson quite closely.[9]

In Britain a strong contemporary voice is that of C. H. Waddington, first in *Science and Ethics* (1942), and more recently in *The Ethical Animal* (1960), one of the fullest and most sophisticated treatments of the subject ever written by a professional scientist. Waddington occupies a position somewhere between Huxley and Simpson, but nearer, on most points, to Huxley. His basic premise is that man is an "ethicizing" animal by his innermost nature, a being who progresses by rendering ethical judgments, which in turn make possible evolution "in the socio-genetic mode." The cultural or socio-genetic system of evolution in man performs precisely the same functions as the genetic systems: it passes on information from one generation to the next, but through education rather than genetic processes. Yet when man ethicizes, he is only doing what he is meant to do, obeying biological imperatives no less natural than the instinct that drives a bird to build a nest. In this way, the hypothetical wall between fact and value crumbles, since the duty to ethicize is for man not something that he can escape, but a fact or condition of his existence.

Waddington goes further to propose that ethical judgment in man can be seen as the tendency of the universe made conscious.

> Observation of the world of living things reveals a general evolutionary direction, which has a philosophical status similar to that of healthy growth, in that both are manifestations of the immanent properties of the objective world. . . . I conclude that any particular set of ethical beliefs . . . can be meaningfully judged according to their efficacy in furthering this general evolutionary direction. . . . The function of ethicizing is to mediate the progress of human evolution.

Concretely, progressive change in evolution consists of increased independence of the external environment and greater capacity for feeling and awareness, or, put another way, increased richness of individual experience. Far from being arbitrary or accidental, the direction of progress in evolution arises "as a result of the general structure of the universe; that is, it is not merely a direction in which progress happens to have occurred, but, in some of its aspects at least, it has the character of an inevitable consequence of the nature of the evolutionary process." [10]

Waddington does not maintain, of course, that man is under precisely the same sort of *Diktat* that prompts nest-building in birds. Although knowledge can influence choice, it does not literally compel. But he criticizes Simpson and Dobzhansky for suggesting that man's choices should not be guided by his rational knowledge of the evolutionary process as a whole. Subjective or irrational decisions made without consulting the social and biological context of one's acts may seem closer to "nature," but in fact they are the most unnatural of all, since they make no use of man's most human faculties. "If we can . . . see mankind as at present the most advanced phase in a process of progressive or anagenetic evolution in which the whole living kingdom is involved, it would seem to follow . . . that it is man's duty, not only to mankind but to the living world as a whole, to use his special faculties of reason and social organization to ensure that his own future evolution carries forward the same general trend." [11]

Ridiculing the idea of progress, Waddington adds, has lately become all too fashionable in certain quarters. "Anyone who is bold enough to assert that it has occurred, or even that the word has a definite meaning, is likely to be dismissed as merely naive and unsophisticated." But nothing could be more wrong-headed. Man— the ethical animal—has made progress all through his history on earth; individual experience has been substantially enriched since prehistoric times. Moreover, although many Western intellectuals sneeringly deny it, the rate of progress has greatly accelerated in the last two hundred years, a fact well understood by the Indians and the Chinese, who do not doubt "that to die at eighty after a healthy life using inanimate sources of power is, in some real and undeniable sense, better than to die at forty after a life of back-breaking labour, hunger and sickness." [12]

In some respects, perhaps, progress has gone too far, notably in the development of social superegos or belief systems so sweeping and fanatical that they call to mind "the extensive body-armour of the later dinosaurs, or the finicky adaptation of certain parasites which fits them to live on only one host." Human societies need such belief systems, but they must not become too overbearing. Those of the twentieth century have provoked devastating wars that threaten to shut off the possibility of further progress. Part of the solution must be heavier reliance on the scientific attitude, with its devotion to reason and intellectual prudence. In any case, Waddington confesses to "a moderate optimism." Despite the numerous outbreaks of fratricidal irrationalism in recent history, he sees "potentialities for a future far preferable to the past," in a world of humanized technology, fulfilling leisure, well-planned cities, freedom from war, and the rise of the under-industrialized parts of the planet to the same level of affluence enjoyed in the West.[13]

Specialists in the other natural sciences and related disciplines— physicists, chemists, mathematicians, engineers—have less occasion than biologists to ponder processes involving time, evolution, and history, and less to say for or against the idea of general human progress. The most socially conscious minds in such fields typically direct their thoughts to what science and engineering can offer mankind in the way of improved conditions of life, and delve no further.

A somewhat broader view was taken by the British physicist J. D. Bernal—perhaps because he was also a fully committed Marxist. His books on the history and social application of science united a strong concern for the relevance of science to human needs with the historical consciousness of Marxism. He recognized that in the non-communist countries of the Western world a reaction of great intensity had set in against the traditional claims of science and the hopes aroused by science in previous centuries. Thinking people no longer automatically assumed that scientific progress results in human progress. Nonetheless, science remained the chief agent of change in society, and one must, in Bernal's judgment, discriminate carefully between "the necessary effects of science and its abuse under capitalism." The harnessing of science for the profit of entrepreneurs introduced an "enormous distortion in the direction in which scientific results are applied." In a sane society, the only purpose of scientific research would be the service of the welfare of all mankind.[14]

The importance of science to progress was not easily overestimated. Bernal divided history so far into three eras, inaugurated by three great events: the founding of societies, the founding of civilizations based on agriculture, and the Scientific Revolution. Each era represented an advance on what came before, but from the modern perspective the decisive event was the last. The prescientific civilizations rose and fell over a period of thousands of years without any substantial progress taking place at all. Not until the arrival of modern science during the Renaissance did humanity begin to move forward rapidly once again. Now, directly the immediate social and political problems of man were resolved by the adoption of world-wide socialism, "we have every chance of realizing a world so different from anything we have had before that the transition is greater than any which has occurred since the first appearance of humanity." In place of today's divided world, with its poverty and cruelty, Bernal anticipated "an age of abundance and leisure." [15] With the perfection of a complete science of society, the freedom of old-fashioned liberalism would yield to scientific planning and "the freedom of necessity." "Each man will be free in so far as he realizes that he is taking a conscious and determinate part in a common enterprise."[16]

Any Marxist scientist could be expected to agree with most of Bernal's views. Another tradition in nineteenth-century thought—Positivism—was revived in the 1920s and 1930s by Alfred Korzybski. In *Manhood of Humanity* (1921) and *Science and Sanity* (1933), Korzybski failed to acknowledge Comte as a forerunner, but the two thinkers were obviously spiritual cousins. Both championed the scientific outlook over the "empty verbalism" and "mythology" of earlier thought, and both applied a technocratic approach to the question of mankind's social salvation. Both constructed their world-views on the foundation of a "scientific" idea of progress. Both acquired a cultic following in later life.

We need not be concerned here with Korzybski's work as the founder of the Institute of General Semantics in Chicago in 1938, and with the details of his "non-Aristotelian" philosophy of language, logic, mathematics, and psychophysiology. But the centrality of a belief in progress to all his thought is quite clear. *Manhood of Humanity* expounded most of his ideas on the subject. He ventured a definition of mankind as the "time-binding" class of organisms. Vegetable life was the "chemistry-binding" class, subhuman animal life the "space-binding." But only men possessed "the capacity to summarise, digest and appropriate the labors and experiences of the past . . . to employ as instruments of increasing power the accumulated achievements of the all-precious lives of the past generations . . . to conduct their lives in the ever increasing light of inherited wisdom." In brief, progress was cumulative and relentless; and progressiveness was man's most distinctive quality, that which alone sufficed to set him apart from all other forms of life. Even progress itself progressed logarithmically. "We humans," Korzybski wrote, "are, unlike animals, naturally qualified not only to progress, but to progress more and more rapidly, with an always *accelerating acceleration,* as the generations pass." [17] Man's duty as a time-binder, therefore, consisted not merely in thriving or moving about but also in engaging in creative and productive work, at an exponential rate of improvement.

Yet progress could not be described as self-sustaining. Man's freedom of will gave him the opportunity to defy the laws of his own nature. Moreover, progress often failed to unfold harmoniously; along some lines of advance progress tended to lag; along others, to occur much more rapidly. The so-called ethical and social sciences were especially slow in developing, just as the natural sciences were especially swift. To this single fact could be attributed all the wars and revolutions of history. Society was repeatedly forced to catch up with its own technical progress by resorting to violent cataclysmic forward leaps. To ensure peaceful, uninterrupted progress, Korzybski called for the perfection of a "science and art of human engineering," whereby man's quintessentially human power of time-binding could be brought to its fullest possible development. Through human engineering, obsolete speculative and mythological systems of thought would be replaced by a scientific view of man, society, and nature; man's philosophy and man's social institutions would at last stay abreast of his technology; and the welfare of the race would be managed efficiently. Experts in science and engineering would supplant politicians and lawyers. Production, instead of politics, would become the chief business of humanity. The World War had disclosed once and for all the idiocy of the old order of things: henceforth, man would enter into his manhood, forsaking the childish ways of the past. "And so I say that these days, despite their fear and gloom, are the beginning of a new order in human affairs— the order of permanent peace and swift advancement of human weal." [18]

Korzybski's later efforts to develop a "general semantics" that would provide the absolute clarity of thought and language necessary for sane human engineering occupied most of his attention in the years after the publication of *Manhood of Humanity*. But he continued to believe fervently in progress. Several other noteworthy physicists, mathematicians, and engineers since the interwar decades have emulated Bernal and Korzybski in their Marxist or positivist conceptions of general human progress. But, as in biology, a somewhat more restrained spirit seems to inform the greater part of very recent thinking on progress from specialists in the "exact" sciences. Perhaps the overwhelming fact of the uranium and hydrogen bombs has had a daunting effect. One thinks, for example, of the mathematician and cyberneticist Norbert Wiener's unambiguous censure of "the simple faith in progress" (although he accepted a somewhat subtler form of it) in the second chapter of *The Human Use of Human Beings* (1950), a chapter that also revived the late nineteenth-century horror of the second law of thermodynamics.

Caution is not necessarily hopelessness, and not all physical scientists are as cautious as Wiener. The writings of Buckminster Fuller, for example, convey an atmosphere of almost psychedelic confidence in man as engineer. In Britain C. P. Snow has chided the literati for their pessimistic *Weltanschauungen,* for their "facile despair" and "intellectual treachery." Moral sensitivity, Snow writes, has increased under our very noses, not to mention knowledge, health, and longevity, just since the nineteenth century, and he sees no technical obstacle to the achievement of all the good things of our Western "enclaves of progress" by the less advanced societies of the world. "As a race, we have scarcely begun to live." [19]

In a recent and well-received book, the American physicist John Rader Platt finds the future coming increasingly under man's ability to control, through scientific prediction and the power of invention. "There is an incredible amount of social engineering to be done, to make social structures that will give us freedom and yet keep us from killing each other; but it is now clear . . . that it can be done." He rejoices in the infinite number of "possible social forms" within man's power to plan and execute, the human adventures that "have just begun," the sheer "fun" of hope and creative action. Although man may yet perish in a thermonuclear war, the period of maximum peril cannot go on forever. If he survives the next few decades, it will be due to his realization that the world "has now become too dangerous for anything less than Utopia." He will have no choice but to take his evolution into his own hands, as he has already begun to do. For man is at last "emerging into Man. . . . It is a quantum jump. It is a new state of matter. The act of saving ourselves, if it succeeds, will make us participants in the most incredible event in evolution. It is the step to Man." [20]

Even though science fiction has often been a medium for prophecies of disaster,* it too provides illustrations of the optimism of many sectors of the scientific and engineering community. H. G. Wells's *Men Like Gods* (1923) and *The Shape of Things to Come* (1933) adapted the techniques of modern science fiction to the service of utopography in the interwar decades, inspiring many imitators. Another important work of the same years, Olaf Stapledon's *Last and First Men,* first published in 1931, stupefied its readers with a chronicle that surveyed the history of man for the next two billion years. Seventeen different species of the genus *Homo* appeared in succession, several far more advanced in every respect than *Homo sapiens.* Mankind twice changed its home planet, from earth to Venus, and from Venus to Neptune, before the end of all life in the solar system.

More recently we have had the stories of Robert Heinlein—especially his utopographic *Stranger in a Strange Land* (1961)—and the novels and prophetic books of Arthur C. Clarke, a British writer-scientist who has several times touched on the theme of the evolution of a higher race from present-day humanity, as in *Childhood's End* (1953) and in *2001: A Space Odyssey* (1968), based on his screenplay of the same title. Clarke believes that space flight will reinvigorate ailing, earthbound man, just as the voyages and conquests of the Renaissance brought new life to the sick society of the late Middle Ages. It may also be true, writes Clarke, that the opening of the space frontier will give man a new sense of boundlessness, enabling him to "break free from the ancient cycle of war and peace." The first men to go forth will be the explorers, then the scientists and engineers.

> Later will come the colonists, laying the foundations of cultures which in time may surpass those of the mother world. . . . Could the builders of Ur and Babylon—once the wonders of the world—have pictured London or New York? Nor can we imagine the citadels that our descendants may one day build beneath the blistering sun of Mercury or under the stars of the cold Plutonian wastes. And beyond the planets, though ages still ahead of us in time, lies the unknown and infinite promise of the stellar universe.[21] **

---

* See above, pp. 229–31.
** Optimistic views of the future may also be found in the writings of several philosophers of science who envisage world cultural synthesis mediated by science. See, for example, F. S. C. Northrop, *The Meeting of East and West,* (New York, 1946); Northrop, *The Taming of the Nations* (New York, 1952); Oliver L. Reiser,

The Promise of Scientific Humanism (New York, 1940); Lancelot Law Whyte, *The Next Development in Man* (London, 1944); and Leo J. Baranski, *Scientific Basis for World Civilization: Unitary Field Theory* (Boston, 1960). Even astrology —perhaps the first science—offers hope for the future of mankind. Quite apart from the popular image of the coming "Age of Aquarius" as a utopian era, see the interweaving of cyclical and progressivist themes in Dane Rudhyar, *Birth Patterns for a New Humanity* (Wassenaar, The Netherlands, 1969), especially ch. 10.

16

The Social Sciences

THE SOCIAL SCIENCES have also struggled to recover some of their nineteenth-century hopefulness, especially since the Second World War. An inevitable interdisciplinary rebellion has erupted against the ethical neutralism and anti-evolutionist functionalism ascendent in all the sciences of man earlier in the century. The rebels do not necessarily embrace a doctrine of progress, but some of them display a receptivity to progressivism absent in the generation that dominated the social sciences in the 1920s and 1930s. As in the case of biology and physics, throughout the century the social sciences have also had their share of Marxists predisposed to expound a belief in progress.

A further ground for hope is provided by the efforts of the technologically backward countries to free themselves from centuries of Western control and political and economic stagnation. Many sympathetic Western scholars take a kind of vicarious pleasure in the "rise of the East" that compensates in some ways for apprehensions about the "decline of the West." The struggle for modernization in Asia, Africa, and Latin America reminds social scientists that cross-sectional and functional analysis alone cannot explain processes of radical sociocultural change. Notions of "evolution," "development," and even "progress" have earned new respectability in the minds of many younger anthropologists, economists, and sociologists. At the same time, they seldom attempt to work out systematic conceptions of general human progress. The building of theoretical structures lags behind empirical research, and scholarship tends to be intensively specialized.

The longest views are those taken by anthropologists and archaeologists. Some have managed to surmount the modern professional bias against large-scale surveys of social evolution and the rendering of value judgments, and have done something to rehabilitate the idea of progress in the bargain. Stanley Casson's *Progress and Catastrophe,* though full of foreboding about the immediate future, gave readers in the period just before the Second World War a cautious defense of the concept of progress from the perspective of a humane and distinguished archaeologist.[1] More emphatic in their support of the belief in progress were two other British scholars of evolutionist persuasions whose careers roughly coincided with Casson's, Robert Briffault and V. Gordon Childe. Both were caught up in the same enthusiasm for Marxism that also captured Haldane, Needham, and Bernal, but Marxism was not by any means the only influence on their doctrines of progress.

Briffault was born in 1876, and turned to productive scholarship only after he had practiced medicine for more than twenty years in New Zealand. He made his mark as an anthropologist in 1927 with an analysis of matriarchy in primitive society, *The Mothers.* His major contribution to the literature of the idea of progress, *The Making of Humanity,* first appeared in 1919 and was subsequently rewritten in 1930 with the title *Rational Evolution.* Whether Briffault's book deserves to be ranked as the last important classic of late nineteenth-century polemical rationalism or among the first examples of twentieth-century anthropological "neo-evolutionism" is an open question. In any event, it spared no rhetorical effect to condemn traditional religious belief and to show mankind painfully but surely advancing through the ages to steadily higher levels of material and moral culture, in spite of the obstacles set in mankind's path by its own irrationality.

All human progress, Briffault contended, was social, occurring through the evolution of culture rather than improvement in the individual or in his genetic material. All progress also sprang ultimately from man's capacity for critical and rational thought, even the progress of morality, which Briffault saw as man's greatest achievement. From the first civilizations of antiquity through the golden ages of Greece and Rome to the Moorish culture of the

Middle Ages and the revolutions of the modern world, man had advanced to ever-wider sympathy, social consciousness, and altruism, by virtue of the constant expansion of his rational faculties. Absurd traditions, superstitions, and elaborate apologies for social injustice on the part of ruling classes had fallen one by one to the superior strength of the inquiring mind. Even the contemporary disenchantment with the idea of progress was itself further evidence of moral progress, since in most cases the disenchantment issued from an unwillingness to rest on laurels already won, and in particular to accept the far from perfect sociality of nineteenth-century industrial civilization. The very impulses that drove critics to despair "are products of that same human world from which they recoil as from a thing unclean. . . . It is that same humanity which has through the strifes and struggles of its long evolution brought into being those ideals which lift them upwards." [2]

An Australian by birth, V. Gordon Childe held chairs in prehistoric archaeology at Edinburgh and London. In his most familiar work for lay consumption, *Man Makes Himself,* first published in 1936, he concurred with Briffault that the belief in progress had entered upon evil days since the First World War. But, by adopting a strictly scientific viewpoint, it was quite a simple matter to "vindicate the idea of progress against sentimentalists and mystics." Following the same strategy developed by Carl Becker in *Progress and Power,* Childe urged acceptance of an "impersonal" and ostensibly value-free definition of progress. In such a definition, "progress becomes what has actually happened—the content of history." [3] In *Man Makes Himself,* the sole criterion of improvement was survival and multiplication of the species. When human populations thrived and increased, one could speak of progress; when they failed and diminished, one could speak of decline. The three greatest advances made by mankind were, therefore, the agricultural revolution, in which savage man had become barbarian man; the urban revolution, which had transformed barbarian man into civilized man; and the industrial revolution. As each revolution took place, the total human population leaped forward dramatically: the proof of progress.

If this were all that Childe had said, we would be well-advised to exclude him from the roll of the believers in progress. In our use of the term no idea of progress can be axiologically neutral; all ideas of progress embody ideas of the good. If Childe meant by "progress" nothing more than "change" or "development," as he appeared to do, his insistence that only demographic growth constituted progress, i.e., change, is obviously nonsensical. If he meant by "progress" the "improvement" of the species, then in spite of himself he must have made an extra-scientific judgment as to the "goodness" of demographic growth. Yet no thinker of Childe's acutely developed social conscience could have believed that an increase in the numbers of human beings was in itself good, let alone the single admissible criterion of racial improvement.

Looking more closely into *Man Makes Himself* and others of Childe's books, we find that demographic growth is more or less incidental to what in fact engaged his sympathy and admiration. In *Progress and Archaeology,* for example, he included among the "results" of progress greater longevity and greater richness and diversity of life—put another way, more freedom of choice for the individual.[4] Throughout *Man Makes Himself* and its sequel, *What Happened in History,* he also affirmed that the source of progress was the inventiveness of scientists, engineers, and other self-disciplined creative men, whose work gave the species new ways of controlling its environment and expanding its knowledge of the world. The enemies of progress, on the other hand, were the exploitative ruling classes, the kings and landlords, and their clerks and priests, under whose aegis millions of men were enslaved and wealth was accumulated unproductively in the coffers of the few.

Childe projected a spiraliform view of progress, with ages of productivity and invention, making life richer, succeeded by ages of stagnation and decline, in which ruling elites discouraged free inquiry and greedily consolidated their power and their fortunes. It was "childish to ask why man did not progress straight from the squalor of a 'pre-class' society to the glories of a classless paradise," since the "dialectics of progress" clearly required conflicts and contradictions *en route,* just as an unsightly scaffolding was essential "to the erection of a lovely building." "The upward curve," he concluded, "resolves itself into a series of troughs and crests. But . . . no trough ever declines to the low level of the preceding one; each crest out-tops its last precursor." [5] All of which suggests a more or less conventional and surely not value-free idea of progress, defined as the growth of freedom, science and knowledge, and human mastery of nature.

In anthropology proper, the most articulate proponents of evolutionism in recent years have been Leslie A. White, Elman R. Service, and Marshall D. Sahlins, all members of the anthropology department at the University of Michigan. White's books, *The Science of Culture* (1949) and *The Evolution of Culture* (1959), have led the way, with loyal support from the symposium edited by Sahlins and Service, *Evolution and Culture,* published in 1960. These American anthropologists, however, have been rather more consistent than Childe in adhering to the idea of an objective and scientific concept of "progress." They make quite clear that, for them, progress means only development, not improvement.

The revival of evolutionism, then, does not necessarily involve the revival of progressivism, even when the word "progress" is freely used. If anything, White's thinking harks back to Rousseauian primitivism. But it would be a mistake to look for concepts of progress in the ranks of the evolutionists alone. As scholars or simply as human beings with an interest in the fate of man, several anthropologists opposed to formal evolutionism or not strongly identified with it have nonetheless spoken out on behalf of a belief in the general improvement of mankind.

One case in point is the Anglo-Polish anthropologist and founder of the functionalist school, Bronislaw Malinowski. Shortly before his death in 1942 he wrote a study of freedom that assailed the "sophistication and relativism" of his colleagues and reaffirmed "the existence of certain values and principles which are indispensable to the very process of maintaining and advancing culture." [6] In the face of the challenge of totalitarianism, Malinowski insisted that freedom and democracy were necessary to the progress of civilization, defining progress itself as the growth of thought and knowledge. Freedom and democracy were not merely moral ideas but scientific ideas as well. A world of free, democratic societies would agree to abolish war forever, and civilization could resume its forward march unhindered by the threat of suicidal mass-destruction.

Even Franz Boas, as we have seen, was provoked by Nazi racism to issue a defense of Western democratic ideals far removed from the spirit of his earlier work.* Since the Second World War, younger anthropologists of White's generation and beyond have been deeply impressed by the energetic efforts of the so-called backward peoples of the world to build new societies in the Western image. In the

* See above, pp. 191–92.

314

light of the obvious enthusiasm of non-Western peoples for Western cultural values, the self-flagellating cultural relativism of the Boas tradition in anthropology has tended to lose ground. The rapid and voluntary Westernization of many sectors of life in nearly every non-Western society on earth is a fact too overwhelming to be ignored.

Thus, a scholar such as Margaret Mead—once a disciple of Boas and Benedict—has become an unblushing spokesman in the name of progress for the swift transformation of primitive societies into integrally modern societies. After revisiting the New Guinea village where she had conducted intensive field work in 1928–29, she writes in *New Lives for Old* that anthropologists were once so fearful of ethnocentrism that they would do everything they could to "protect" primitive societies against Western encroachment. But what if the less developed peoples actually wanted to enter the modern world? What if their Western benefactors begrudged them what they needed to make this entry as rapidly and painlessly as possible?

In the instance of Mead's New Guinea village, progress had occurred at a headlong pace. In twenty-five years circumstances had totally changed for the villagers; they had uprooted their old way of life—and it had all been for the better! The people were happier, more affectionate, healthier, more fulfilled than under their traditional life pattern. What we need today, writes Mead, "is imagination, imagination free from sickly nostalgia, free from a terror of machines bred by mediaeval fantasies or from the blind and weatherbound dependence of the peasant or the fisherman." Inveighing against the pessimism of the "Old World," she affirms her belief that

American civilization is not simply the last flower to bloom on the outmoded tree of European history . . . but something new and different . . . because it has come to rest on a philosophy of production and plenty instead of saving and scarcity, and . . . because the men who built it have themselves incorporated the ability to change and change swiftly as need arises.

She looks forward to an open world, fertilized by the genius of American civilization, a world of people not constrained by custom but flexible and exuberant, "the age of the air, when the world becomes one great highway, and in any inn along the way there must be room and welcome for each and every guest." [7]

The Chicago anthropologist Robert Redfield confronted the problem of cultural relativism and ethical neutralism with remarkable candor in his book *The Primitive World and Its Transformations*. He also came down unequivocally on the side of "progress" and "civilization." Like Mead, Redfield had had the experience of studying at first hand a "village that chose progress." In two field trips to a small Mayan community in Yucatan in 1931 and 1948 he had been astonished by the speed and efficiency with which a primitive village had transformed itself into something like a Spanish town.[8] He now questioned the tradition of objectivity in anthropology that forbade the rendering of judgments of value. Could one man investigating the lives of other men avoid judging them?

It was easy enough, Redfield noted, to be benevolently impartial about all sorts of value systems "so long as the values were those of unimportant little people remote from our own concerns. But the equal benevolence is harder to maintain when one is asked to anthropologize the Nazis, or to help a Point Four administrator decide what to do for those people he is committed to help." To help, one had to institute change. To institute change, one had to have a scale of values. Belief in the possibility of improvement amounted to a belief in that much maligned concept, "progress." "When the anthropologists helped modern people to see that the nineteenth-century belief in progress was a faith, not a proven fact," Redfield added, "they threw out the baby with the bath so far that its persisting cry to be heard could not reach their ears."[9] Yet having values and believing in progress was entirely natural for modern man, even for modern men who happened to be scientists.

Redfield did not hesitate to adhere to a doctrine of progress himself. In his reading of history, mankind in the era of civilization had become adult. It had abandoned the childish ignorance and cruelty of precivilized life. "On the whole the human race has come to develop a more decent and humane measure of goodness," and in this sense "there has been a transformation of ethical judgment which makes us look at noncivilized peoples, not as equals, but as people on a different level of human experience. I find it impossible to regret that the human race has tended to grow up." To those who would find Redfield guilty of ethnocentricity, he had only one reply. He was guilty, and proudly so. Although he had taken the pledge of objectivity as a scientist, "somehow the broken pledge—if it is broken—sits lightly on my conscience."[10]

In the other social sciences—economics, sociology, and political science—the inclination of nearly all trained specialists and even most popularizers in recent decades has been to concentrate on analyses of narrowly defined current social problems, or, less often, on prediction and problem solving with reference to the immediate future. Economists, for example, have been just as fascinated as cultural anthropologists with the penetration of the underdeveloped peoples into the modern industrial age, and much theoretical as well as empirical work has been done in the "economics of development" by scholars of the caliber of Colin Clark, A. K. Cairncross, J. K. Galbraith, Alexander Gerschenkron, Gunnar Myrdal, and W. W. Rostow. But little in the way of a concept of general human progress emerges from their studies, although they have acted as a powerful stimulant to hopefulness in the postwar mental climate.

One exception, perhaps, is the French economist Jean Fourastié. He has little patience with "our French intellectuals of today who, in unison, express their resentment against *machinisme* and expose the profound decadence of our civilization, who prophesy the end of Christianity and the end of the world." Although two hundred years ago these same intellectuals would probably not even have learned to read, they conveniently ignore the "passage of the popular masses from a vegetal life," the rise of universal education, the disappearance of famines, and the emancipation of man by his own techniques of production from servile labor, disease, and early death. In the most general terms, "the maintenance and growth of the capacity for innovation (a capacity so feeble in traditional humanity and so much threatened even in humanity today), is the key to human progress." Technical innovation is not all of life nor all of progress, but it alone has the power of initiative to continue transforming the state of the world from age-old mass misery to the universal affluence in which all men, if they choose, can actualize their potentialities as human beings.[11] *

* Economic thought is seen by Sidney Pollard, Head of the Department of Economic History at the University of Sheffield, as the most significant expression of man's renewed belief in progress since 1945. Pollard contends in his recent book, as we noticed earlier (see p. 242 above), that faith in progress is back again "as strongly as ever." Rapid economic growth has relieved political tensions and promoted a more humane attitude toward hitherto submerged groups within societies. Economists, once the apostles of gloom, "have plunged fully into the stream of progress. . . . Most of them have absorbed in their work . . . the unspoken assumption that growing wealth and the social changes which are its preconditions,

Among sociologists, two German scholars first deserve our attention, both émigrés to Anglo-Saxon countries in the 1930s, both liberal socialists by persuasion who sought perhaps unsuccessfully to fuse elements of Western liberal-idealist thought with Marxism. The older of the two, Franz Oppenheimer, practiced medicine for some years, but then became convinced that he could benefit mankind more as a social scientist. He returned to his studies and earned a degree in economics and sociology at the University of Berlin. In 1908 he published a short book on the sociology and history of the state, which was well-received in Germany and abroad. He devoted himself during the Weimar years to a complete *System of Sociology,* which appeared in several volumes between 1922 and 1935. Oppenheimer's sociology is very much in the tradition of Comte and Marx, and it creates the same interpretative problems as Marx's for the historian of the idea of progress. An almost idyllic picture of precivilized man is juxtaposed with a view of the history of civilization as the tragic but necessary record of class struggle, and transcending both is the prophecy of a blissful future world order of small states, small towns, and small proprietors.[12]

Better known and more influential in academic circles than Oppenheimer was another sociologist of the German-speaking world, Karl Mannheim.* He was born in 1893 in Budapest and launched his career as a professional sociologist at the University of Heidelberg. In 1930 he was appointed to the chair in sociology at the University of Frankfurt am Main, which he lost when the Nazis came to power. From 1933 until his death in 1947 he lectured in sociology and education at the University of London. Although he expressed complete sympathy with historicism, he deplored its use by scholars such as Max Weber to support an ethically neutral and nontemporal theory of social processes. The fact that all ideas and events were conditioned by their times did not render impossible a

will themselves ensure improvements in the other aspects of social life which together make up the 'progress' of humanity." There are still reasons for doubting that all will necessarily work out for the best, Pollard concedes, but "for those who are in the van of humanity, groping forward into the dark, the belief that they are moving in an upward direction is . . . a necessity. Today, the only possible alternative to the belief in progress would be total despair." Sidney Pollard, *The Idea of Progress* (London, 1968), pp. 185 and 203. See also David Hamilton, *Evolutionary Economics: A Study of Change in Economic Thought* (Albuquerque, N.M., 1970).

* See the discussion of Mannheim above, p. 198, in connection with the rise of historicism.

rational understanding of the way in which the elements of culture progressed as a whole through history, nor did it invalidate the attempt to make value comparisons between different epochs and stages of culture. Mannheim's own "relationist" theory of knowledge, not unlike Ortega y Gasset's perspectivism, proposed the possibility of steadily higher approximations to truth (including historical and ethical truth) through progressively more comprehensive and therefore more complete visions of the whole.

As applied to the history of mankind, Mannheim's epistemology persuaded him to divide the elements of culture into three classes: those that progressed in a rectilinear fashion, such as science and technology; those that progressed dialectically, such as philosophy, by repeatedly starting over again from different points of departure but on the whole rising to higher and higher levels of comprehensiveness; and those that did not progress at all, because they had nothing to do with reason, such as art and religion.[13] Through the progress that had occurred thus far in human history, mankind had reached a higher plane of existence than ever before, a plane of self-awareness and potentiality for mastery of its own future development. At the same time, the modern period was also an era of disintegration. In modern man's economic, political, social, psychological, and moral-religious life, he found himself in something very much like chaos. His faith in progress, as mentioned before, had crumbled. How could one explain the fact of power and the sense of powerlessness, both coming at the same moment in history?

Mannheim's answer was scarcely original, although he clothed it in the rather arcane language typical of modern sociologists: change had come too fast to be assimilated by society. In particular, scientific and technical progress had accelerated out of all proportion to social progress, and the progress of the natural sciences out of all proportion to the social sciences. In ethical behavior, actual regression had taken place in recent decades. A new mode of thought was needed, a higher mode of problem solving, which would supersede "invention," as "invention" had long ago superseded "chance discovery." This higher mode could be termed "planning," the level of thought achieved "when man and society advance from the deliberate invention of single objects or institutions to the deliberate regulation and intelligent mastery of the relationships between these objects."[14]

The key to future progress, therefore, was "integrative thinking and behavior," "democratic planning for freedom," the engineering through an improved social psychology of democratic human co-operation. Mannheim looked with special favor in his last years on the emergence of the democratic welfare state in Western Europe; he also came to feel that social and behavioral scientists would have to work hand in hand with churchmen to help provide the ethico-spiritual integration indispensable to social harmony, a suggestion that recalls the later thought of Auguste Comte.[15]

The British sociologist most concerned with the belief in progress in the middle decades of the twentieth century has been Morris Ginsberg, a disciple of L. T. Hobhouse and his successor at the London School of Economics. Ginsberg devoted his address as chairman of the first annual general meeting of the British Socio-logical Association in 1952 to a survey of the idea of progress, later published in book form, and he has also written papers on progress for the recent symposia of humanist thought edited by Sir Julian Huxley (*The Humanist Frame*) and A. J. Ayer (*The Humanist Outlook*).[16]

Like his *maître* Hobhouse, Ginsberg focuses on the progress of morals and reason, and takes encouragement from both. The essen-tial point in the theory of progress remains true, he finds, "namely, that in the course of historical development man is slowly ration-alized and that man is moralized in proportion as he becomes more rational." "Is it not fair to say," he adds, "despite the difficulty of balancing gains and losses, that in no previous age has so much been done to relieve suffering, and to abolish poverty, disease and ignorance in all parts of the world? . . . On the whole, the evi-dence suggests, I think, that while morality is still far from domi-nant in human affairs it is gaining in strength." A unified world community, grounded in a rational and humane ethical philosophy, lies within man's power to build, although only man himself can build it, and no sociological law assures him success.[17]

In the United States, where the work of Hobhouse and Ginsberg is almost unknown, sociologists since the 1920s have been curiously silent on the question of general progress. To be sure, elements of the classical faith survive. The confident scientism of George A. Lundberg furnishes one instance. The "New Left" sociology of C. Wright Mills and many others is progressivist in spirit, although it seldom deals in grand-scale theories.[18] Noteworthy, also, is the vogue in the 1960s and 1970s of "futurology," promoted by social scientists such as Daniel Bell, Burnham Beckwith, and John McHale, and—with no little public fanfare—by Herman Kahn and the Hudson Institute. Futurology is often a more or less dispassionate endeavor to forecast the future "scientifically," but the findings of the futurologists tend to be optimistic. Beckwith's ingenuous book of forecasts, *The Next 500 Years,* anticipates—mainly by extrapolation—the progress of education, social control, rationalization of social policies, equalization of incomes, personal freedom, cultural homogenization, humanitarianism, and eugenics. A professed admirer of Condorcet, most of whose predictions, he notes, "have already been largely verified," Beckwith does admit the probability of a nuclear world war, but even this prospect holds few terrors for him: a nuclear war would only hurry along the trends visible on the horizon, and help persuade humanity to create a world government.[19]

We turn finally to the one social science not treated so far: political science—but political science broadly interpreted to include the ideologies and normative political theory. It is impossible to disagree with Alfred Cobban's lament that political theory has sadly deteriorated in its intellectual quality during the twentieth century.[20] Even the ideologies operate at a noticeably lower level of discourse than in the nineteenth century. There is some danger that discourse will eventually cease to matter at all in politics. Television, the new medium of politics, is not primarily a verbal medium; already elections are won or lost on the strength of a fresh shaven jaw or a pleasant smile or an earnest eye.

Nonetheless, in what there is of it, twentieth-century political thought provides its share of new expressions of faith in progress. The socialists have, as during *la belle époque,* made the greatest contribution, but even fascist thought does not lie entirely outside our jurisdiction. John T. Marcus argues in a recent book that all totalitarian political thinkers in the twentieth century have used the historical sense only as a way of escaping history, as a route to "secular immortalization" by promising a future classless utopia, an everlasting empire, or some other transcendental redemptive goal. In totalitarian—as also in conservative—thought, "the concept of progress could not arise."[21] There is certainly merit in Marcus' contention, especially as it relates to the varieties of fascist thought, all of which fiercely attack both the liberal and the socialist ideas of progress inherited from the nineteenth century.

But substantial fragments of the old belief in progress survive in fascist thought, which was nothing if not eclectic. Italian and German fascism from 1920 to 1945 drew heavily on the socialist tradition. Both harbored radical-rightist elements who saw the improvement of the lot of the peasant and working man as a major objective of fascist social policy. The economist Werner Sombart, in his important book *German Socialism* (1934), challenged the Germans of the Third Reich to free themselves "entirely from the fatal belief in progress." Many things could not progress—including art, religion, and philosophy. "But we do believe," he went on, "that there are conditions of collective life that are more favorable for the fulfillment of man's mission on earth than those which have been set up by the economic age, which permit the better sides of human life to develop, conditions under which the individual will be able to develop his inclinations and capacities more uniformly and thereby contribute more to the good of his community and the service of God." Sombart anticipated "a great, creative period of the human race in which the spirit of unity shall bring the individuals into a more thoughtful whole, wherein each individual, in living the most complete life, thereby renders service to the community." [22]

Still more significant, perhaps, at least in the case of German fascism, was its crude but compulsive adherence to the Darwinist notion of racial struggle as the rule of life and the agency of progress. Hitler's *Mein Kampf* is studded with references to progress, even to the progress of humanity. In Nazi theory, as developed by Hitler and by Alfred Rosenberg, "everything we admire on this earth today—science and art, technology and inventions—is only the creative product of a few peoples and originally perhaps of *one* race." The higher culture that now spanned the entire globe had been created almost exclusively by members of the Aryan race, who had succeeded by obeying "Nature's stern and rigid laws," by work, struggle, conquest, and avoidance of miscegenation. Lower races had played their part, too, by serving in early times as the slaves and serfs of higher men. He who refused to acknowledge the sovereignty of the iron laws of nature sought to thwart "the triumphal march of the best race and hence also the precondition for all human progress," and was doomed to "the animal realm of helpless misery." [23]

Yet in due course, "when the highest type of man has . . . conquered and subjected the world" and state eugenic policies had cleansed and improved the existing Aryan racial stock, the world would at last enjoy what Germany should long ago have given it: "a peace, supported not by the palm branches of tearful, pacifist female mourners, but based on the victorious sword of a master people, putting the world into the service of a higher culture." In that future era of world peace, humanity would not stagnate, but would confront "problems which only a highest race, become master people and supported by the means and possibilities of an entire globe, will be equipped to overcome." [24] In a psychological sense, perhaps Hitler's Thousand-Year Reich was a promise of "secular immortality," as Marcus would have it, a substitute for heaven. But this does not exhaust its meaning, and something of the old faith in progress as a continuous unfolding of ever-higher life and truth lingered on.

In the liberal West, political scientists have tended toward a relatively value-free and "realistic" study of contemporary politics, leaving the discussion of the question of progress to professional politicians, who generally accept progress in some sense as part of their official programs; to scholar-journalists, such as Raymond Aron and Walter Lippmann, who have eminent academic qualifications, but feel less reluctance in dealing with questions of value than the typical political scientist of the academy; and to the socialist intelligentsia, who have continued on the whole to accept the gospel of progress, though not without doubts and a measure of honest anxiety, as we observed earlier.*

Non-socialist liberal politics in the period since 1914, for all the broad phrases of the politicians, can hardly be described as exuberant. Many liberals have turned to pessimism.** Even the optimists have been cautious, and far from satisfied with the drift of the age toward collectivism and mass democracy—much like their late Victorian precursors. A great weathervane in this regard has been Lippmann, who was graduated from Harvard in the last years of the Progressive Era. In his twenties he thought of himself as a moderate socialist, but he gradually moved to a more conservative posture, finally opposing collectivism in any form and tempering his hopes for progress with a concern for the "universal laws of the rational order" and the "great tradition of the public philosophy." The West, he wrote in 1955, has entered into a phase of decline,

* See above, pp. 234-35.
** See above, p. 233.

marked by barbarization and a taste for violence and tyranny. Leaders have abandoned their calling as leaders to appeal to mankind's lowest impulses; the traditions of civility, the rational laws and the order of the "good society," are universally discarded in favor of a frenzied search for "one great public massive, collective redemption." [25]

Nonetheless, the essential Lippmann is a true-believer in progress, who prays that "men may find again the conviction of their forefathers that progress comes through emancipation from—not the restoration of—privilege, power, coercion, and authority." In *The Good Society,* the fullest statement of his mature thought, he blames the domestic unrest and total warfare of the twentieth century on collectivism, including the gradualist variety that pretends to believe in "the dictatorship of transient majorities" but reduces in the end to the rule of pressure groups. Collectivism is the worship of power, and it promotes an endless and barbarizing struggle for power within and among nations. But all progress in history, in man's ascent from savagery, has occurred through liberation, "by the removal of constraints" and "by the disestablishment of privilege." Progress has meant the humbling of power and the transforming of slaves into "free men inviolate in the ways of the spirit." How else, he asks, "can the human race advance except by the emancipation of more and more individuals in ever-widening circles of activity?" All the great movements of the modern world, from the Renaissance and Reformation to the Industrial Revolution were "movements to disestablish authority. It was the energy released by this progressive emancipation which invented, wrought, and made available to mankind all that it counts as good in modern civilization." [26]

The duty of government, therefore, is not to plan, organize, or dictate, but to ensure liberty under common law. Despite the public clamor for a planned society, and the vogue of fascism and communism, Lippmann expresses confidence that in the long run liberty will once more prevail, and progress resume. Men are not ants: they are born to be free. The energy of freedom has through the ages

> moved men to rise above themselves, to feel a divine discontent with their condition, to invent, to labor, to reason with one another, to imagine the good life and to desire it. This energy must be mighty. For it has overcome the inertia of the primordial savage. Against this mighty

energy the heresies of an epoch will not prevail. For the will to be free is perpetually renewed in every individual who uses his faculties and affirms his manhood.[27]

It is to socialist thinkers—academicians, publicists, politicians— that we must repair for the strongest affirmations of the doctrine of progress in twentieth-century political thought. The historical optimism of contemporary Marxist scholars in physics, biology, anthropology, and sociology has already been amply illustrated, but we have not yet considered those who have been most directly responsible for shaping socialist (and communist) thought and policies since the First World War.

In the West leadership in non-communist socialist thought after the passing of Bernstein and Kautsky fell, *faute de mieux* perhaps, to British socialists. Britain alone, of the major Western democracies, has had a traditionally powerful socialist party throughout the century with frequent opportunities to govern; and Britain's relatively smaller share in the suffering of the century has helped to give British thought in general a somewhat more hopeful tone than is typically encountered on the Continent, even among socialists.* Few twentieth-century French socialists, for example, have been as invincibly optimistic as Léon Blum, whose confession of faith in progress, *For All Mankind* [*A l'Echelle humaine*], was writen in 1941 and smuggled from his prison cell by friends. "This moment will pass," he promised his readers, "the dictatorships that now hold Europe in their grip will pass, present sufferings and ills will pass, and the eternal truths will remain. . . . Who can be sure that a century or two hence, when philosophers can regard the events of our time with complete detachment, they will not conclude that even Nazism and Fascism had some share in the providential march of progress?"[28]

For that matter, one might ask if even in Great Britain there have been many democratic socialist writers as optimistic as Blum in his wartime manifesto. The generation of R. H. Tawney, G. D. H. and Margaret Cole, Harold J. Laski, George Orwell, and John Strachey, which succeeded that of Shaw, Wallas, the Webbs, and Wells, has believed—at times passionately—in the ultimate triumph

---

* It cannot be a coincidence that most of the writers since 1914 who have devoted books to the history and anaylsis of the idea of progress are British: J. B. Bury, Dean Inge, Christopher Dawson, John Baillie, Morris Ginsberg, Vincent Brome, R. V. Sampson, Sidney Pollard, Leslie Sklair.

of socialism, but it has also shared fully in the anxieties of the century. Most of its members became disillusioned with the Soviet experiment at some point, most have been deeply troubled by the tendency toward hyper-bureaucratization in both communist and non-communist states, most lived to view the post-1945 nuclear arms race with horrified concern.

Of those British socialists who came to prominence in the years after 1914, G. D. H. Cole and Laski were the most prolific and perhaps the most influential. Both belonged to the academy, Cole as a lecturer in economics and political theory at Oxford, Laski as a political scientist at the London School of Economics. Both were active from time to time in the higher echelons of the Fabian Society, and Cole served twice as its chairman. In later years Laski wielded considerable power in the British Labour party. Neither shrank from public utterance: between them they wrote nearly one hundred volumes of history, politics, and economics.

Cole did not produce what could be called a systematic theory of progress, nor even, although he wrote a great deal of political history, an original theory of history. He possessed none of Laski's oracular style. In his most representative books, such as *The Simple Case for Socialism,* what stands out is his surefooted Anglo-Saxon common sense and his concern that socialism should serve humane values, rather than production, efficiency, or the laws of history. He returned to the old Benthamite formula, so influential in Fabianism, of the greatest happiness of the greatest number. We are socialists, he affirmed, purely and simply because we believe that socialism will make the mass of mankind happy, "and because that above all else is what we want." Socialism called upon men "to live up to a higher ethical standard than capitalism," but it did not ask the impossible. Most men were fundamentally decent. By organizing society for equality rather than for competition "we shall liberate pent-up human energy and goodwill to a degree that will before long triumphantly refute the sceptics." [29]

Even under capitalism, mankind had already made spectacular progress. The most predatory capitalist was less of a brute than the feudal robber baron, and in modern times the conditions of the masses had improved palpably, not only in a material sense, but also through the progressive conquest in the Western societies of basic civil, political, and social rights. Cole expressed faith, in his last book, that these rights, once won, could never be permanently lost and would serve as "a stepping-stone to further advances," to the development of "the conditions of a classless society." [30]

Despite his general optimism, Cole grew increasingly unhappy with the major forms of socialism in his own time. With the detachment of the academic idealist, he found both Western democratic socialism and Soviet communism blighted by the bureaucratic spirit, and looked rather to guild socialism, syndicalism, the thought of Proudhon, and the example of the *kibbutzim* of Israel for alternatives to the excessive reliance on state power of twentieth-century socialist politics. In this respect, as in others, he stood rather far from his fellow socialist Harold J. Laski. Several years younger, Laski was more of a liberal pragmatist than a socialist during the first part of his career, but when he turned decisively to Marxism during the Depression era, he abandoned his earlier scruples and became a self-appointed apologist of the Soviet Union and Leninist Machiavellism, a role he continued to play until his death in 1950. At the same time, he remained active in the inner circles of the Labour party in his own country and had a part in the planning of the welfare programs of the Attlee government. Although he never aspired personally to the parliamentary or ministerial career that lay well within his reach, Laski was too impatient and too passionate to be satisfied with the chair of the remote spectator-critic. He applauded success, and he had the true politician's respect for power effectually wielded.

The Laski of the years after 1929 shared Marx's apocalyptic conception of history. He looked with febrile gloom on his own epoch as a time of decay, as the predestined winding-up of an exhausted civilization, but he retained enough of his earlier liberalism to agree with Cole that in its time and place, capitalist society had been eminently progressive, and had not met with irreparable disaster until the Great Depression. The triumph of the liberal philosophy of history, politics, and society, which was nothing less than the ideology of capitalism, had represented "a real and profound progress." It had raised the average level of material life, nurtured the sciences, heightened regard for personal dignity, and created a new zeal for truth and experiment, "all parts of a social heritage which would have been infinitely poorer without them." Liberalism had not failed to exact a tragic price for its blessings, but the price had been lower than the benefits, and net progress had resulted.[31]

Now, in the 1930s and 1940s, the old liberal civilization had demonstrated its impotence to generate further progress. The Depression ruled out any prospect of continued steady gains in material well-being for the working class. Hitler and Mussolini were "the *condottieri* of big business," and their wars, as well as Churchill's manufacture of a "Cold War" against Soviet Russia after 1945, represented quite clearly a massive counterrevolutionary thrust on the part of embattled capitalism to save itself from ruin. In his last books Laski referred repeatedly to the contemporary age as a time of "darkness" and of "winter," lit only by the beacon of Soviet socialism, which, for all its errors, showed the way to the future of mankind and had accomplished "a transvaluation of all values, the birth of a new civilization." He sometimes let himself be convinced that a violent revolution might not be necessary in the West, as it had been in Russia, but of the suffering and struggle that lay ahead, he had no doubts.* It was unrealistic to hope for the salvation of humanity "in a society where the exploited are to look to their exploiters for redemption." If Lenin "was surely right when the end he sought for was to build his heaven upon earth," he was also right in recognizing "that the prelude to peace is a war." "Our generation, certainly, dare not hope to enter the promised land; it must be enough for us if we labour to set the feet of our children on the road at whose end it lies." [32]

Laski's generation did not, to be sure, live to see the "promised land." Laski and Orwell both died in middle age in 1950, Cole passed on in 1959, Tawney in 1962, Strachey in 1963. In the 1960s their place in Western democratic socialist thought was taken, if by anyone, by a much younger group of thinkers, the so-called New Left, men born chiefly in the interwar decades, who have not yet fully proved themselves as social theorists, or as prophets of progress. The motherland of the New Left is—surprisingly—the United States, rather than Great Britain, but the movement finds exponents in all the Western countries.

* *Cf.* John Strachey, in *The Coming Struggle for Power* (New York, 1935), p. 395: "For if men hesitate before the task of achieving a new civilization; if they draw back because no new order of society can be born without violent conflict, they will not achieve an epoch of peaceful stability. The alternative to the violence entailed by the lifting of human life to a new level is the violence entailed by the decline of human society, the break-up of such world civilization as exists, the dawn of a new dark age of perpetual conflict. It is not given to men to stand still upon the path of history. Forward or back is their only alternative."

New Left thought, like Old Left thought, concentrates on radical social criticism; it shares the apocalyptic consciousness of Laski and the rejection of Soviet totalitarianism of Orwell. But many of its anxieties are new. More than world war, dictatorship, depression, and genocide, it fears what Michael Harrington terms the "gentle apocalypse" of impersonal, collectivized, bureaucratized, megalopolitan decadence. It bemoans the failure of ideology and the triumph of the Organization Man. It finds the world in the pocket of a small number of giant American private corporations and their Soviet counterparts. It discovers new proletariats in the racial minorities of the United States and the millions of Asians, Africans, and Latin Americans exploited by white Western economic imperialism, which has resorted to subtleties not foreseen even by Lenin. It fears the misuse of affluence and leisure as much as the older Left feared unemployment and underconsumption.

Without expounding a full-fledged theory of historical progress, New Left thought rejects the melancholy temper of earlier decades for a program of ideological *risorgimento,* resistance (if not revolution), and ebullient optimism. Its most typical formula for progress is Harrington's: not the abolition of technology and bureaucracy, but their submission to "democratic social control." Mastery over the chaotic trends of the century must be regained, and power restored to "the people." [33] That "the people" themselves have been thoroughly corrupted and bewitched by "the Establishment" or "the military-industrial complex" or "the power elite" is a fact often conceded, but the spell—it is thought—can be broken by political education. Socialism is seen not only as a device for redistributing wealth or collectivizing the means of production, but, even more, as a philosophy of equality, brotherhood, peace, and liberation.

Just as the New Left has obviously called back to life most of the constructive ideas of traditional socialism, in spite of its redefinition of socialism's enemies, so it has experienced all over again the divisions of the older socialism on questions of strategy and tactics. While some New Leftists confine themselves to persuasion and non-violent resistance, more radicalized segments of the New Left insist on neo-Maoist street warfare, the seizure and democratization of universities, and public demonstrations aimed at social disruption rather than mere protest. But all New Leftists, from Michael Harrington and Staughton Lynd to the urban guerrillas of Paris in 1968 and the liberation committees of Berkeley in 1969, share a revolutionary euphoria that suggests the possibility of a new era of socialist activism and new expositions of the doctrine of progress.*

In Soviet Russia and the parties of the Third International, the belief in progress has retained, at least officially, its binding authority as an article of Marxist-Leninist dogma, and pervades every department of Soviet culture. The decisive influences are those of Marx, Engels, and Lenin, but others are notable: in anthropology, for example, the thought of L. H. Morgan, whose evolutionism was endorsed by Engels; in psychology, the behaviorism of Ivan Pavlov, and in biology, the theories of Michurin and Lysenko, all of which suggest the plasticity of human nature, and the possibility of bringing it under control through behavioral and social engineering. Both of Lenin's heirs, Trotsky and Stalin, were fully committed to the belief in progress; nor did Trotsky let his bitter disappointment with the course of Soviet history after Stalin's rise to power rob him of his confidence in the ultimate triumph of world communism and the salvation of mankind.[34] **

A representative Soviet statement on progress was offered in 1961 in the *Journal of the History of World Culture*. In a lengthy article on the meaning of history, N. I. Konrad argues that history so far has been a record of general improvement, achieved dialectically through contest with nature and the forces of reaction. The future will be unambiguously splendid, once all mankind has succeeded in making the painful transitions from capitalism to socialism to pure communism.[35] Soviet educators have repeatedly voiced their confidence that human nature is malleable, and that a "new man" can be constructed, a man who will work for the sheer love of labor and of his fellow man, rather than for private gain. In the new Soviet man "moral purity will become an inherent quality, and voluntary observance of the rules of living in a communist society will be a matter of conscience for every member of society. . . . As for vices, these will inevitably die away once the social conditions that have produced them shall have been removed."[36] Another excellent source for Soviet thinking on the future quality of life under communism is Soviet science fiction. In Ivan Efremov's novel *Androm-*

---

* One New Left philosophy of progress of some note has already appeared, although its proponent is by no means young. See the discussion of Herbert Marcuse's thought below, pp. 343–48.

** But see also above, p. 235 and fn.

eda, for example, we visit civilization three thousand years from now, an affluent world of happy, productive, profoundly communized men and women, whose children are reared by public authorities after weaning, and who have time to pursue five or six different careers in a life-span lasting as long as three hundred years.[37]

Communists in the Western countries have tended to follow closely the lead of Soviet writers on the subject of progress, as on all other subjects—at least until very recently.[38] But the 1960s witnessed new stirrings in communist thought throughout the world, and most dramatically in the Eastern European people's democracies. Writers return to Marx, as the Protestant Reformers returned to the New Testament, to seek out the democratic humanism especially of his earliest works, which they find irreconcilably opposed in letter and spirit to the bureaucracy and totalitarianism of certain modern communist regimes.[39] Another, quite different, indication of the rise of an Eastern New Left is the remarkable essay by the Soviet physicist Andrei D. Sakharov, *Progress, Coexistence and Intellectual Freedom,* circulated privately in the U.S.S.R. and published as a short book in the United States in 1968. Sakharov dismisses the notion of a necessary struggle to the finish between East and West, and foresees an eventual convergence of the socialist and capitalist worlds, an immense program of Soviet-American assistance to the underdeveloped countries, and the creation of a world government.[40]

Two substantial thinkers, apart from Marx himself, have served as mentors of the Eastern New Left, Georg Lukács in Hungary and Ernst Bloch in Germany, both born in 1885. Lukács earned his reputation chiefly as a literary critic, but Bloch is above all a prophet of hope and progress, who has made one of the few original contributions to speculative philosophy by a Marxist thinker in our century. Bloch devoted much of his attention as a writer in the pre-Hitler days to the exploration of utopian thought, and wrote an important study of Thomas Münzer, the Anabaptist revolutionary, whose career had also inspired a book by Engels. In 1933 he left Germany, spending ten of the next fifteen years in the United States (without, however, becoming Americanized). He returned to Germany after the war to teach philosophy at the University of Leipzig in the German Democratic Republic. In 1956, the year of the Polish "October" and the Hungarian Revolution, his heterodox views forced him into retirement, but shortly thereafter he accepted a call to the University of Tübingen in the Federal Republic, where he has remained ever since. He is without doubt one of the most widely discussed intellectual figures in Central Europe today. In the Anglo-Saxon countries his influence thus far has been limited to the current crop of avant-garde theologians, who are already comparing his impact on theology to that of Heidegger.[41] Such younger German theologians as Jürgen Moltmann and Wolfhart Pannenberg are clearly in his debt.

Bloch's most ambitious work, *The Hope Principle,* does indeed provide a revealing contrast to Heidegger's *Being and Time,* or for that matter to Spengler's *The Decline of the West.* All three men were born in the 1880s, but Spengler and Heidegger, as analysts of decline and death, and Bloch, as an analyst of hope and futurity, speak to different generations. It is not surprising that Bloch should have found his public only in the last ten or fifteen years, whereas Spengler and Heidegger found theirs three or four decades ago.

For Bloch, the universe has a distinctively Bergsonian quality. The ground of all our being and becoming is matter, as Marx posited, but matter should not be understood in the usual way, as dead, inert, changeless stuff, foreclosing the future by its obduracy. Rather, it is dynamic, fruitful of endless possibility, generating the whole world out of itself. The nature of the real universe of matter is to unfold by moving forward, and in man its forwardness takes form as the conscious hope for better things, the hope of the not-yet (*das Noch-Nicht*), symbolized in folklore, religion, art, literature, and music in innumerable ways, but most pointedly in the dream of utopia, as in the Christian image of the kingdom of heaven, which is only a mythopoetic anticipation of the Marxist vision of the kingdom of freedom. Every great revolution in history, too, such as the French Revolution, represents a fresh upsurge of creative and liberating hope, no matter how far short the revolutionists fall of their goals, or how limited their goals may have been from the start. The Jacobins' hope "of a vastly improved kind of *polis,* their sense of human progress as within the bounds of historical possibility . . . endowed their cause with so much greater moral grandeur than any mere emancipation of the Third Estate. . . . There was already more than a little red in the old tricolor, introduced by the so-called Fourth Estate—the red of irreversible progress."[42]

Marxism, then, so often misunderstood by both Marx's enemies and would-be orthodox apologists, is a philosophy of hope and freedom, not a deterministic dogma that makes the future subject to irresistible laws and leaves no place for man and his dreaming. It is utopian through and through. It catches sight of a possible—but not an inevitable—future, in which men are happy, free, and at home. Bloch's thought culminates in nostalgia for the "homeland" (*Heimat*) that waits for man in the future, if only he will strive to reach it. "The root of history is working, creating man, man who transforms and outstrips the conditions of his existence. Let him achieve self-comprehension and ground his life in real democracy, without renunciation and estrangement, and then something will arise in the world that all men see in childhood, a place where no one has yet lived: homeland." [43]

17

The Meta-Psychology of Progress

I HAVE SAVED until last what some students of twentieth-century thought would regard as the most prestigious and seminal of all the sciences of our time. Mental illness fascinates twentieth-century man more than physical. Every contemporary figure in the public eye is subjected to ruthless psychoanalysis in the press. Historians write psychoanalytical history. Freud and Jung have exerted more influence on *belles-lettres* and the fine arts than any philosopher. In the academy, psychology has advanced from a minor discipline hardly recognized as a discipline at all by nineteenth-century universities, to a major field of research, with numberless applications in medicine, education, and law.

It is not surprising that we have had no occasion thus far to discuss psychological thought, except for a few words about Jung.* Most psychologists have confined themselves to studies of individual development and behavior, giving them no chance to consider the problem of general human progress. Even the majority of social psychologists are more concerned with the dynamics of specific kinds of group behavior than with the whole sweep of history. For that matter, psychologists are usually not interested in history at all. The psychoanalytical approach to history often results in the total dehistoricization of the historical subject.

But professional psychologists, psychoanalysts, and philosophers with a special interest in psychology have sometimes produced theories of progress in our century. In nearly every case, their theories

* See above, p. 222.

are based not on psychological thought alone, but on a combination of psychological themes with the perspectives of some other discipline or ideology. Thus, Freud's idea of progress was steeped in the rationalism of the Enlightenment and in nineteenth-century scientism. B. F. Skinner's hope for progress is also strongly scientistic. The work of Erich Fromm and Herbert Marcuse is inconceivable without the influence of Marxism. In their hands, psychology becomes meta-psychology; it clothes itself in a *Weltanschauung* and is transformed into a doctrine of progress.

Even religious ideas may help convert psychological into meta-psychological thought. Jung's social philosophy underwent such a transformation, although for him the result was more a theory of anti-progress. In the writings of the Anglo-American philosopher Gerald Heard, on the other hand, religious inspiration worked to the opposite effect. He interpreted history as an ever-rising spiral from lower to higher forms of spirituality and consciousness, and predicted, in *The Ascent of Humanity* (1929) and later works, the eventual transcendence of both the tribal and the personal psyche in "the superconsciousness of a purely psychologically satisfying state." [1]

Heard is scarcely a representative figure in modern psychology, however. More typical is the Harvard psychologist B. F. Skinner, a prominent spokesman of the behaviorist movement. Behaviorism takes its inspiration from nineteenth-century scientism and nowadays dominates many American psychology departments—not to mention those in the Soviet Union, where (in the Pavlovian version) it is the only officially sanctioned school of psychological theory.

In his utopian novel, *Walden Two* (1948), and elsewhere, Skinner submits that human behavior, individually and collectively, has always been subject to outside influences, and always will be. Man must choose whether he wishes to be controlled, as in the past, by historical accidents, or by his own scientific knowledge of himself, rationally implemented by behavioral engineers. "The slow growth of the method of science, now for the first time being applied to human affairs," Skinner urges,

> *may* mean a new and exciting phase of human life to which historical analogies will not apply. . . . If we are worthy of our democratic heritage we shall, of course, be ready to resist any tyrannical use of science for immediate or selfish purposes. But if we value the achievements and goals of democracy we must not refuse to apply science to the

design and construction of cultural patterns, even though we may then find ourselves in some sense in the position of controllers.

Blind fears of intelligent social planning for "a better way of life" must be conquered—and in the very act of conquest, "we shall become more mature and better organized and shall, thus, more fully actualize ourselves as human beings." [2]

Although most critics equate *Walden Two* with *Nineteen Eighty-Four,* Skinner himself is very much in earnest: man, he assures us, can scientifically determine what he wants and needs, learn how to design a social order that automatically supplies his wants and ministers to his needs through psychological conditioning, and then set about transferring his knowledge from the drawing board to the real world. One can almost see the abbé de St.-Pierre nodding his head in fatherly approval!

In the Western world, at least, the chief rival of behaviorism, and by far the stronger influence on medical practice, is the thought of Sigmund Freud. The place of Freud in literature, the arts, philosophy of religion, social thought, and the Western image of man is no less secure than his place in psychology and psychiatry. There is no way to escape him. In a book such as this, it would be a "Freudian slip" if we even tried.

But Freud, like Nietzsche, is a mind not easily classified. On the question of progress, his meta-psychological views can be interpreted as theories of both progress and anti-progress. The most pertinent documents for either interpretation are the two brief books he published in his seventies, *The Future of an Illusion* and *Civilization and Its Discontents.* Yet virtually all his work has influenced Western thinking on progress. On the one hand, it has given rise to a sense of liberation from old fears and feelings of guilt, and prescribed rational therapies for psychic illness in the individual and in society alike. On the other hand, Freud's thought has affirmed in a new mode the Judeo-Christian doctrine of original sin; it has rediscovered the "beast in man," resurrected Schopenhauer's intuition that man can never be truly happy, and demonstrated the stubborn irrationality of *Homo sapiens* not only in his "natural" state but in the "highest" stages of civilization as well. On balance, Freud has probably contributed more to the pessimism of the century than to its optimism. Yet the man himself was from first to last a votary of enlightenment through science and reason, a figure essentially of the nineteenth and even eighteenth century; or, as Abraham Kaplan writes, "a rationalist, following in the Jewish tradition of Maimonides, Spinoza, and Einstein . . . not so much a pessimist as a realist . . . always a yea-sayer to life."[3]

Freud's explicit statements on progress contradict one another. In *The Future of an Illusion* he noted that "while mankind has made continual advances in its control over nature and may expect to make still greater ones, it is not possible to establish with certainty that a similar advance has been made in the management of human affairs." Yet some progress even in the management of human affairs had clearly occurred. Civilization was nothing if not a system for repressing human aggressiveness, which it accomplished by conditioning men to love one another and by arousing in them feelings of guilt and unworthiness. In *Civilization and Its Discontents* Freud spoke of an actual "loss of happiness" as the price for civilization's services; yet this seems to contradict his own rigorously unsentimental account of prehistoric times in the same book. Only the head of the primal family, he observed, had enjoyed the pleasures of an unrepressed instinctual life; the rest were his slaves; and even for the primal paterfamilias, happiness was likely to be of short duration.[4] One would have to retreat to a time before the establishment of the first patriarchal families, to the subhuman stage of man's evolution, to encounter a nonrepressive life order. What Freud appears to have meant was that since repression produced psychic suffering, which more or less negated the pleasures of security, civilization had brought as yet no real increase in happiness; but when this idea is coupled with his judgment that civilization had given mankind civil peace, control of nature, and great advances in knowledge, the result, arguably, is a theory of net progress. Freud was a sometimes disappointed lover of civilization: but a lover no less.

The same difficulties arise in considering Freud's thoughts on the future of civilization. On one side, we have his well-known dictum that the program of the pleasure principle is "at loggerheads with the whole world, with the macrocosm as much as with the microcosm. There is no possibility at all of its being carried through; all the regulations of the universe run counter to it." Social controls notwithstanding, each human being still bore, as T. H. Huxley had long since pointed out, the indelible mark of the wolf. A man's neighbor was not only someone to love but also a tempting object of aggression, a victim to be humiliated, injured, exploited, even tortured and killed. *"Homo homini lupus.* [Man is a wolf toward his fellow man.] Who, in the face of all his experience of life and of history, will have the courage to dispute the assertion?" Society, therefore, would always be forced to chain up the beast in man, and a noncoercive, nonrepressive ordering of human affairs was forever out of reach.[5]

On the other side, we also have Freud's hopes for at least modest and gradual improvement in the human condition. The fact that his hopes were usually accompanied by sighs of regret that one could no longer reasonably hope for perfection does not disqualify him as a believer in progress. His recipe for reform was to extend the empire of reason in human affairs: to understand things as they are, without illusions, to work for a better balance between the demands of society and of the individual, to replace the religious sanctions of social discipline with rational ones, to reduce the number and severity of social restrictions where reason permits, and in the end perhaps to convert the majority of men from an attitude of hostility toward civilization to an attitude of willing acceptance. In place of the fierce father-god of religion, Freud proposed to deify *Logos,* the spirit of reason. "The primacy of the intellect lies, it is true, in a distant, distant future," he wrote, "but probably not in an *infinitely* distant one." Its voice was soft, but it "does not rest till it has gained a hearing. Finally, after a countless succession of rebuffs, it succeeds." Those who would scold science for not having already rescued mankind from its misery "forget how young she is, how difficult her beginnings were and how infinitesimally small is the period of time since the human intellect has been strong enough for the tasks she sets." [6]

In the end, we are left with almost the positivist theory of progress, despite Philip Rieff's caveat that "nothing was more foreign to [Freud's] mind than the optimistic temper in which the three-stage positivist theory was regularly set forth." [7] But three stages of progress emerge in Freud's historical vision just the same: the age of primal man; the age of religious civilization; and the dawning age of scientific civilization, the best of all possible, if not of all imaginable, worlds. Freud could not resist the image of the child growing to manhood. In the age of religion men had played the part of children, but in the nascent age of science and reason, "they will be in the same position as a child who has left the parental house where he was so warm and comfortable. But surely infantilism is destined to be surmounted. Men cannot remain children for ever." The idea forced itself upon the psychologist "that religion is comparable to a childhood neurosis, and he is optimistic enough to suppose that mankind will surmount this neurotic phase." [8]

In reply to sceptics who might doubt that mankind could bear the thought of its aloneness in the cosmos, Freud could only hope for the best.

> It is something, at any rate, to know that one is thrown upon one's own resources. One learns then to make a proper use of them. And men are not entirely without assistance. Their scientific knowledge has taught them much since the days of the Deluge, and it will increase their power still further. . . . By withdrawing their expectations from the other world and concentrating all their liberated energies into their life on earth, they will probably succeed in achieving a state of things in which life will become tolerable for everyone and civilization no longer oppressive to anyone.

Then, like the poet Heine, they could say without regret: "We leave heaven to the angels and the sparrows." [9]

A further index to the fertile meliorism of Freud's thought is the number of eminent thinkers of the next generation who have acknowledged its power, and who have moved on from certain Freudian premises and concepts to doctrines of progress still more hopeful than his own. How Freud's basic concepts could be applied almost unchanged to the formation of a doctrine of progress far less hobbled by doubts and misgivings than his own was displayed with uncompromising integrity by the British psychologist J. C. Flugel in his *Man, Morals, and Society,* first published in 1945. Two other noteworthy Freudians, Erich Fromm and Herbert Marcuse, blend their Freudianism with Marxism. Refugees from the Third Reich who came to the United States in the 1930s, Fromm and Marcuse have emerged as two of the most influential minds of the postwar world and as leading apostles of the radical optimism of the 1960s. In the thought of both, Freud's presence is strongly felt, but they are not so much Freudian as post-Freudian thinkers—almost in the sense that Hegel was post-Kantian and Marx, post-Hegelian.

Each has a distinctive prophetic style. Whereas Marcuse attracts the tough-minded, the young, and *soi-disant* revolutionists, Fromm finds his public among the tender-minded, the middle-aged, and liberal churchmen and humanists. To Marcuse and his followers, Fromm is dull and naive; to Fromm and his followers, Marcuse is cynical and irresponsible.

Erich Fromm was born in Frankfurt in 1900. A practicing psychoanalyst, he worked with Karen Horney and Harry Stack Sullivan in the 1930s, and published his first book, *Escape from Freedom,* in 1941. Since that time he has written extensively in the fields of psychology, religion, and social philosophy; at least one of his books, *The Art of Loving* (1956), has become something of a best seller. Despite deep respect for Freud as a pioneer, Fromm is unable to accept the greater part of Freud's social thought. Indeed, one of his chief criticisms of Freud is that he was all but oblivious of the sociohistorical dimension of human existence, preferring to adopt the nineteenth-century liberal view of mankind as composed of atomic individuals, more or less self-contained and subject to the rule of instincts and psychic forces that vary little through all the eras of history. Although Freud's research shed much light on the behavior and development of individuals, Fromm urges that the thought of a man such as Karl Marx is needed to explain the evolution and prospects of human society. Yet, in the final analysis, Freud and Marx shared the same goal. "They have in common an uncompromising will to liberate man, an equally uncompromising faith in truth as the instrument of liberation and the belief that the condition for this liberation lies in man's capacity to break the chain of illusion." [10] The illusions of supernatural religion, of the impotence of reason, of the inevitability of exploitation, of the permanence of man's enslavement to established power—these are the illusions, the obstacles to human progress, that Freud and Marx, one or both, sought to demolish.

Fromm interprets history not as a struggle between Eros and Thanatos, but as the progress of freedom, as man's upward striving to extricate himself from the anonymity of primeval nature, a process that began in earnest some four thousand years ago. The quest for freedom and personality, however, is not an easy one. At each new step forward, man fears the loss of his security, and experiences a temptation to regress to the womb-like unfreedom of his former oneness with nature. "In the history of the individual, and of the race, the progressive tendency has proven to be stronger, yet the phenomena of mental illness and the regression of the human race to positions apparently relinquished generations ago, show the intense struggle which accompanies each new act of birth." In this perpetual tension between the desire for freedom and the fear of freedom, Fromm sees "the true kernel in Freud's hypothesis of the existence of a life and death instinct." The "death-wish" is in fact nothing but man's fear of the unknown; and in any event, "normally, the forward-going life instinct is stronger and increases in relative strength the more it grows." [11]

Progress has also occurred in two other areas: in moral thought and in control and knowledge of nature. Most of the former took place more than two thousand years ago. The first civilizations attempted to create "a new and truly human home to take the place of the irretrievably lost home in nature." A society bound by the bonds of brotherly love, justice, and truth became the goal of progressive man. Then, some five hundred years before Christ "the idea of the unity of mankind and of a unifying spiritual principle underlying all reality assumed new and more developed expressions" in the teachings of Lao-tse, Buddha, Isaiah, Heraclitus, and Socrates. The great social prophets of recent times, including Marx, have done no more than apply old wisdom to the concrete historical situation of civilized man in the industrial age. Fromm has even devoted a book to the messianic tradition of the Old Testament, which he identifies as a major source of modern secular radicalism. [12]

In science and technology, he notes, progress has continued with few interruptions since earliest times, and in the last few centuries at a pace that defies all precedent. Such progress "is truly awe-inspiring." Man has "developed his reason to a point where he is solving the riddles of nature, and has emancipated himself from the blind power of the natural forces." The best of art, literature, and music have been brought to every member of society, and productive forces have been released, "which will permit everybody to have a dignified material existence, and reduce work to such dimensions that it will fill only a fraction of man's day." Yet material progress must not be overrated: "The most valuable achievement of human culture" is our respect for, and our cultivation of, "the uniqueness of the self." [13]

Fromm finds the self in peril in the twentieth century. The freedom and power it has won in modern times have raised it to dizzy heights, but it is precisely this enormous progress that has so powerfully reawakened the ancient fear of freedom in the breast of modern man. He looks for routes of retreat from the responsibilities that are now his, and, as in the past, turns regressively to the security of enslavement. In some societies, he has built superstates that denude the self of its freedom and vest all authority in a *Führer,* a *Duce,* or a Generalissimo. Such a fate has not been escaped even by the world's first officially "Marxist" country, Soviet Russia. In other societies, such as the United States, the flight from freedom is accomplished more subtly by surrendering one's autonomy to the impersonal requirements of the mass market, bureaucratic government, and uncontrolled technological progress. In either case, the end is the same: robotization. The robot-man is an alienated man, and at any time his alienation may plunge him into aggressive or self-destructive acts, such as wars, which put his very survival in question, given the power of modern technics. The problem of alienation was well-defined by Marx, but even he did not foresee that under twentieth-century conditions it would envelop men of all classes: not workers alone, but professional men, managers, all men who must make their livelihood by conforming to the demands of the superorganization, the bureaucracy, and the machine.

Nevertheless, Fromm refuses to despair. He writes of a "revolution of hope" that can sweep the world and restore to man control over his own destiny. The very vigor of our science, not least our psychological thought, suggests that, whatever else it might be, ours is not a decadent society, foredoomed to exhaustion and death. Hope itself can become the antidote to fear, and our growing knowledge of our nature and our needs can be applied to the work of saving civilization from its own folly. As Marx believed, man is naturally both good and sociable. He finds his deepest satisfaction in the spontaneous unfolding of his powers, in freely willed labor and love. He needs to be both free and united in freedom with his fellows and with nature, to feel at home in the world without being swallowed up in it or reduced to slavery. Fromm submits that a social order can be constructed that offers both personal freedom and the security of fellowship—not, at least in the West, through revolutionary violence, but through a more fully participatory democracy, experiments in communitarian socialism, the gradual decentralization of power, and the humanizing of economic production and consumption.

Fromm also anticipates the possible emergence of a new secular religion, which "would embrace the humanistic teachings common to all great religions of the East and of the West," creating its own rituals and art forms "conducive to the spirit of reverence toward life and the solidarity of man." He agrees with Jung, as against Freud, that man's desire for religious faith is legitimate, expressive of his need for meaning in life, although he cannot accept authoritarian religion or the literal existence of a supernatural world order. Man must first have faith in man, and his one hope for the future is the creation of

> a sane society which conforms with the needs of man. . . . Building such a society means taking the next step; it means the end of "humanoid" history, the phase in which man had not become fully human. . . . We are in reach of achieving a state of humanity which corresponds to the vision of our great teachers; yet we are in danger of the destruction of all civilization, or of robotization. A small tribe was told thousands of years ago: "I put before you life and death, blessing and curse—and you chose life." This is our choice too.[14]

But to some post-Freudian thinkers, such as Herbert Marcuse, the social philosophy of Fromm fails to enter into mortal combat with the forces ranged against freedom. Its ideals are sound, as far as they go, but its critique of history and society lacks muscle; its account of human nature glosses over the deep instinctual bases of behavior; it affirms, without attending to the equally essential task of negation. The realism of Marx and Freud is exchanged for a somewhat sentimental optimism that plays into the hands of the established social order.[15]

Whether his judgment of Fromm is fair or not, Marcuse is the philosopher *par excellence* of the new radicalism. His books have given the program of the New Left meaningful roots in the radical tradition in European philosophy, set its dissent from bourgeois society in the framework of a totalistic world-view, and articulated its hopes for the future progress of mankind. Like Bloch in Germany, Marcuse in his old age has become the prophet of what is basically a movement of young men. He was born in Berlin in 1898, and emigrated to the United States shortly after the Nazis came to power. From 1941 to 1950 he was employed by the Office of Strategic Services and the U.S. State Department. Since 1954 he has taught political science and philosophy, first at Brandeis University and then at the University of California at San Diego. His first major scholarly works analyzed Hegel and Soviet Marxism. In 1955 he published *Eros and Civilization,* "a philosophical inquiry into Freud," which contains the quintessence of his thinking as a prophet of revolutionary utopism; further reflections on progress and utopia appeared in *One-Dimensional Man* (1964) and *An Essay on Liberation* (1969).

As one might expect of a disciple of both Marx and Freud who is also well-versed in Hegel, Marcuse does not subscribe to a rectilinear theory of past progress. History is not the simple unfolding of man's creative powers, but a record of repression. Marcuse adopts the most pessimistic aspects of Freud's doctrine of civilization as a repressive order: the more man has "progressed," the more he has become un-free, the victim of a social system that organizes his life more and more minutely, until—with the triumphs of modern technology—every hour of every man's day has fallen into bondage. A minimal amount of instinctual repression has been absolutely necessary be-cause of the scarcity of the means of subsistence, but what Freud failed to bring out clearly was the prevalence in civilization—beginning even with the primal horde—of "surplus-repression," repression necessary to permit the domination of man by man. Cer-tain men, certain groups, certain classes take more than their share—as in Marx's theory of surplus value—and thus gratify their own instincts at the expense of those whom they deny and repress. At the same time they impose on the societies they rule a special historical form of Freud's "reality principle," which Marcuse terms the "per-formance principle," dictating the stratification of society according to the competitive economic performance of its members. The result is alienated labor, required by a society founded on the logic of domination. Instead of making one's own accommodation to the hard facts of reality, one does what he is told, and leads a socially preestablished life.

History, therefore, contains a painful and irrational paradox. Al-though civilization, with the help of reason, has steadily increased man's control of nature, making repression rationally less impera-tive, repression nonetheless continues and indeed intensifies. But more and more the repression that continues is "surplus-repression," the denial of gratification and freedom for the sake of exploitation. As the apparatus of repression becomes more efficient and all-pervasive, it becomes an end in itself, and we arrive in the twentieth century at a radically depersonalized, totalitarianized social order that lumbers on, like a juggernaut, engulfing everyone in its path. Marcuse would hardly wish to be thought of as another prophet of "culture lag," but in effect he is. The prevailing social system and value-system lags behind the progress of technology, and must be replaced, he warns, from top to bottom.

Marcuse's criticism of contemporary civilization leaves nothing unscathed. The hostility of civilization to the pleasure principle, which carries so much further than reason alone would require, and its insistence upon vast amounts of alienated labor, generates aggressiveness on the part of its victims that can be discharged only against supposed "enemies" of the society—whether the Jew, the Negro, the Vietnamese peasant, the Communist, the "traitor" within, the pervert, the political "deviationist," or what you will. The result is a century made unspeakably horrible by world wars, genocide, and persecution beyond all measuring. Against the "enemy" technological society cannot exert itself too much.

> He is everywhere at all times; he represents hidden and sinister forces, and his omnipresence requires total mobilization. The difference between war and peace, between civilian and military populations, between truth and propaganda, is blotted out. There is regression to historical stages that had been passed long ago, and this regression reactivates the sado-masochistic phase on a national and international scale.[16]

Aggressiveness manifests itself in a war against nature as well, in heedless reproduction, spoliation of natural resources, the wiping out of wild life, the rape of the earth. But the advanced industrial society of the twentieth century also has its "benevolent" side. Although it permits "legitimate" outlets for the aggressiveness that it breeds, it spares no effort to subdue opposition from within by buying off its masses with a higher material standard of life, by manipulating their leisure time with its entertainment "industry," and by controlling their thoughts with its corporate and government propaganda, mass advertising, public education, and puppet media. Never has there been so much "happiness" for so many people—but the happiness is a plastic commodity, mass-produced by the Establishment for the better subjugation of its servile classes. The majority of people still must perform meaningless and alienated labor "full time," their instincts are still carefully repressed and managed by society, and they have no experience of true joy or freedom.

But Marcuse cannot accept Freud's thesis that civilization and repression are fatally intertwined, or even his hope for improvement of social relations through the enthronement of Logos. Although

345

Freud's basic categories, his inventory of the instincts, his anatomy of consciousness, and his critique of civilization are all fundamentally correct when supplemented by Marxian insights into the nature of social domination, he failed to transcend his limitations as a member of the European liberal bourgeoisie. He failed to appreciate the degree to which even science and reason can be prostituted for exploitative ends. Yet his analysis of the profound irrationality of civilization as an agency of perpetual repression helps to create the possibility that men will—with opened eyes—reject that civilization and refuse to serve as its victims any longer. His disclosure of the secret contents of the unconscious opened up to view a world of sensuous primeval truth denied by civilization. Promises and potentialities came to light "which are betrayed and even outlawed by the mature, civilized individual, but which had once been fulfilled in his dim past and which are never entirely forgotten. . . . Regression assumes a progressive function. The rediscovered past yields critical standards which are tabooed by the present. . . . The *recherche du temps perdu* becomes the vehicle of future liberation." [17]

In assessing the chances for "future liberation," Marcuse proposes that history, if one probes more deeply, reveals itself to be more than an endless process of enslavement, of revolutions betrayed, of misery and frustration. With Marx, he suggests that man's martyrdom has also paved the way for man's liberation. The mechanism is quite simple. Civilization, with its fierce discipline and enforced sacrifice, has very nearly succeeded in abolishing the scarcity that has always, in the past, doomed mankind to a certain measure of deprivation and pain even under the least repressive social orders. In the twentieth century, automation promises to eliminate the need for human labor almost entirely. With the progress of technology, the rational case for continued repression rapidly vanishes, until civilization as we know it becomes obsolete, immoral, and unjust. It is a scaffolding which must be dismantled, now that the fortress of abundance that protects mankind against natural scarcity has at last been built—at a terrible, and often needless, cost in human suffering.

Of course the Establishment will argue that it remains indispensable; the bureaucracy, the industrial complex, the military wish to preserve themselves in power. In order to maintain their authority, they will continue to provide a copious flow of material goods and an inexhaustible supply of skillfully fabricated foreign and domestic "enemies" against whom the oppressed citizenry can unleash its resentment. But Marcuse rejects the bureaucratic definition of progress as a steadily rising standard of material living. With Hegel and Marx, he insists that the only true progress is progress in freedom: in Freudian terms, liberation from natural or artificial fetters on the pleasure principle. True freedom is possible only in an economy of abundance, but true freedom for this very reason is possible here and now for the first time in human history. At last Eros can be set free, not only to seek but also to find gratification. The visions of childhood, of utopian speculation, of the primeval ape-man enter into congruence with the reality principle, and the pleasures that civilization has been forced, in some measure, to deny, can now be sought without shame or anxiety. Life can be lived not for production or for work, but for joy—as Nietzsche foresaw in his fantasy of the *Übermensch.*

Marcuse's picture of a civilization without repression attacks some of the most deeply rooted taboos of civilized morality and has for this reason been subjected to savage criticism; one can be sure that Marx and Freud themselves would have censured parts of it rather severely. For example, the unshackling of Eros means a regression from exclusively genital sexuality and the institution of the monogamous patriarchal family to a more "primitive" polymorphous sexuality that takes the whole body as an instrument for pleasure. In the coming nonrepressive civilization many of the so-called sexual perversions will be freely practiced. But Marcuse envisages no "explosion" of the libido, and certainly none of the violent excesses that mark the behavior of civilized man when he is temporarily released from sexual restraint. Such excesses are an index to how powerfully he is repressed under "normal" circumstances. In a free society, sexuality will actually assume less importance than it has now, and much erotic energy will be freely sublimated by diffusion into other kinds of social relationships.

Such relationships will be characterized by nonsexual eroticism. People and animals will be loved, not treated as objects for profitable exploitation. At the same time, most labor will be abolished or made pleasurable; private property will disappear; disease will be conquered; war, deliberate waste, and commercial advertising will cease; and the world will become, as for the Arapesh of New Guinea described by Margaret Mead, "a garden that must be tilled, not for one's self, not in pride and boasting, not for hoarding and usury, but that the yams and the dogs and the pigs and most of all the children

347

may grow." In many ways, Marcuse's free civilization also revives the utopism of Charles Fourier—a point that has not escaped Marcuse himself.[18]

In the end, then, Marcuse provides us with yet another tableau of history as progress. Human evolution takes the form of a three-stage dialectical movement, from prehistoric freedom circumscribed by scarcity, to civilized repression during which scarcity is progressively eliminated, to the coming nonrepressive civilization of freedom fulfilled by abundance. The second stage may be regretted, but it is not absolutely evil, and certainly not dispensable: through the sufferings of the second stage, mankind arrives at a bliss inconceivable to his prehistoric ancestors except in myths and dreams.

But for Marcuse the idea of a nonrepressive civilization is in no sense mythical: it is real, it is possible, it is rational, and it is necessary, although not inevitable, given the destructive power now wielded by advanced industrial society, which cannot be wished away by mere words or good will. Marcuse prefers not to take anything for granted. He aligns himself with the forces of anti-bureaucratic revolutionary radicalism in the contemporary world, with Castro's Cuba, with the spirit of the Red Guard movement, with the Viet Cong in its struggle against American "neocolonialism," with the Paris rebels of 1968 and the campus radicals throughout the Western world. But he admits that recent events "refute all optimism." How and when a nonrepressive civilization can be born catastrophically, as it must be born, out of the womb of the present order, he does not know. He can do no better than quote the lines of Walter Benjamin: "It is only for the sake of those without hope that hope is given to us." [19]

From Marx and Darwin to Fromm and Marcuse we have traveled less than a hundred years in the spiritual history of Western man. If the journey has nevertheless been long, it is a tribute to the still unfailing springs of Western man's profound concern for the meaning of history, and to the depths of his caring and hoping. Even pessimism is a form of aspiration, for the authentically hopeless culture lives without fears or hopes, in a silence of resignation that Western man has never known.

At the same time, we cannot help being struck—indeed, overwhelmed—by the fecundity of the modern Western mind, and by the fantastic confusion into which it has plunged our culture. If the historian's highest calling is to discern a few clear, simple lines of development, to cut through morasses of data with a sword of deathly sureness, then *Good Tidings* has failed. I have done little more than survey and map the morasses, but I would answer that modern Western culture itself has failed: not in genius, but in the power to educe from its genius an organic order of spiritual life. It is a culture of glorious fragments. These fragments may form a new and higher unity in some future age, perhaps joining forces with what survives of the traditional cultures of Asia and Africa. Or mankind may be compelled to start over again, from new beginnings unforeseeable in the 1970s. Whatever happens, there is no spiritual center in our civilization today, no axis around which our thoughts revolve. We hope; we fear; ideas of progress and anti-progress mingle in fathomless confusion.

EPILOGUE

The Great Explosion

S IX YEARS HAVE PASSED since I set down the promise in the first
chapter to bring this book to a close with some thoughts of my
own on the question of human progress. These years have not been
empty, for me or the world. All the great social issues that tormented
mankind in the early 1960s, when I began my study of the idea of
progress, remain unsolved, but the climate of opinion has altered
significantly. Hope has returned as a fashionable category of Western
thought. Radicalism has become possible again for the very young—
if for no one else. The freezing spiritual temperatures of my under-
graduate days (1949 to 1953) no longer prevail, although the weather
is not yet tropical. By no means!

What is left to say about progress, after hearing the voices of so
many prophets? The easiest points to make are the negative ones. Let
us be brief and speak the truth. Every idea of progress is the result
of a private judgment of value for which no claim of objectivity has
the slightest merit. Prophets who make use of accurate and comprehensive empirical data proceed in a more rational manner than
prophets who know less of the phenomenal universe, but there is no
way to test the "truth-value" of their normative judgments. The best-
informed mind in the world might stray as far from moral "rectitude" as the least. The criterion of what is good is the moral insight
of the valuing self, or the valuing community, or whoever values:
not a standard external to the valuing subject.

Even if it were possible (which it is not) to establish objective
criteria for progress, could we measure anything as vast as the gen-

350

eral progress of mankind since the beginning of human time? One must answer very plainly: no! Too many facts are lost beyond recall. Take the question of whether there has been "improvement" in man's somatic and genetic material since *Pithecanthropus erectus.* Nearly all the knowledge we possess of the tissues of men who died before the present century must be derived by conjecture from miscellaneous skeletal remains; their behavior must be reconstructed from fragmentary historical records and artifacts. It is not enough. If one recalls that a precise knowledge of the physiology of intelligence and feeling still eludes science, what chance could there be of making accurate comparisons between the living and the dead? In any event, cultural circumstances change, and who can pretend to be able to compare objectively the acts or desires of ancient Gallic cave dwellers with the acts or desires of modern Parisians or Muscovites? Who would venture to measure and compare the sum of happiness experienced by representative individuals of Mousterian and Soviet culture? And when all the experiences and qualities of all men in all periods and civilizations must be measured, what hope can there be of answers acceptable to any science?

The apostles and opponents of the doctrines of progress and retrogression do no more than guess on the basis of prejudice and intuition. It is amusing, therefore, to encounter the claim that "gains" outweigh "losses" or vice versa, or that every "gain" is matched by a corresponding "loss." As if one could keep accurate accounts! It is no less amusing to be advised that "more" does or does not mean "better." If one assumes that a human life is absolutely precious, the needless loss of two lives will involve twice as much loss as the loss of one. If one assumes that a certain type of pleasurable sensation is absolutely good, twice as much will be twice as good. But it all depends on what one assumes as a valuing subject. The science of progress, if it means anything at all, requires bookkeeping—but on such a colossal scale, and with so many arbitrary judgments of value, that only a god could perform the tasks demanded. Men have a right to try their hands, but from our present perspective, no right to believe they can succeed.

A further limitation is the opacity of the future. Who has ever foreseen what will happen? Mankind perpetually wishes to know, and listens to self-styled prophets, but prophets invariably fail. Yet many ideas of progress depend on the thinker's presumed power to project old trends or predict new ones. Even if one is not certain, he speaks of "possibilities." But who knows what is possible, as a matter of cognition? Who can quote accurate odds? Bookmaking is no less hazardous than bookkeeping, when the destiny of all mankind is at stake.

Quite obviously, ideas of progress and retrogression are also relative to particular historical situations. They are conditioned by the age, childhood, education, and temperament of the thinker. They are conditioned by his milieu in time and space. Whether his mind was formed in the 1680s or 1790s or 1920s will make a tangible difference. Whether he was born a North American, a Frenchman, a German, or a non-Westerner will make a tangible difference. Certain periods encourage hope, others despair. What a man thinks progress consists of or does not consist of, and how progress can be accelerated or retarded is further conditioned by his training and his field of special competence. Biologists tend to propound biological theories of progress, historians focus on history, political scientists on politics. All this is exactly as it should be, so long as minds must content themselves with knowing and caring more about some aspects of human experience than others. But again, the result is something less than godlike objectivity.

One's choice of a scale of observation matters, too. The observer who studies only the last (or the next) few hundred years is likely to gain a quite different view of the curve of human evolution from that of the observer who carries his work through hundreds of millennia. Who can say how long progress must take place or continue to take place, in order to constitute the general progress of mankind?

Yet each thinker must plunge into the icy waters for himself, and swim as best he can. My scepticism about progress as an object of cognition is matched only by my willingness to speak of progress as an exercise in moral intuition. I am not alone in readiness to speak: every thinker studied in *Good Tidings* has shared my conceit. But for a convinced sceptic to render judgments about progress does seem to require a certain intellectual schizophrenia. If I cannot pretend to know, why should I dare to believe?

I can only answer that contemporary man's existential situation unequivocally demands such schizophrenia. As scholars, we know that we do not know whether progress has occurred or will occur. But as human beings, each of us must try to discover what history means to him. We must also have some sense of the human prospect, even though it may lack any cognitive significance whatsoever, if we are to attempt to bring our future under conscious control. We may please neither reason nor intellect by the judgments we make, but when we make none at all, we abandon our humanity.

With these massive reservations, I must still affirm my belief in the progress of mankind since the first appearance on earth of the genus *Homo*. I think it is even necessary to accuse certain modern nonbelievers of a kind of moral infantilism. We are all born weak and helpless; a feeling of inferiority to parents, ancestors, natural forces, and established cultures and institutions marks most men and most societies throughout history. But I am stubborn enough to suspect, without for a moment supposing I can prove, that such deference to the past in contemporary men is morally dishonest.

How has *Homo sapiens* progressed? Bodies and cerebral cortices and psyches have probably not improved, but I do venture to believe that we have progressed in knowledge of the world and ourselves, and in technical mastery of our environment. We have progressed in material wealth, personal comfort, security of life and limb, longevity, freedom from pain, and powers of perception, reasoning, and sensual enjoyment. We have progressed in individuation, self-awareness, and freedom of personal choice and thought. We have progressed in the richness and variety of our cultures, in the scope and sensitivity and quantity of our art, music, literature, philosophy, scholarship, and religion. We have progressed toward world unity and community. We have progressed toward equality of status and opportunity. We have progressed in the ideals and practice of peace, nonviolence, and brotherhood beyond the family and the clan.

I do not mean, of course, that every thought and deed of every twentieth-century man excels every thought and deed of every nineteenth-century or first-century or Paleolithic man; or that perfection has been reached in any department of human endeavor; or that radical change is not demanded if progress is to continue. Perfection has never been reached, and change is always demanded. But the level of human achievement, I believe, steadily rises. The total aesthetic culture, let us say, of modern Europe outstrips the total aesthetic culture of classical Greece and Rome, even though certain individual Greeks or Romans accomplished as much in their way as certain individual modern Europeans. The total aesthetic culture of Greece and Rome, in turn, outstripped the total aesthetic culture of the Bronze Age. The longer the spans of time considered, the easier it is to see progress. The shorter the spans, the easier it is to become confused by interludes of stagnation or decay. The smaller the geographical field of inquiry, the easier it is to be misled by the falling of whole regions into more or less permanent petrifaction. Yet the race advances: so it seems to me. *Eppur si muove*.

Again, I must insist that these are matters of personal belief, not of cognition. Knowledge has nothing to do with it. One believes that the music of Mozart, Mahler, and Stravinsky surpasses the music of the Middle Ages or ancient Greece, or he does not. One believes that modern men have more highly developed powers of empathy and a higher valuation of human life than earlier men, or he does not. One believes that the majority of persons enjoy greater opportunities for self-fulfillment in the twentieth century than they did in the tenth, or he does not. One believes that a longer expectancy of life or the harnessing of electric power or the progress of medicine and mathematics are good, or he does not.

Even those who are unprepared to rank the modern world above the medieval or classical or ancient worlds, or the modern West above the traditional East, may still accept the belief that progress has occurred on the ampler time scale of geology. Because rates of progress may vary along different lines of development, and because there are periods of retrogression or lateral movement without appreciable change for the better, it is sometimes difficult to convince oneself that the human race has experienced net progress over the course of only a few centuries. Yet no one but the most deeply entrenched primitivist can argue that mankind has not progressed since the founding of the great riverine civilizations of early antiquity, or since the close of the Paleolithic Age, or since the biological emergence of *Homo sapiens*.

We must also reckon with those modern thinkers who have resurrected the lament of the Renaissance and the Middle Ages that modernity not only fails to improve on antiquity, but even represents calculable decline, for one reason or another. Many good minds insist that demons are loose among us, whose crimes more than compensate for all the obvious good things of modern civilization. Who dares to write of progress in the century of Sarajevo, Auschwitz, Hiroshima, and Vietnam? Evil has never been so well-armed, and life and freedom have never been held more cheaply than by some authentic modern monsters. But for that matter, what of the atrocities of the past, the millennia of human slavery, the wars of religion, the massacres decreed by barbarian warlords and conquering emperors, the burnings, rapes, colosseum spectacles, persecutions, and ritual murders? Even the pages of the Bible and the Koran stink with the corpses of men, women, and children slain for the greater glory of Yahweh and Allah. Why not sweep all of history away, past as well as present, and strive for the annihilation of the will in timeless nothingness?

I am not convinced that any increase in human savagery has taken place in the twentieth century. In all likelihood it has diminished sharply, despite the immense power for destruction that technology and political circumstance have put at the disposal of a few men such as Hitler, Stalin, and Mao Tse-tung. That evil should be better armed is no proof that evil itself has grown more virulent or that more men have abandoned their humanity. Nor can I embrace the belief that modern men are more enslaved or crushed by conformism than their ancestors. Men have always tended to obey and conform. The tribe and the village are more efficient censors of freedom than the giant corporation and the bureaucratic national state. What dismays us, and justly so, is the degree to which freedom is still circumscribed, in new forms and by new agencies, despite the passing away of so many ancient wrongs. There has been progress, but there could and should have been much more. The very fact that modern man so vigorously lashes himself for his shortcomings is in itself evidence of the progress of moral sensibility.

What has happened in modern history is not a sudden plague of demons, but something far simpler and more humanly understandable. Even Hitler built splendid highways and rescued a stricken economy, even Stalin directed a vast program of industrialization and a great war of national defense, even Mao brought to his people a new and badly needed sense of social discipline and historical idealism. It is not a question of demons, but of unintended and unforeseeable growth too rapid and too complex to be assimilated by mortal men in the time allowed. The twentieth century is the age of the Great Explosion of mankind, when everything has happened and everything has been demanded and everything has been unveiled and unleashed all at once, in a matter of one or two generations. Beliefs and institutions have not kept pace with the Great Explosion of knowledge, power, commodities, people, desires, and wastes. Whether they can catch up in time often seems madly improbable. We are perhaps simultaneously at the zenith of history and on the brink of total and irreparable catastrophe.

But let us not resort to the contemptible *Schadenfreude* of the neo-Augustinian theologians, the obscurantists, and all the pious and aesthetic and mystical refugees from progress who detect a fatal moral insufficiency in the heart of man. For thousands of years we have listened to apologies for God, to theodicies that explain why God made the world so unlike the paradise that a benevolent deity could easily have bestowed upon his hapless creatures. There is no shame in turning the tables on these tiresome apologists and contriving anthropodicies that explain why a fallible and predatory race of recently evolved apes has not yet succeeded in building its own paradise.

Why should we not acknowledge that we have even done rather well, we who are not gods?

Notes
Index

NOTES

PART ONE

1. *Definitions*

1. Arthur O. Lovejoy, *The Great Chain of Being* (Cambridge, Mass., 1936), Lecture I.
2. J. B. Bury, *The Idea of Progress* (New York, 1932), p. 2.
3. Morris Ginsberg, *The Idea of Progress: A Revaluation* (London, 1953), p. 3; John Baillie, *The Belief in Progress* (London, 1950), pp. 2–3.
4. Bury, p. 5.
5. Ginsberg, pp. 42 and 71; Baillie, p. 40.
6. Ginsberg, p. 68; Baillie, pp. 155–56.
7. Bury, pp. 20–29; Ginsberg, pp. 7–8; and Baillie, pp. 94–96.
8. Bury, p. 4.
9. See Jules Delvaille, *Essai sur l'idée de progrès jusqu'à la fin du XVIIIe siècle* (Paris, 1910); R. V. Sampson, *Progress in the Age of Reason* (Cambridge, Mass., 1956); and Frank E. Manuel, *The Prophets of Paris* (Cambridge, Mass., 1962). A brief bibliographical guide to the history of the idea of progress is provided in Manuel, pp. 319–20.

2. *Origins*

1. See Frank E. Manuel, *Shapes of Philosophical History* (Stanford, Cal., 1965).
2. Thucydides, *The Peloponnesian War*, tr. Rex Warner (Baltimore, 1954), p. 24.
3. J. B. Bury, *The Idea of Progress* (New York, 1932), p. 6.
4. In addition to John Baillie, *The Belief in Progress* (London, 1950), see Karl Löwith, *Meaning in History* (Chicago, 1949): Reinhold Niebuhr, *Faith and History* (New York, 1949); Ernest Lee Tuveson, *Millennium and Utopia: A Study in the Background of the Idea of Progress* (Berkeley and Los Angeles, 1949); Eric Voegelin, *The New Science of Politics* (Chicago, 1952); Emil Brunner, *Eternal Hope*, tr. Harold Knight (London, 1954); Joseph Needham, *Time and Eastern Man* ([London], 1965); and Ludwig Edelstein, *The Idea of Progress in Classical Antiquity* (Baltimore, 1967).

5. See Theodor E. Mommsen, "St. Augustine and the Christian Idea of Progress," *Journal of the History of Ideas,* XII (June 1951), 346–74; also the discussion of Eusebius and Lactantius in C. N. Cochrane, *Christianity and Classical Culture* (New York, 1944), pp. 183–86 and 191–97.

6. See especially Baillie, p. 95; and Brunner, p. 10.

7. Löwith, p. 84.

8. See W. Warren Wagar, "Modern Views of the Origins of the Idea of Progress," *Journal of the History of Ideas,* XXVIII (January–March 1967), 55–70.

9. Fontenelle, "Digression sur les Anciens et les Modernes," in *Oeuvres de Fontenelle* (Paris, 1790), reprinted in Wagar, ed., *The Idea of Progress since the Renaissance* (New York, 1969), tr. Wagar, p. 49.

10. R. V. Sampson, *Progress in the Age of Reason* (Cambridge, Mass., 1956), ch. 2.

11. Peter Gay, *The Enlightenment: An Interpretation* (New York, 1966–69), II:100.

12. Cf. Henry Vyverberg, *Historical Pessimism in the French Enlightenment* (Cambridge, Mass., 1958); and Charles Vereker, *Eighteenth-Century Optimism* (Liverpool, 1967).

PART TWO

3. La Belle Époque

1. Gerhard Masur, *Prophets of Yesterday: Studies in European Culture, 1890–1914* (New York, 1961), p. 36.

2. Friedrich Nietzsche, *The Joyful Wisdom,* tr. Thomas Common, in Oscar Levy, ed., *The Complete Works of Friedrich Nietzsche* (Edinburgh, 1909–13), X:275–76. H. G. Wells used the same image to much the same effect in the closing pages of his novel *Tono-Bungay* (1909).

3. J. M. Robertson, *A History of Freethought in the Nineteenth Century* (London, 1929), II:391. See also Franklin L. Baumer, *Religion and the Rise of Scepticism* (New York, 1960), ch. 3.

4. H. Stuart Hughes, *Consciousness and Society: The Reorientation of European Social Thought, 1890–1930* (New York, 1958), especially ch. 2.

5. See Léon Bloy, *Au Seuil de l'apocalypse* (Paris, 1916).

4. The Cult of Science

1. Karl Pearson, *The Ethic of Freethought* (London, 1888), pp. 22–23, 25, and 32.

2. Two useful books on the Positivist movement are John Edwin McGee,

A Crusade for Humanity: The History of Organized Positivism in England (London, 1931); and W. M. Simon, *European Positivism in the Nineteenth Century* (Ithaca, N.Y., 1963).

3. Pierre Laffitte, *Cours de philosophie première* (Paris, 1889–94).

4. Laffitte, II:181, 189–90, and 193.

5. Laffitte, *The Positive Science of Morals,* tr. J. Carey Hall (London, 1908), pp. 9, 110, and 157.

6. Emile Corra, *La Philosophie positive* (Paris, 1904), pp. 13 and 91. See also Corra, *Les Enseignements philosophiques de la guerre* (Paris, 1915).

7. Frederic Harrison, *The Positive Evolution of Religion: Its Moral and Social Reaction* (London, 1913), p. 3; Harrison, *The Creed of a Layman: Apologia pro Fide Mea* (London, 1907), p. 54.

8. Harrison, *The Philosophy of Common Sense* (London, 1907), pp. 63–65 and 396.

9. See the examples published in *The Creed of a Layman*. Other important books produced by British Positivists include J. Cotter Morison, *The Service of Man: An Essay towards the Religion of the Future* (London, 1887); and F. S. Marvin, *The Living Past: A Sketch of Western Progress* (Oxford, 1913).

10. Simon, *European Positivism,* p. 263. See his ch. 9 for an excellent account of the reception of Positivism in Germany.

11. Ernst Haeckel, *The Riddle of the Universe,* tr. Joseph McCabe (New York, 1900), p. 14.

12. Arnold Zweig, *The Case of Sergeant Grischa,* tr. Eric Sutton (New York, 1928), p. 285.

13. Wilhelm Ostwald, "Der energetische Imperativ," *Annalen der Naturphilosophie,* X (1911), 114.

14. See Ostwald, *Natural Philosophy,* tr. Thomas Seltzer (London, 1911).

15. Max Nordau, *Degeneration* (New York, 1895), pp. 541–43.

16. Nordau, *The Interpretation of History,* tr. M. A. Hamilton (New York, 1911), especially chs. 8–10.

17. See Elie Metchnikoff, *The Nature of Man: Studies in Optimistic Philosophy,* ed. P. Chalmers Mitchell (London, 1904); and Metchnikoff, *The Prolongation of Life: Optimistic Studies,* ed. P. Chalmers Mitchell (London, 1907).

18. Alfred Russel Wallace, *Social Environment and Moral Progress* (London, 1913), p. 132.

19. See Wallace, *The World of Life: A Manifestation of Creative Power, Directive Mind and Ultimate Purpose* (London, 1910).

20. Arthur James Todd, *Theories of Social Progress* (New York, 1918), p. vii.

21. W. K. Clifford, "Cosmic Emotion," in *Lectures and Essays,* ed. Leslie Stephen and Frederick Pollock, second edition (London, 1886), p. 417.

22. Guillaume De Greef, *Le Transformisme social* (Paris, 1895), pp. 8–306.

23. Ibid., p. 513; and De Greef, *Problèmes de philosophie positive* (Paris, 1900), p. x. See also De Greef, *La Structure générale des sociétés* (Paris, 1908).

24. Julius Lippert, *The Evolution of Culture* [1886–87], tr. George Peter Murdock (London, 1931), p. 2.

25. Sir James Frazer, *The Golden Bough,* abridged edition (New York, 1922), p. 712.

26. Franz Müller-Lyer, *The History of Social Development,* tr. Elizabeth Coote Lake and H. A. Lake (London, 1920), pp. 310 and 320.

27. Ibid., pp. 355–56. See also Müller-Lyer, *Die Entwicklungsstufen der Menschheit* (Munich, 1923–24).

28. Ludwig Gumplowicz, *La Lutte des races,* tr. Charles Beye (Paris, 1893), p. 346.

29. Jacques Novicow, *La Politique internationale* (Paris, 1886), p. 365.

30. Howard Becker in Becker and Harry Elmer Barnes, *Social Thought from Lore to Science,* third edition (New York, 1961), II:719–21; Don Martindale, *The Nature and Types of Sociological Theory* (Boston, 1960), pp. 69–72; and Richard Hofstadter, *Social Darwinism in American Thought,* revised edition (New York, 1959), p. 78.

31. Lester Ward, *Pure Sociology* (New York, 1903), p. 91.

32. Ibid., pp. 204 and 238–40.

33. Ward, *The Psychic Factors of Civilization* (Boston, 1893), p. 135.

34. *Pure Sociology,* pp. 469–71.

35. Ibid., p. 450. Cf. Charles Letourneau, *L'Evolution de la morale* (Paris, 1894).

36. Ward, *Dynamic Sociology* (New York, 1910), II:249–50.

37. *Pure Sociology,* p. 135.

38. Thomas Henry Huxley, *Evolution and Ethics* (London, 1894), p. 83.

39. Huxley, *Essays upon Some Controverted Questions* (London, 1892), pp. 235–36.

40. *Evolution and Ethics,* p. 36 fn.

41. Ibid., p. 44; *Essays upon Some Controverted Questions,* pp. 48–49.

42. *Evolution and Ethics,* pp. 44 and 85–86.

43. *Essays upon Some Controverted Questions,* p. 371.

44. H. G. Wells, Preface, *Seven Famous Novels* (New York, 1934), p. ix.

45. For a fuller treatment of Wells's thinking on progress, see W. Warren Wagar, *H. G. Wells and the World State* (New Haven, Conn., 1961), pp. 76–87 and passim.

46. L. T. Hobhouse, *Development and Purpose: An Essay towards a Philosophy of Evolution* (London, 1913), pp. xv–xvi.

47. Ibid., p. 208.

48. Hobhouse, *Mind in Evolution,* second edition, London, 1915, p. 442.

49. Hobhouse, *Morals in Evolution,* revised edition, London, 1915, p. 637.

50. Ibid., p. 634.

51. *Development and Purpose,* p. xxix.

52. Ibid., pp. 371–72.

53. Hobhouse, *Social Development: Its Nature and Conditions* (London, 1924), pp. 336–37 and 339–40.

5. *The Will to Power*

1. Alfred Fouillée, *La Liberté et le déterminisme* (Paris, 1872), p. 190.
2. Walter Kaufmann, *Nietzsche: Philosopher, Psychologist, Antichrist* (Princeton, N.J., 1950), p. 7.
3. Eric Fenby, *Delius As I Knew Him,* revised edition (London, 1966), p. 171.
4. See, in addition to Kaufmann, Karl Jaspers, *Nietzsche: An Introduction to the Understanding of His Philosophical Activity,* tr. Charles F. Wallraff and Frederick J. Schmitz (Tucson, 1965); Martin Heidegger, *Nietzsche* (Pfullingen, 1961); and F. A. Lea, *The Tragic Philosopher: A Study of Friedrich Nietzsche* (London, 1957). An interpretation stressing Nietzsche's "aesthetic pessimism" is Charles Andler's *Nietzsche: sa vie et sa pensée* (Paris, 1958). For Nietzsche and the Nazis, consult Fritz Stern, *The Politics of Cultural Despair: A Study in the Rise of the Germanic Ideology* (Garden City, N.Y., 1965), pp. 344–50.
5. Nietzsche, *The Antichrist,* tr. Anthony M. Ludovici, in Oscar Levy, ed., *The Complete Works of Friedrich Nietzsche* (Edinburgh, 1909–13), XVI:129. For Kaufmann's argument see his *Nietzsche,* pp. 270–92.
6. George Allen Morgan, *What Nietzsche Means* (Cambridge, Mass., 1941), ch. 13.
7. See Walter Kaufmann's new edition of *The Will to Power,* tr. Kaufmann and B. J. Hollingdale (New York, 1967).
8. Nietzsche, *Beyond Good and Evil,* tr. Helen Zimmern, in *Complete Works,* XII:234.
9. *The Antichrist,* in *Complete Works,* XVI:129.
10. See Morgan, ch. 12.
11. Nietzsche, *The Will to Power,* tr. Anthony M. Ludovici, in *Complete Works,* XV:317 and 378.
12. *Beyond Good and Evil,* in *Complete Works,* XII:129–30.
13. Wilhelm Wundt, *Ethics,* tr. Julia Gulliver, Edward Bradford Titchener, and Margaret Floy Washburn (London, 1897–1901), III:90. See also Wundt, *System der Philosophie* (Leipzig, 1889), pp. 591–626, on the collective will and the meaning of history. Eucken's philosophy of progress may be studied in several of his books, including *Life's Basis and Life's Ideal,* tr. Alban G. Widgery (London, 1912); *Main Currents of Modern Thought,* tr. Meyrick Booth (London, 1912); *Knowledge and Life,* tr. W. Tudor Jones (London, 1913); and *The Truth of Religion,* tr. W. Tudor Jones (London, 1913).
14. See Hermann Cohen, *Ethik des reinen Willens* (Berlin, 1904); and *Die Religion der Vernunft aus den Quellen des Judentums* (Leipzig, 1919).
15. Fouillée, *Les Eléments sociologiques de la morale* (Paris, 1905), p. 339.
16. Emile Boutroux, *Science and Religion in Contemporary Philosophy,* tr. Jonathan Nield (London, 1912), pp. 373 and 400. See also his lecture against pessimism in *Education and Ethics,* tr. Fred Rothwell (London, 1913), pp. 80–113.

17. Jean-Marie Guyau, *The Non-Religion of the Future* (London, 1897), pp. 11 and 13.

18. Ibid., pp. 16 and 425. Cf. Guyau, *A Sketch of Morality Independent of Obligation or Sanction,* tr. Gertrude Kapteyn (London, 1898), p. 214.

19. *A Sketch of Morality,* p. 215.

20. Ibid., p. 211.

21. Ibid., pp. 211–12 and 87.

22. *The Non-Religion of the Future,* pp. 476, 211, 499, and 533.

23. Henri Bergson, *Creative Evolution,* tr. Arthur Mitchell (New York, 1911), p. 100.

24. Ibid., p. 271.

25. Gerhard Masur, *Prophets of Yesterday: Studies in European Culture, 1890–1914* (New York, 1961), pp. 263–64.

26. Bergson, *The Two Sources of Morality and Religion,* tr. R. Ashley Audra and Cloudesley Brereton (Garden City, N.Y., 1956), pp. 205, 213, and 255–56.

27. Ibid., p. 317.

28. Léon Brunschvicg, *Le Progrès de la conscience dans la philosophie occidentale* (Paris, 1927), ch. 21.

29. Samuel Alexander, *Space, Time, and Deity* (London, 1920), II:346 and 429. Cf. C. Lloyd Morgan's 1922–23 Gifford Lectures, published in two volumes: *Emergent Evolution* (London, 1923), and *Life, Mind, and Spirit* (London, 1926).

30. J. C. Smuts, *Holism and Evolution* (New York, 1926), pp. 318 and 345. Cf. Herbert F. Standing, *Spirit in Evolution: From Amoeba to Saint* (London, 1930).

31. Bernard Shaw, *Three Plays for Puritans* (London, 1901), p. 202.

32. Shaw, *Back to Methuselah: A Metabiological Pentateuch* (London, 1921), p. xxxvii.

33. Shaw, *Man and Superman* (London, 1903), pp. 105, 129, and 133.

34. Ibid., pp. 134 and 104.

35. *Back to Methuselah,* p. 8.

36. Ibid., pp. 266–67.

37. William Jennings Bryan and Mary Baird Bryan, *The Memoirs of William Jennings Bryan* (Philadelphia, 1925), p. 501.

38. C. S. Peirce, "Evolutionary Love," in Justus Buchler, ed., *Philosophical Writings of Peirce* (New York, 1955), p. 364.

39. William James, *The Will to Believe* (New York, 1897), pp. 100 and 103.

40. James, *A Pluralistic Universe* (New York, 1909), p. 124.

41. James, *Talks to Teachers on Psychology* (New York, 1899), p. 300.

42. John Dewey, "Progress," in *Characters and Events* (New York, 1929), II:822–23.

43. Ibid., II:824, 826, and 823.

44. Dewey, *Human Nature and Conduct* (New York, 1930), pp. 285 and 288–89.

6. The Theology of Progress

1. H. G. Wells, *God the Invisible King* (New York, 1917), pp. 64–65.

2. James P. Martin, *The Last Judgment in Protestant Theology from Orthodoxy to Ritschl* (Edinburgh, 1963), p. 201. On Ritschl, see also A. E. Garvie, *The Ritschlian Theology* (Edinburgh, 1899).

3. Wilhelm Herrmann, *Die evangelische Glaube und die Theologie Albrecht Ritschls* (Marburg, 1890), p. 20. Herrmann's major works include *The Communion of the Christian with God,* tr. J. S. Stanyon (London, 1895); and *Faith and Morals,* tr. D. Matheson and R. W. Stewart (London, 1904).

4. Adolf von Harnack, *What Is Christianity?,* tr. Thomas Bailey Saunders (New York, 1901), p. 8.

5. Ibid., pp. 108 and 122–23. See also Harnack and Herrmann, *Essays on the Social Gospel,* tr. G. M. Craik, ed. M. A. Canney (London, 1903).

6. See Bernard Holland, ed., *Selected Letters of Baron Friedrich von Hügel, 1896–1924* (London, 1927), pp. 15–16.

7. Alfred Loisy, *The Gospel and the Church,* tr. Christopher Home (London, 1903), pp. 170–71.

8. Loisy, *La Morale humaine* (Paris, 1923), p. 219.

9. Loisy, *Religion et humanité* (Paris, 1926), p. 251.

10. Loisy to Maude Petre, October 23, 1917, in Petre, *Alfred Loisy: His Religious Significance* (Cambridge, 1944), p. 119.

11. See Maurice Blondel, *L'Action* (Paris, 1893).

12. Blondel, *Lutte pour la civilisation et philosophie de la paix* (Paris, 1939), pp. 164, 198, and 303.

13. George Tyrrell, *Christianity at the Crossroads* (London, 1910), pp. 41–42.

14. Ibid., pp. xxi, 177, 166, 50, and 70.

15. Ibid., pp. 121, 124, and 157.

16. Ibid., pp. 134, 197, and 235–36.

17. Tyrrell, *Through Scylla and Charybdis: The Old Theology and the New* (London, 1907), p. 181.

18. Henry Drummond, *The Ascent of Man* (New York, 1894), pp. 335 and 340.

19. Ibid., pp. 341, 343, and 345–46.

20. R. J. Campbell, *The New Theology* (London, 1907), pp. 15 and 23–24.

21. Ibid., pp. 61–63.

22. Campbell, *Christianity and the Social Order* (London, 1907), p. 149.

23. Campbell, *A Spiritual Pilgrimage* (London, 1916), p. 277.

24. Quoted in Arthur S. Bolster, Jr., *James Freeman Clarke* (Boston, 1954), p. x.

25. Quoted in David B. Parke, ed., *The Epic of Unitarianism* (Boston, 1957), p. 124.

26. Lyman Abbott, *The Theology of an Evolutionist* (New York, 1897), pp. iii, 9, 13, and 191.

27. Francis Howe Johnson, *What Is Reality?* (Boston, 1891), p. 362.

28. Johnson, *God in Evolution* (New York, 1911), pp. v, 3, and 260–61.

29. Ibid., pp. 37, 95, and 91–92.

30. Shailer Mathews, *The Spiritual Interpretation of History* (Cambridge, Mass., 1916), pp. 216–17.

31. Mathews, *The Faith of Modernism* (New York, 1924), p. 167. Cf. Mathews, *The Gospel and the Modern Man* (New York, 1910), p. 327.

32. Walter Rauschenbusch, *Christianity and the Social Crisis* (New York, 1907), pp. 41, 59, and 60–61.

33. Ibid., pp. 142 and 422.

34. Rauschenbusch, *A Theology for the Social Gospel* (New York, 1919), pp. 142–43, 224, and 227.

35. Harry Emerson Fosdick, *Christianity and Progress* (New York, 1922), pp. 246–47.

7. Progress and Politics

1. Gustave de Molinari, *Les Problèmes du XXe siècle* (Paris, 1901), p. 302. See also W. E. H. Lecky, *Democracy and Liberty* (London, 1896); and Sir Henry Maine, *Popular Government* (London, 1885). Late Victorian suspicions of democracy and socialism took another form in the writings of W. H. Mallock, who attributed all human progress to the work of "great men." See his *Aristocracy and Evolution* (London, 1898).

2. Gertrude Himmelfarb in Lord Acton, *Essays on Freedom and Power,* ed. Himmelfarb (Boston, 1948), pp. xxxiii and xlv–xlvi. See also Himmelfarb, *Lord Acton* (London, 1952), pp. 237–41.

3. Acton, p. 25.

4. Ibid., pp. 193 and 32, from essays originally published in 1862 and 1877.

5. Ibid., pp. 12–13, 15, 7, and 29.

6. Ibid., p. 159.

7. Ibid., pp. 193–95.

8. Heinrich von Treitschke, *Politics,* tr. Blanche Dugdale and Torben de Bille, ed. Hans Kohn (New York, 1963), pp. xxiii–xxv.

9. Ibid., pp. 40 and 45.

10. Ibid., pp. 293 and 55–56.

11. Ibid., pp. 12, 14, and 16. See also p. 40.

12. Friedrich Naumann, *Geist und Glaube* (Berlin, 1911), p. 60.

13. Naumann, *Das blaue Buch von Vaterland und Freiheit* (Königstein im Taunus and Leipzig, 1913), p. 84.

14. Ibid., pp. 253–54 and 262.

15. Naumann, *Central Europe* [*Mitteleuropa*], tr. Christabel M. Meredith (London, 1916), pp. 194–95 and 204.

16. Ibid., p. 179.

17. Léon Bourgeois, *Solidarité* (Paris, 1902), p. 99.

18. Ibid., pp. 100–101 and 123.

19. Ibid., pp. 129–30.

20. L. T. Hobhouse, *Liberalism,* London, [1911], p. 137.

21. See Hobson's tribute to Hobhouse in J. A. Hobson, *Confessions of an Economic Heretic* (London, 1938), pp. 75–79.

22. Hobson, *Imperialism,* third edition (London, 1938), p. 367.

23. Ibid., pp. 184 and 189.

24. Ibid., p. 169.

25. Ibid., p. 185.

26. Ibid., pp. 237 and 230.

27. *Confessions of an Economic Heretic,* pp. 160–61 and 163.

28. See David W. Noble, "*The New Republic* and the Idea of Progress, 1914–1920," *Mississippi Valley Historical Review,* XXXVIII (December 1951), 387–402.

29. Herbert Croly, *The Promise of American Life* (New York, 1909), pp. 15–16.

30. Ibid., p. 284.

31. Ibid., p. 454.

32. Walter Weyl, *The New Democracy* (New York, 1912), pp. 351 and 354–55.

33. Ibid., p. 41. Cf. Croly, pp. 21–22.

34. Edward P. Cheyney, *Law in History* (New York, 1927), pp. 18, 20, 23–24.

35. See James Harvey Robinson, *The Mind in the Making* (New York, 1921).

36. Lord Milner, *The Nation and the Empire* (London, 1913), p. 496. See also J. A. Cramb, *The Origins and Destiny of Imperial Britain* (London, 1900).

37. John W. Burgess, *Political Science and Comparative Constitutional Law* (Boston, 1890), I:39 and 44.

38. Ibid., I:46.

39. Ibid., I:85 and 89.

40. Josiah Strong, *Our Country* (New York, 1885), p. 6.

41. Ibid., pp. 161 and 165.

42. Ibid., pp. 175 and 177.

43. Ibid., pp. 177–78 and 180.

44. Karl Kautsky, *Ethics and the Materialist Conception of History,* tr. John B. Askew (Chicago, 1914), p. 120.

45. Kautsky, *Die materialistische Geschichtsauffassung* (Berlin, 1927), II:835.

46. Kautsky, *The Labour Revolution,* tr. H. J. Stenning (London, 1925), p. 282.

47. *Die materialistische Geschichtsauffassung,* II:837–38.

48. The fullest statement of Kautsky's thoughts on progress is to be found in ibid., II:741–845. See also the discussion of "progress and adaptation" in I:401–406.

49. Eduard Bernstein, *Evolutionary Socialism,* tr. Edith C. Harvey (London, 1909), pp. 13–14.

50. Bernstein, *Wie ist wissenschaftlicher Sozialismus möglich?* (Berlin, 1901), p. 35.

51. *Evolutionary Socialism,* pp. viii–ix.

52. Peter Gay, *The Dilemma of Democratic Socialism: Eduard Bernstein's Challenge to Marx* (New York, 1952), p. 136.

53. Jean Jaurès, *Pages choisies* (Paris, 1922), pp. 226 and 236.

54. See especially "L'Idéalisme de l'histoire," ibid., pp. 358–74.

55. Jaurès, *Studies in Socialism,* tr. Mildred Minturn (New York, 1906), pp. 160–61.

56. Ibid., pp. 168–69.

57. Edward R. Pease, *The History of the Fabian Society,* revised edition (New York, 1926), pp. 90–91.

58. Sydney Olivier, "The Moral Basis of Socialism," in G. Bernard Shaw, ed., *Fabian Essays in Socialism* (Boston, 1908), p. 116.

59. Sidney Webb, "The Historic Basis of Socialism," *Fabian Essays in Socialism,* p. 27.

60. Henry George, *Progress and Poverty* (Garden City, N.Y., 1926), pp. 476 and 478.

61. Ibid., pp. 504–505.

62. Ibid., p. 525.

63. Ibid., p. 482.

64. Ibid., pp. 526–27.

65. Ibid., p. 540.

66. Ibid., p. 549. Cf. p. 327.

67. Georges Sorel, *Les Illusions du progrès,* second edition (Paris, 1911), p. 265.

68. Syndicalist utopography is well illustrated by Emile Pataud and Emile Pouget, *Syndicalism and the Co-operative Commonwealth,* tr. Charlotte and Frederic Charles (Oxford, 1913).

69. G. V. Plekhanov, *The Role of the Individual in History* [1898] (New York, 1940), pp. 58–59.

70. Plekhanov, *Fundamental Problems of Marxism* [1908], tr. Eden and Cedar Paul (New York, n.d.), p. 93.

71. See Plekhanov, *The Development of the Monist View of History* [1895], tr. Andrew Rothstein, in Plekhanov, *Selected Philosophical Works* (Moscow, n.d.), I:625–30.

72. Alfred G. Meyer, *Leninism* (New York, 1962), p. 247 fn.

73. V. I. Lenin, *The State and Revolution* [1918], in Lenin, *Selected Works* (New York, 1943), VII:80–81.

74. Ibid., VII:75 and 88.

PART THREE

8. *The Decline of Hope*

1. William Barrett, *Irrational Man: A Study in Existential Philosophy* (Garden City, N.Y., 1957), p. 28.
2. Carl Gustav Jung, "The Spiritual Problem of Modern Man" (1928), in *Civilization in Transition,* tr. R. F. C. Hull (New York, 1964), p. 77.
3. Sigrid Undset to Stanley J. Kunitz and Howard Haycraft, in Kunitz and Haycraft, eds., *Twentieth Century Authors* (New York, 1942), p. 1433.
4. H. Stuart Hughes, *Oswald Spengler,* revised edition (New York, 1962), p. 165.
5. Emil Brunner, *Eternal Hope,* tr. Harold Knight (London, 1954), p. 10.

9. *Romanticism, Positivism, and Despair*

1. Recent scholarly studies of nineteenth-century pessimism include Judith N. Shklar, *After Utopia: The Decline of Political Faith* (Princeton, N.J., 1957), chs. 1–3; Hans J. Schoeps, *Vorläufer Spenglers: Studien zum Geschichtspessimismus im 19. Jahrhundert* (Leiden, 1953); A. E. Carter, *The Idea of Decadence in French Literature, 1830–1900* (Toronto, 1958); and Koenraad W. Swart, *The Sense of Decadence in Nineteenth-Century France* (The Hague, 1964). See also Fritz Stern, *The Politics of Cultural Despair: A Study in the Rise of the Germanic Ideology* (Berkeley and Los Angeles, 1961), especially the Introduction and Conclusion.
2. See Jacques Barzun's discussion of romanticism as a movement of reconstruction in his *Classic, Romantic and Modern* (Boston, 1961). Cf. Eugene N. Anderson, "German Romanticism as an Ideology of Cultural Crisis," *Journal of the History of Ideas,* II (June, 1941), 301–17.
3. Joseph de Maistre, *The Saint Petersburg Dialogues* (1821), in Jack Lively, ed. and tr., *The Works of Joseph de Maistre* (New York, 1965), p. 253.
4. Prince Richard Metternich, ed., *Memoirs of Prince Metternich, 1773–1835,* tr. Mrs. Alexander Napier (London, 1880–82), III:456–57. For a view of Metternich as the "last champion" of the Enlightenment, see Henry A. Kissinger, "The Conservative Dilemma: Reflections on the Political Thought of Metternich," *American Political Science Association Review,* XLVIII (December 1954), 1017–30.
5. Metternich, in *Aus Metternich's nachgelassenen Papieren* (Vienna, 1880–89), III:347, as quoted in Kissinger, p. 1029.
6. Arthur J. Balfour, "A Fragment on Progress," in *Essays and Addresses* (Edinburgh, 1893), pp. 241 and 244.

7. Ibid., p. 279; and "The Religion of Humanity," in *Essays and Addresses,* p. 292.

8. See Mario Praz, *The Romantic Agony,* tr. Angus Davidson, second edition (London, 1951).

9. Théophile Gautier, *Mademoiselle de Maupin* (Paris, 1928), pp. 28–29.

10. Charles Baudelaire, "Exposition universelle de 1855," in *Oeuvres complètes* (Paris, n.d.), II:218–19.

11. J.-K. Huysmans, *Against the Grain,* tr. John Howard (New York, 1931), pp. 338–39.

12. See Hans Kohn, *The Mind of Germany* (New York, 1960), pp. 244–51.

13. See Morse Peckham, *Beyond the Tragic Vision: The Quest for Identity in the Nineteenth Century* (New York, 1962), p. 173.

14. Arthur Schopenhauer, *The World as Will and Idea,* tr. R. B. Haldane and J. Kemp, eighth edition (London, n.d.), III:227.

15. Edgar Evertson Saltus, *The Philosophy of Disenchantment* (Boston, 1885), p. 160.

16. See Eduard von Hartmann, *Philosophy of the Unconscious,* tr. William Chatterton Coupland, second edition (London, 1893), especially III:120–42. A more popular introduction to Hartmann's thought is his essay, "The Comforts of Pessimism," in *The Sexes Compared, and Other Essays,* tr. A. Kenner (London, 1895), pp. 68–101.

17. Sören Kierkegaard, *Concluding Unscientific Postscript,* tr. David F. Swenson and Walter Lowrie (Princeton, N.J., 1941), p. 141. For a good example of Kierkegaard's thoughts on progress, see his sarcastic "Eulogy upon the Human Race," in *Kierkegaard's Attack upon Christendom, 1854–1855,* tr. Walter Lowrie (Princeton, N.J., 1944), pp. 105–106.

18. Jacob Burckhardt, *Force and Freedom: Reflections on History,* tr. Mary D. Hottinger, ed. James Hastings Nichols (New York, 1943), p. 352.

19. See ibid., especially pp. 359–63.

20. Ibid., pp. 149–50.

21. Burckhardt, *On History and Historians,* tr. Harry Zohn (New York, 1965), p. 3.

22. See Oswald Spengler, "Nietzsche und sein Jahrhundert," in *Reden und Aufsätze* (Munich, 1937), pp. 110–24; and Eberhard Gauhe, *Spengler und die Romantik* (Berlin, 1937).

23. See Spengler, *The Decline of the West,* tr. Charles Francis Atkinson (New York, 1926–28), especially the tables following I:428; and the convenient summaries in Hughes, *Oswald Spengler,* ch. 5, and in Albert William Levi, *Philosophy and the Modern World* (Bloomington, Ind., 1959), pp. 117–26.

24. *The Decline of the West,* I:40–41 and II:507.

25. Danilevsky's book has never been translated into English. A good introduction to his thought may be found in Pitirim A. Sorokin, *Social Philosophies of an Age of Crisis* (Boston, 1950), pp. 49–71.

26. Brooks Adams, *The Law of Civilization and Decay* (New York, 1943), p. 59.

27. Ibid., p. 60.

28. Ibid., p. 349.

29. Brooks Adams, "Introductory Note" and "The Heritage of Henry Adams," in Brooks and Henry Adams, *The Degradation of the Democratic Dogma* (New York, 1919), pp. vii and 86.

30. Ernest Millard, *Une Loi historique* (Brussels, 1903–08), I:9. For another pre-1914 cyclical theory of history, see W. M. F. Petrie, *The Revolutions of Civilization* (London, 1912).

31. See the excellent discussion of this theme in J. H. Randall, Jr., *The Making of the Modern Mind,* revised edition (Boston, 1940), ch. 21.

32. Marquis de Sade, *Juliette,* tr. Austryn Wainhouse (New York, 1968), p. 401.

33. Sade, *La nouvelle Justine* (Sceaux, 1953), I:152.

34. See Loren Eiseley, *Darwin's Century* (Garden City, N.Y., 1958), ch. 9.

35. H. G. Wells, *The Time Machine,* in *Three Prophetic Novels,* ed. E. F. Bleiler (New York, 1960), p. 335.

36. Wells, *Mind at the End of Its Tether,* in G. P. Wells, ed., *The Last Books of H. G. Wells* (London, 1968), pp. 73 and 77. See Anthony West, "The Dark World of H. G. Wells," *Harper's,* CCXIV (May 1957), 68–73.

37. Anatole France, *The Garden of Epicurus* (New York, n.d.), pp. 61, 16, and 18.

38. See Carter Jefferson, *Anatole France: The Politics of Skepticism* (New Brunswick, N.J., 1965). For an example of the *fin-du-monde* theme in the popular scientific literature of *la belle époque,* see Camille Flammarion, *Astronomie populaire,* first published in 1879. Flammarion's description of mankind's final years closely resembles that of France. It is quoted at length in translation by Henry Adams in *The Degradation of the Democratic Dogma,* pp. 182–84.

39. See *The Degradation of the Democratic Dogma,* especially p. 308.

40. Ibid., pp. 157–58.

41. Roger B. Salomon, *Twain and the Image of History* (New Haven, Conn., 1961).

42. Madison Grant, *The Passing of the Great Race, or The Racial Basis of European History* (New York, 1916), p. 228.

43. Charles H. Pearson, *National Life and Character: A Forecast,* second edition (London, 1894), pp. 275, 355, and 363.

44. Gustave Flaubert, *Bouvard and Pécuchet,* tr. T. W. Earp and G. W. Stonier (Norfolk, Conn., 1954), pp. 344–46.

10. *The Relativity of Values*

1. Charles Frankel, *The Love of Anxiety* (New York, 1965), p. 73. Cf. Franklin L. Baumer, in Baumer, ed., *Main Currents of Western Thought* (New York, 1952), pp. 577–85.

2. Erich Kahler, *The Meaning of History* (New York, 1964), p. 175; Georg G. Iggers, "The Idea of Progress: A Critical Reassessment," *American Historical Review,* LXXI (October 1965), 8–9. Iggers' quotation is from

Ernst Troeltsch, *Die Absolutheit des Christentums und die Religionsgeschichte* (Tübingen, 1902), p. 49. Both Kahler and Iggers prefer to call the earlier *Historismus* "historism" and the more recent variety "historicism."

3. Leopold von Ranke, *Weltgeschichte* (Leipzig, 1888–1902), IX, tr. Moltke S. Gram, in Alan Donagan and Barbara Donagan, eds., *Philosophy of History* (New York, 1965), pp. 72–75.

4. See R. G. Collingwood, *The Idea of History* (Oxford, 1946), Part IV; Karl Heussi, *Die Krisis des Historismus* (Tübingen, 1932); Carlo Antoni, *From History to Sociology: The Transition in German Historical Thinking,* tr. Hayden V. White (Detroit, 1959); Pietro Rossi, *Lo storicismo tedesco contemporaneo* (Turin, 1956); Georg G. Iggers, "The Dissolution of German Historism," in Richard Herr and Harold T. Parker, eds., *Ideas in History* (Durham, N.C., 1965), pp. 288–329; and Iggers, *The German Conception of History* (Middletown, Conn., 1968), especially chs. 6–7.

5. Wilhelm Dilthey, *Pattern and Meaning in History,* ed. and tr. H. P. Rickman (New York, 1962), pp. 167–68. Cf. Dilthey's lecture, "Traum," in his *Gesammelte Schriften* (Leipzig, 1914–36), VIII:218–24, available in English in William Kluback, *Wilhelm Dilthey's Philosophy of History* (New York, 1956), pp. 103–109.

6. Rickman, General Introduction, in *Pattern and Meaning in History,* p. 26, and see pp. 56–59.

7. Quoted in Antoni, *From History to Sociology,* p. 75.

8. Ernst Troeltsch, *The Social Teaching of the Christian Churches,* tr. Olive Wyon (London, 1931), II:1012–13.

9. Troeltsch, *Christian Thought: Its History and Application,* ed. Friedrich von Hügel (London, 1923), p. 26.

10. H. Stuart Hughes, *Consciousness and Society* (New York, 1958), pp. 240–41. Cf. the similar judgment in Antoni, pp. 74–77.

11. Friedrich Meinecke, "Kausalitäten und Werte in der Geschichte," in *Staat und Persönlichkeit* (Berlin, 1933), tr. Julian H. Franklin in Fritz Stern, ed., *The Varieties of History* (New York, 1956), pp. 288 and 411.

12. See his last testament, *The German Catastrophe: Reflections and Recollections,* tr. Sidney B. Fay (Cambridge, Mass., 1950).

13. Benedetto Croce, *History as the Story of Liberty,* tr. Sylvia Sprigge (New York, 1941), p. 35.

14. Croce, *History: Its Theory and Practice,* tr. Douglas Ainslie (New York, 1960), p. 59.

15. Ibid., pp. 84–85 and 90.

16. *History as the Story of Liberty,* pp. 278–79.

17. Ibid., p. 52, and see also ch. 8, "Historiography as Liberation from History."

18. *History: Its Theory and Practice,* p. 85.

19. See Part II, "Concerning the History of Historiography," ibid.; Croce, *History of Europe in the Nineteenth Century,* tr. Henry Furst (New York, 1933); and A. Robert Caponigri, *History and Liberty: The Historical Writings of Benedetto Croce* (London, 1955), Part III.

20. *History as the Story of Liberty,* pp. 59 and 61.

21. Hughes, *Consciousness and Society,* p. 225.

22. Ibid., p. 226.

23. See Collingwood, "Progress as Created by Historical Thinking," in *The Idea of History,* pp. 321–24.

24. Karl R. Popper, *The Poverty of Historicism* (Boston, 1957), p. 3. See Dwight E. Lee and Robert N. Beck, "The Meaning of 'Historicism'," *American Historical Review,* LIX (April 1954), 568–77; and Hans Meyerhoff's rebuke of Popper in Meyerhoff, ed., *The Philosophy of History in Our Time* (Garden City, N.Y.), 1959, pp. 299–300.

25. Popper, *The Open Society and Its Enemies* (Princeton, N.J., 1950), pp. 462–63.

26. Ibid., p. 463.

27. *The Poverty of Historicism,* pp. 150–51.

28. H. A. L. Fisher, *A History of Europe,* revised edition (Boston, 1939), p. ix.

29. G. R. Elton, *The Practice of History* (New York, 1968), p. 46.

30. J. B. Bury, *The Idea of Progress* (New York, 1932), p. 352. Bury's Epilogue is quoted and similar conclusions are reached by Carl L. Becker in his article on "Progress" in the *Encyclopaedia of the Social Sciences* (New York, 1933), XII:499.

31. Edward Westermarck, *Ethical Relativity* (New York, 1932), p. 215.

32. H. R. Hays, *From Ape to Angel: An Informal History of Social Anthropology* (New York, 1958), p. 261.

33. Franz Boas, *Race, Language, and Culture* (New York, 1940), p. 268.

34. Ibid., pp. 286 and 310–11.

35. Boas, *The Mind of Primitive Man* (New York, 1911), p. 207.

36. Ibid.

37. Boas, *Race and Democratic Society* (New York, 1945), pp. 170–71 and 143.

38. Boas, "Anthropology," in *Encyclopaedia of the Social Sciences* (New York, 1930), II:103.

39. Ruth Benedict, *Patterns of Culture* (Boston, 1934), p. 278; but cf. the same author's *Race: Science and Politics* (New York, 1940), where her relativism is rather less pure, for the same reasons that affected Boas' judgment.

40. Melville J. Herskovits, *The Economic Life of Primitive Peoples* (New York, 1940), pp. 7–8; and "The Problem of Cultural Relativism," in *Man and His Works: The Science of Cultural Anthropology* (New York, 1948), especially p. 69.

41. Alfred L. Kroeber, *Configurations of Culture Growth* (Berkeley and Los Angeles, 1944), pp. 821–22.

42. Kroeber, *Anthropology* (New York, 1948), p. 304.

43. Julian H. Steward, *Theory of Culture Change: The Methodology of Multilinear Evolution* (Urbana, Ill., 1955), pp. 13–14.

44. E. E. Evans-Pritchard, *Social Anthropology* (London, 1951), pp. 40–41.

45. See Ferdinand Tönnies, *Fundamental Concepts of Sociology,* tr. Charles P. Loomis (New York, 1940).

46. Max Weber, "The Meaning of 'Ethical Neutrality' in Sociology and

Economics," in *The Methodology of the Social Sciences,* tr. Edward A. Shils and Henry A. Finch (Glencoe, Ill., 1949), p. 5. See also his lecture, "Science as a Vocation," in *From Max Weber: Essays in Sociology,* tr. H. H. Gerth and C. Wright Mills (New York, 1958), pp. 129–56.

47. *The Methodology of the Social Sciences,* p. 5.

48. Ibid., pp. 38–39.

49. Raymond Aron, *German Sociology,* tr. Mary and Thomas Bottomore (New York, 1964), pp. 105–106.

50. See the argument for this point of view in S. E. Finer's Introduction to Vilfredo Pareto, *Sociological Writings,* ed. Finer (New York, 1966), pp. 65–66.

51. Pareto, *The Mind and Society* [*Trattato di sociologia generale*], tr. Andrew Bongiorno and Arthur Livingston (New York, 1935), IV:1726.

52. Florian Znaniecki, *Cultural Reality* (Chicago, 1919), pp. 21–22 and 16.

53. See Znaniecki, *Cultural Sciences: Their Origin and Development* (Urbana, Ill., 1952).

54. The ambiguities of his position are elucidated in Aron, *German Sociology,* pp. 55–62.

55. See Alfred Weber, *Kulturgeschichte als Kultursoziologie,* second edition (Munich, 1950); Robert M. MacIver, *Society: A Textbook of Sociology* (New York, 1937), chs. 14 and 28; and MacIver, *The Challenge of the Passing Years: My Encounter with Time* (New York, 1962), ch. 15. On Alfred Weber, consult also Aron, pp. 43–51.

56. See William F. Ogburn and Meyer F. Nimkoff, *Sociology* (Boston, 1940), pp. 904–909. For another discussion of sociology and disbelief in progress, see Robert A. Bailey, III, *Sociology Faces Pessimism: A Study in European Sociological Thought Amidst a Fading Optimism* (The Hague, 1958).

57. Arnold J. Toynbee, *A Study of History,* second edition (New York, 1962), I:193 and 159.

58. Ibid., I:147–81.

59. Ibid., VI:312–21.

60. Ibid., VI:167. For the whole argument, VI:149–75 and 365–69.

61. Pitirim A. Sorokin, *Social and Cultural Dynamics,* revised edition (Boston, 1957), pp. 667–69. For an explanation of immanent change and limit see chs. 38–39.

62. For Sorokin's "integral" theory of truth, see ibid., pp. 683–92.

63. Ibid., pp. 628 and 702. See also Sorokin's *The Crisis of Our Age* (New York, 1941); *The Reconstruction of Humanity* (Boston, 1948); and *Social Philosophies of an Age of Crisis* (Boston, 1950).

64. See Philip Bagby, *Culture and History: Prolegomena to the Comparative Study of Civilizations* (Berkeley and Los Angeles, 1963); and Carroll Quigley, *The Evolution of Civilizations: An Introduction to Historical Analysis* (New York, 1961).

65. Charles Galton Darwin, *The Next Million Years* (Garden City, N.Y., 1953). The author was a grandson of Charles Darwin. For an economist's view, see Shepard B. Clough, *The Rise and Fall of Civilization: An Inquiry*

into the Relationship between Economic Development and Civilization (New York, 1951).

11. *The Age of Anxiety*

1. Albert Camus, *The Rebel,* tr. Anthony Bower (New York, 1956), p. 225.
2. The terrain surveyed in this chapter has also been covered by Judith N. Shklar, *After Utopia: The Decline of Political Faith* (Princeton, N.J., 1957), chs. 4–6. Cf. Franklin L. Baumer, "Twentieth-Century Version of the Apocalypse," *Journal of World History,* I (January 1954), 623–40, reprinted in W. Warren Wagar, ed., *European Intellectual History since Darwin and Marx* (New York, 1967), pp. 110–34.
3. The most readable introduction to the new eschatological theology before 1914 is Albert Schweitzer, *The Quest of the Historical Jesus,* tr. W. Montgomery (London, 1910), especially chs. 15–20.
4. Karl Barth, *The Epistle to the Romans,* tr. Edwyn C. Hoskyns (London, 1933), p. 169.
5. Barth, *The Word of God and the Word of Man,* tr. Douglas Horton (Boston, 1928), pp. 17 and 166.
6. Barth, *Church Dogmatics: The Doctrine of Reconciliation,* Part 3, tr. G. W. Bromiley (Edinburgh, 1961–62), II:939.
7. Barth, *Against the Stream: Shorter Post-War Writings, 1946–1952,* ed. Ronald Gregor Smith (New York, 1954), p. 35.
8. See T. F. Torrance, *Karl Barth: An Introduction to His Early Theology, 1910–1931* (London, 1962).
9. Emil Brunner, *The Theology of Crisis* (New York, 1929), p. 60.
10. Brunner, *Eternal Hope* (London, 1954), pp. 10–11.
11. Ibid., pp. 23, 78, and 80.
12. Reinhold Niebuhr, *Faith and History: A Comparison of Christian and Modern Views of History* (New York, 1949), pp. 6–7.
13. Niebuhr, *The Nature and Destiny of Man* (New York, 1941–43), II:318.
14. Ibid., II:207.
15. *Faith and History,* pp. 232–33.
16. See Nicolas Berdyaev, *The Fate of Man in the Modern World,* tr. Donald A. Lowrie (Ann Arbor, Mich., 1961), pp. 38–39. The first edition was published in 1934.
17. *The Meaning of History* [1923], as quoted in Matthew Spinka, *Nicolas Berdyaev: Captive of Freedom* (Philadelphia, 1950), p. 150. Spinka provides a good introduction to Berdyaev's philosophy of history and his eschatology. See especially chs. 7–10.
18. *The Fate of Man in the Modern World,* p. 130.
19. Berdyaev, *The Destiny of Man,* tr. Natalie Duddington (London, 1937), pp. 374–75.

20. Ibid., pp. 367–68.

21. Rudolf Bultmann, *History and Eschatology* (Edinburgh, 1957), pp. 126, 143–44, and 155.

22. Bultmann, *Jesus Christ and Mythology* (New York, 1958), p. 66.

23. See Paul Tillich, *The Interpretation of History,* tr. N. A. Rasetzki and Elsa L. Talmey (New York, 1936); and *Systematic Theology* (Chicago, 1951–63), III, Part 5. Tillich has also provided an explicit analysis of modern cultural despair as a function of the loss of faith in *The Courage to Be* (New Haven, Conn., 1952).

24. See especially the essays by C. H. Dodd, Edwyn Bevan, and Heinz-Dietrich Wendland in H. G. Wood et al., *The Kingdom of God and History* (Chicago, 1938). Dodd addresses himself to the question of progress on pp. 36–37, Bevan on pp. 59 and 70–71, and Wendland on pp. 162–64. Tillich was also among the contributors, and his essay—perhaps the best of the lot—appears on pp. 107–41.

25. Karl Löwith, *Meaning in History* (Chicago, 1949), p. 3; and Erich Frank, *Philosophical Understanding and Religious Truth,* tr. Mrs. Ludwig Edelstein (New York, 1945), p. 128.

26. See W. R. Inge, *The Idea of Progress* (Oxford, 1920); Paul Althaus, *Die letzten Dinge* (Gütersloh, 1922); Dodd, *History and the Gospel* (New York, 1938); Eric Voegelin, *The New Science of Politics* (Chicago, 1952), especially pp. 119–21; and Josef Pieper, *The End of Time,* tr. Michael Bullock (London, 1954). Cf. Eric Rust, *Towards a Theological Understanding of History* (New York, 1963).

27. Mircea Eliade, *The Myth of the Eternal Return,* tr. Willard R. Trask (New York, 1954), ch. 4; and Thomas J. J. Altizer, *Mircea Eliade and the Dialectic of the Sacred* (Philadelphia, 1963), p. 200. The last two clauses are adapted from Nietzsche's *Thus Spake Zarathustra.*

28. Cf. the contrast, borrowed from Heidegger, between "human time" and "world-time" in John Wild, *The Challenge of Existentialism* (Bloomington, Ind., 1955), pp. 242–49.

29. See Martin Heidegger, *Being and Time,* tr. John Macquarrie and Edward Robinson (New York, 1962), especially sections 72–83, pp. 424–88. Dilthey is examined in section 77, Hegel in section 82.

30. See Albert Rabil, Jr., *Merleau-Ponty: Existentialist of the Social World* (New York, 1967), pp. 148–53.

31. Jean-Paul Sartre, *Existentialism and Humanism,* tr. Philip Mairet (London, 1948), p. 33.

32. Wilfrid Desan, *The Marxism of Jean-Paul Sartre* (Garden City, N.Y., 1966), p. 261. Sartre expounds his view of Marxism in *Critique de la raison dialectique* (Paris, 1960); see pp. 640–43 in particular.

33. Albert Camus, *The Myth of Sisyphus and Other Essays,* tr. Justin O'Brien (New York, 1955), pp. 60 and 92.

34. Camus, *The Rebel,* p. 179.

35. Camus, *The Plague,* tr. Stuart Gilbert (New York, 1948), p. 278.

36. *The Rebel,* p. 303. See also Gabriel Marcel, *Man Against Mass Society,* tr. G. S. Fraser (Chicago, 1952).

37. Paul Valéry, *History and Politics,* tr. Denise Folliot and Jackson Mathews (New York, 1962), p. 23.

38. A perceptive study of the theme of hostility to time in twentieth-century thought is available in Douglas K. Wood's doctoral dissertation in history at Yale University, "Men Against Time," 1967. I had the pleasure of serving on Wood's dissertation committee and although the thinking that follows is entirely my own, we may owe our common interest in this theme to the fact that we were both students of F. L. Baumer. Wood gives special consideration to the thought of Jung, A. Huxley, Eliot, and Berdyaev. See also A. A. Mendilow, *Time and the Novel* (London, 1952); Hans Meyerhoff, *Time in Literature* (Berkeley and Los Angeles, 1955); and Georges Poulet, *Studies in Human Time,* tr. Elliott Coleman (Baltimore, 1956), and *The Metamorphoses of the Circle,* tr. Carley Dawson and Elliott Coleman (Baltimore, 1966).

39. Altizer, *Mircea Eliade and the Dialectic of the Sacred,* p. 164.

40. Erich Heller, *The Disinherited Mind* (Cleveland, 1959), p. 219.

41. Samuel Beckett, *Malone Dies,* tr. Beckett, in *Three Novels* (London, 1959), p. 193.

42. Hermann Hesse, *Steppenwolf,* tr. Basil Creighton (New York, 1957), p. 215. *Steppenwolf* originally appeared in 1927.

43. Hesse, *Magister Ludi,* tr. Mervyn Savill (New York, 1949), pp. 37 and 327.

44. Wyndham Lewis, *Time and Western Man* (Boston, 1957), pp. 437–38.

45. For a close analysis of Eliot's doctrine of time, see Staffan Bergsten, *Time and Eternity: A Study in the Structure and Symbolism of T. S. Eliot's Four Quartets* (Stockholm, 1960); cf. Poulet, *The Metamorphoses of the Circle,* pp. 342–47.

46. T. S. Eliot, *Notes Towards a Definition of Culture* (London, 1948), pp. 18–19, 29, and 33–34.

47. Aldous Huxley, *The Perennial Philosophy* (New York, 1945), pp. vii, 20, and 79–80.

48. Ibid., p. 200.

49. E. M. Cioran, *The Temptation to Exist,* tr. Richard Howard (New York, 1968), p. 196. The first French edition was published in 1956.

50. Cioran, *Histoire et utopie* (Paris, 1960), pp. 181, 75, and 171.

51. *The Temptation to Exist,* pp. 100–101.

52. *Histoire et utopie,* pp. 73–74.

53. *The Temptation to Exist,* p. 221.

54. *Histoire et utopie,* pp. 192–94; and *The Temptation to Exist,* p. 199.

55. Huxley, *Ape and Essence* (New York, 1948), pp. 120, 123, and 125–26. For an analysis of counter-utopias from the Christian point of view, see Chad Walsh, *From Utopia to Nightmare* (New York, 1962). Also of interest is George Kateb, *Utopia and Its Enemies* (New York, 1963).

56. Karl Jaspers, *Man in the Modern Age,* tr. Eden and Cedar Paul (Garden City, N.Y., 1957), pp. 226–27.

57. See, e.g., Erich Kahler, *The Tower and the Abyss* (New York, 1957); Friedrich Georg Juenger, *The Failure of Technology,* tr. F. D. Wieck (Hins-

dale, Ill., 1949); C. S. Lewis, *The Abolition of Man* (New York, 1947); Alex Comfort, *Art and Social Responsibility* (London, 1946); Jacques Ellul, *The Technological Society* [1954], tr. John Wilkinson (New York, 1964); Lewis Mumford, *The Transformations of Man* (New York, 1956); Joseph Wood Krutch, *The Measure of Man* (Indianapolis, 1954); and Roderick Seidenberg, *Post-Historic Man* (Chapel Hill, N.C., 1950). Krutch is also the author of *The Modern Temper* (New York, 1929), an early but widely discussed book in the vein of late nineteenth-century naturalistic *Weltschmerz*. Jaspers, Mumford, and Kahler will be examined further in Part IV below.

58. Seidenberg, pp. 27, 107, and 237–38.

59. Ellul, pp. 428 and 434.

60. Gordon Rattray Taylor, *The Biological Time Bomb* (Cleveland, 1968). Cf. Jean Rostand, *Can Man Be Modified?*, tr. Jonathan Griffin (London, 1959).

61. See Shklar, *After Utopia,* ch. 6.

62. Albert Salomon, *The Tyranny of Progress* (New York, 1955), p. 104; and Walter Lippmann, *The Public Philosophy* (Boston, 1955), pp. 178–79. Cf. Friedrich A. Hayek, *The Counter-Revolution of Science: Studies in the Abuse of Reason* (Glencoe, Ill., 1952); and Thomas Molnar, *The Decline of the Intellectual* (Cleveland, 1961), especially ch. 4. Molnar has also mounted an attack on the utopian tradition in his *Utopia: The Perennial Heresy* (New York, 1967).

63. R. H. S. Crossman, "Towards a Philosophy of Socialism," in Crossman, ed., *New Fabian Essays* (New York, 1952), pp. 8–10.

64. Eugene Kamenka, "Marxian Humanism and the Crisis in Socialist Ethics," in Erich Fromm, ed., *Socialist Humanism* (Garden City, N.Y., 1966), p. 128.

65. Shklar, p. 268.

PART FOUR

12. *In Defense of Modern Man*

1. Bernard Delfgaauw, *Geschichte als Fortschritt*, tr. Bruno Loets (Cologne, 1962–66), I: 11–12. There is no English translation as yet.

2. Charles Frankel, *The Case for Modern Man* (New York, 1955); and Erich Fromm, *The Revolution of Hope: Toward a Humanized Technology* (New York, 1968).

3. Adlai Stevenson, as quoted in Clarke A. Chambers, "The Belief in Progress in Twentieth-Century America," *Journal of the History of Ideas,* XIX (April 1958), 221.

4. Nelson A. Rockefeller, as quoted in *Time*, LXXXI (January 11, 1963), 26.

5. See Gustav A. Wetter, *Dialectical Materialism: A Historical and Systematic Survey of Philosophy in the Soviet Union,* tr. Peter Heath (New York, 1958), pp. 329–30.

6. W. Warren Wagar, *The City of Man: Prophecies of a World Civilization in Twentieth-Century Thought,* revised edition (Baltimore, 1967).

7. Sidney Pollard, *The Idea of Progress* (London, 1968), pp. 182–83.

8. Wagar in Wagar, ed., *European Intellectual History since Darwin and Marx* (New York, 1967), p. 10.

13. Holy Worldliness

1. Hans Urs von Balthasar, *A Theology of History,* second edition (New York, 1963), pp. 135–36.

2. Jacques Maritain, *The Peasant of the Garonne: An Old Layman Questions Himself about the Present Time,* tr. Michael Cuddihy and Elizabeth Hughes (New York, 1968).

3. Maritain, *True Humanism,* tr. M. R. Adamson (London, 1938), p. 88.

4. Most of the above will be found in Maritain's *On the Philosophy of History,* ed. Joseph W. Evans (New York, 1957), passim. See also *The Peasant of the Garonne,* pp. 36–37.

5. *On the Philosophy of History,* p. 154.

6. *The Peasant of the Garonne,* p. 4.

7. *True Humanism,* pp. 237–39.

8. Emmanuel Mounier, *Be Not Afraid,* tr. Cynthia Rowland (New York, 1962), pp. 10–11 and 104. Mounier's title was *La Petite peur du XXe siècle.*

9. Ibid., pp. 49 and 102.

10. Jean Daniélou, *The Lord of History: Reflections on the Inner Meaning of History,* tr. Nigel Abercrombie (London, 1958), pp. 10–11.

11. Ibid., pp. 272–73 and 275.

12. Ibid., pp. 93–94.

13. Ibid., p. 126.

14. See Martin C. D'Arcy, *The Meaning and Matter of History: A Christian View* (New York, 1959); and D'Arcy, "Is There a Nascent World Culture?," in A. William Loos, ed., *Religious Faith and World Culture* (New York, 1951), pp. 259–77.

15. Christopher Dawson, *The Historic Reality of Christian Culture: A Way to the Renewal of Human Life* (London, 1960), p. 63; and Dawson, *The Dynamics of World History,* ed. John J. Mulloy (New York, 1956), p. 412.

16. For Fosdick, see above, p. 102. See also E. E. Kresge, *The Church and the Ever-Coming Kingdom of God* (New York, 1922); Sherwood Eddy, *Religion and Social Justice* (New York, 1927); Eddy, *The Kingdom of God and the American Dream* (New York, 1941); and Shirley Jackson Case, *The Christian Philosophy of History* (Chicago, 1943).

17. John Macmurray, *Creative Society: A Study of the Relation of Christianity to Communism* (New York, 1936), pp. 57–58 and 108. See also Mac-

murray, *The Clue to History* (London, 1938); Kenneth Scott Latourette, *The Emergence of a World Christian Community* (New Haven, Conn., 1949); and John Baillie, *The Belief in Progress* (London, 1950).

18. Baillie, pp. 220 and 235.

19. Albert Schweitzer, *The Decay and the Restoration of Civilization,* tr. C. T. Campion and J. P. Naish (London, 1923), pp. 40 and 44.

20. Ibid., p. 87.

21. Schweitzer, *Civilization and Ethics* [1923], third edition, tr. C. T. Campion and Mrs. Charles E. B. Russell (London, 1946), pp. xviii–xix and xxiii.

22. Ibid., pp. 29 and 66.

23. *The Decay and the Restoration of Civilization,* pp. x–xi.

24. *Civilization and Ethics,* p. 273.

25. A good introduction to the new theology is available in John Macquarrie, *God and Secularity* (Philadelphia, 1967).

26. Eric Mascall, *The Secularization of Christianity* (London, 1965), p. 120.

27. See Friedrich Gogarten, *Verhängnis und Hoffnung der Neuzeit: Die Säkularisierung als theologisches Problem* (Stuttgart, 1953); and the discussion of Gogarten in Macquarrie, pp. 36–37.

28. Dietrich Bonhoeffer, *Ethics,* ed. Eberhard Bethge, tr. Neville Horton Smith (London, 1955), pp. 65, 287, and 294.

29. Ibid., pp. 99, 102–103, and 105–106.

30. Bonhoeffer, *Letters and Papers from Prison,* third edition, ed. Eberhard Bethge, tr. Reginald Fuller, Frank Clarke et al. (New York, 1967), pp. 38–39.

31. Harvey Cox, *The Secular City: Secularization and Urbanization in Theological Perspective,* revised edition (New York, 1966), p. 18.

32. Cox, "Afterword," in Daniel Callahan, ed., *The Secular City Debate* (New York, 1966), p. 193. Callahan reprints several pieces on Cox's relationship to the theology of Rauschenbusch. See especially David Little, "The Social Gospel Revisited," and George D. Younger, "Does *The Secular City* Revisit the Social Gospel?" pp. 69–74 and 77–80.

33. Cox, "Afterword," in *The Secular City Debate,* p. 191.

34. See Hendrik Kraemer, *World Cultures and World Religions: The Coming Dialogue* (Philadelphia, 1960).

35. Arend T. van Leeuwen, *Christianity in World History: The Meeting of the Faiths of East and West,* tr. H. H. Hoskins (London, 1964), pp. 18, 408, and 14.

36. Ibid., p. 409.

37. Ibid., p. 431.

38. See Jürgen Moltmann and Jürgen Weissbach, *Two Studies in the Theology of Bonhoeffer,* tr. Reginald H. Fuller and Ilse Fuller (New York, 1967).

39. Moltmann, *Theology of Hope: On the Ground and Implications of a Christian Eschatology,* tr. James W. Leitch (New York, 1967), p. 16.

40. Ibid.

41. Ibid., pp. 329–30. See also Moltmann, *Religion, Revolution, and the Future,* tr. M. Douglas Meeks (New York, 1969).

42. Edward Schillebeeckx, *God the Future of Man,* tr. N. D. Smith (New York, 1968); and Louis Weeks, III, "Can Saint Thomas's *Summa Theologiae* Speak to Moltmann's *Theology of Hope?*" *The Thomist,* XXXIII (April 1969), 215–28.

43. Abraham J. Heschel, *The Insecurity of Freedom* (New York, 1966), p. 146.

44. Heschel, *Israel: An Echo of Eternity* (New York, 1969), pp. 220–21 and 218.

14. *Philosophies of Hope*

1. Bertrand Russell, "A Free Man's Worship," in *Philosophical Essays* (London, 1910), p. 70. Cf. Russell, *New Hopes for a Changing World* (New York, 1952), p. 187: "The more we realize our minuteness and our impotence in the face of cosmic forces, the more astonishing becomes what human beings have achieved."

2. Russell, *Human Society in Ethics and Politics,* London, 1954, pp. 156–57.

3. *New Hopes for a Changing World,* p. 184; *Human Society in Ethics and Politics,* pp. 136–37.

4. See Russell, *The Scientific Outlook* (New York, 1931), chs. 12–17.

5. Russell, *Has Man a Future?* (Harmondsworth, Middlesex, 1961), p. 36.

6. Ibid., pp. 126–27. Cf. Russell, *The Prospects of Industrial Civilization* (with Dora Russell) (New York, 1923), pp. 286–87; Russell, *Portraits from Memory* (New York, 1956), p. 12; and *New Hopes for a Changing World,* ch. 21.

7. H. J. Blackham, *The Human Tradition* (London, 1953), pp. 236, 188, and 237. Blackham has served as secretary of the International Humanist and Ethical Union, and is a founder of the British Humanist Association. See also A. J. Ayer, ed., *The Humanist Outlook* (London, 1968). "The balance of the evidence," writes Ayer, "is on the side of moral progress. . . . The average man is more humane, more pacific and more concerned with social justice than he was a century ago." Ibid., p. 8.

8. Corliss Lamont, *The Philosophy of Humanism,* fourth edition (New York, 1957), pp. 9–10, 228, 11, and 236.

9. See Charles Frankel, *The Faith of Reason: The Idea of Progress in the French Enlightenment* (New York, 1948).

10. Frankel, *The Case for Modern Man* (New York, 1955), p. 44.

11. For his worst fears, see José Ortega y Gasset, *The Revolt of the Masses* (New York, 1932).

12. Ortega, *The Modern Theme* [1923], tr. James Cleugh (New York, 1961), pp. 74 and 69; and Ortega, *History as a System, and Other Essays toward a Philosophy of History,* tr. Helene Weyl, William C. Atkinson, and Eleanor Clark (New York, 1961), pp. 216–17, 219, and 81.

13. *History as a System,* p. 218.

14. See Alfred North Whitehead, *Process and Reality* (New York, 1929), p. 139.

15. Whitehead, *Science and the Modern World* [1925] (New York, 1948), p. 114.

16. Whitehead, *Adventures of Ideas* [1933] (New York, 1955), pp. 32, 90, and 93; *Science and the Modern World,* p. 1. *Adventures of Ideas,* Part I, is the best introduction to Whitehead's conception of human progress.

17. *Science and the Modern World,* p. 205.

18. *Adventures of Ideas,* p. 273. For the thought of J. E. Boodin, see his *Cosmic Evolution* (New York, 1925); *God: A Cosmic Philosophy of Religion* (New York, 1934); and *The Social Mind* (New York, 1939).

19. See Ernst Benz, *Evolution and Christian Hope,* tr. Heinz G. Frank (Garden City, N.Y., 1966), pp. 177–82.

20. Edgar Dacqué, *Die Urgestalt* (Leipzig, 1940), p. 78, as quoted in Benz, p. 181. See also Dacqué, *Urwelt, Sage und Menschheit: Eine naturhistorisch-metaphysische Studie* (Munich, 1924); and Dacqué, *Leben als Symbol: Metaphysik einer Entwicklunglehre* (Munich, 1928).

21. Pierre Lecomte du Noüy, *La Dignité humaine* (New York, 1944), p. 23.

22. Lecomte du Noüy, *Human Destiny* (New York, 1947), pp. 133–34.

23. Ibid., pp. 177, 187, and 225.

24. Benz, p. 226.

25. See, for example, Benz, chs. 12–13; Claude Tresmontant, *Introduction à la pensée de Teilhard de Chardin* (Paris, 1956); Charles E. Raven, *Teilhard de Chardin: Scientist and Seer* (London, 1962); Ignace Lepp, *Die neue Erde: Teilhard de Chardin und das Christentum in der Modernen Welt* (Olten-Freiburg, 1962); Madeleine Barthélemy-Madaule, *Bergson et Teilhard de Chardin* (Paris, 1963); Claude Cuénot, *Teilhard de Chardin: A Biographical Study,* ed. René Hague, tr. Vincent Colimore (London, 1965); Henri de Lubac, *The Faith of Teilhard de Chardin,* tr. René Hague (London, 1965); Emile Rideau, *The Thought of Teilhard de Chardin,* tr. René Hague (New York, 1968); and Jules Chaix-Ruy, *The Superman from Nietzsche to Teilhard de Chardin,* tr. Marina Smyth-Kok (Notre Dame, Ind., 1968).

26. Pierre Teilhard de Chardin, *The Future of Man,* tr. Norman Denny (New York, 1964), pp. 127 and 228.

27. Ibid., p. 127.

28. Ibid., p. 18.

29. Ibid., pp. 137 and 139.

30. Teilhard de Chardin, *The Phenomenon of Man,* tr. Bernard Wall (New York, 1959), p. 254; and *The Future of Man,* pp. 303 and 81.

31. *The Phenomenon of Man,* pp. 273–90.

32. *The Future of Man,* pp. 128 and 122–23.

33. Maritain, *The Peasant of the Garonne,* p. 119.

34. Bernard Delfgaauw, *Geschichte als Fortschritt,* tr. Bruno Loets (Cologne, 1962–66), I: 28.

35. Ibid., II: 182 and 212.

36. Karl Jaspers, *The Origin and Goal of History*, tr. Michael Bullock (New Haven, Conn., 1953), p. 233.

37. Ibid., p. 264.

38. Ibid., p. xv.

39. Ibid., p. 263.

40. Ibid., pp. 2 and 25.

41. Ibid.; p. 48.

42. Ibid., p. 227.

43. See Jaspers, *Man in the Modern Age* [1931], tr. Eden and Cedar Paul (Garden City, N.Y., 1957), pp. 226–28.

44. *The Origin and Goal of History*, pp. 148–49.

45. Lewis Mumford, *The Transformations of Man*, New York, 1956, pp. 248–49.

46. Erich Kahler, *Man the Measure* (New York, 1943), pp. 607–608.

47. Ibid., pp. 606 and 639.

48. Kahler, *The Meaning of History* (New York, 1964), p. 208.

49. Ibid., p. 219. See Kahler, *The Tower and the Abyss: An Inquiry into the Transformation of the Individual* (New York, 1957), ch. 7, for the fullest statement of his vision of world order.

50. Arnold J. Toynbee, *A Study of History* (New York, 1934–61), XII:27.

51. Ibid., VII:374.

52. Ibid., VII:562–64.

53. Toynbee reports that even this list of six is not "exhaustive," ibid., XII:218, fn 4.

54. See Toynbee, *Change and Habit* (New York, 1966), pp. 22–25.

55. *A Study of History*, XII:563–64. See also the discussion of the term "progress," XII:266–68.

56. Ibid., XII:570 and 143. For other twentieth-century philosophies of history endorsing the belief in progress, see Feliks Koneczny, *On the Plurality of Civilisations* [1935] (London, 1962); William Ernest Hocking, *The Coming World Civilization* (New York, 1956); Denis de Rougemont, *Man's Western Quest*, tr. Montgomery Belgion (New York, 1957); and René Silvain, *Les Origines de la pensée moderne* (Paris, 1963).

57. J. H. Plumb in Plumb, ed., *Crisis in the Humanities* (Harmondsworth, Middlesex, 1964), pp. 36 and 34; and E. H. Carr, *What Is History?* (New York, 1962), pp. 149 and 158. Cf. the B.B.C. talk of another Cambridge historian, B. H. G. Wormald, "Progress and Hope," *The Listener*, LXXI (April 9, 1964), 581–83.

58. See Charles Beard, "Introduction," in J. B. Bury, *The Idea of Progress* (New York, 1932), pp. ix–xl; Harry Elmer Barnes, *An Intellectual and Cultural History of the Western World*, third edition (New York, 1965), III: 1310–42; and Carl Becker, *Progress and Power* (Stanford, California, 1936). Cf. Becker's article on "Progress" in *Encyclopaedia of the Social Sciences*, XII (New York, 1934), pp. 495–99. Even in *Progress and Power*, he attempted to define progress in value-free terms as man's gradual acquisition of power

and control over the conditions of life, but by the end of the book this process had definitely assumed a normative character, the strictures of its first chapter notwithstanding.

59. Sidney B. Fay, "The Idea of Progress," *American Historical Review,* LII (January 1947), 246.

60. Crane Brinton, *Ideas and Men: The Story of Western Thought* (Englewood Cliffs, N.J., 1950), p. 492. See also Brinton, John B. Christopher, and Robert Lee Wolff, *A History of Civilization,* second edition (Englewood Cliffs, N.J., 1960), II:685. More recently, Georg G. Iggers has offered a cautious defense of progressivism in "The Idea of Progress: A Critical Reassessment," *American Historical Review,* LXXI (October 1965), 1–17.

61. Vincent Brome, *The Problem of Progress* (London, 1963).

62. Will and Ariel Durant, *The Lessons of History* (New York, 1968), p. 102. See also Herbert J. Muller, *The Uses of the Past* (New York, 1952), pp. 367–74.

15. *Science and the Human Prospect*

1. The image of the bees and the angels is borrowed from H. G. Wells, *The Open Conspiracy: Blue Prints for a World Revolution* (Garden City, N.Y., 1928), pp. 147–48.

2. H. J. Muller, *Out of the Night: A Biologist's View of the Future* (New York, 1935), pp. 124–25. See also Muller, "The Guidance of Human Evolution," in Sol Tax, ed., *Evolution after Darwin* (Chicago, 1960), II:423–62; Muller, "The Human Future," in Sir Julian Huxley, ed., *The Humanist Frame* (New York, 1961), pp. 399–414; Edwin Grant Conklin, *The Direction of Human Evolution* (New York, 1921); Conklin, *Man, Real and Ideal* (New York, 1943); J. B. S. Haldane, *Possible Worlds* (New York, 1928); Haldane, *Science and Human Life* (New York, 1933); Haldane, *Science Advances* (New York, 1947); Joseph Needham, "Integrative Levels: A Revaluation of the Idea of Progress," in *Time: The Refreshing River* (London, 1943), pp. 233–72; and Needham, *History Is on Our Side: A Contribution to Political Religion and Scientific Faith* (New York, 1947), especially chs. 2 and 12.

3. See Julian Huxley, "Progress, Biological and Other," in *Essays of a Biologist* (New York, 1923), pp. 3–65; and Huxley, "Eugenics in Evolutionary Perspective," in *Essays of a Humanist* (New York, 1964), pp. 251–80. In the latter, the Galton Lecture for 1962, Huxley strongly supports what he calls "E.I.D.—eugenic insemination by deliberately preferred donors" as a means of accelerating human progress.

4. The two lectures are reprinted in T. H. Huxley and Julian Huxley, *Touchstone for Ethics* (New York, 1947), pp. 67–166.

5. Huxley, *Religion without Revelation* [1927], revised edition (New York, 1953), p. 217; and Huxley, *Evolution in Action* (New York, 1953), p. 149.

6. *Touchstone for Ethics,* p. 256; and *Religion without Revelation,* p. 217.

7. Huxley in Huxley, ed., *The Humanist Frame,* pp. 26 and 48.

8. *Touchstone for Ethics,* pp. 139, 254, and 140.

9. See George Gaylord Simpson, *The Meaning of Evolution* (New Haven, Conn., 1949); Simpson, *This View of Life: The World of an Evolutionist* (New York, 1964); Theodosius Dobzhansky, *The Biological Basis of Human Freedom* (New York, 1956); Dobzhansky, *Mankind Evolving* (New Haven, Conn., 1962); and Dobzhansky, *The Biology of Ultimate Concern* (New York, 1967). For their views of Teilhard de Chardin contrast ch. 11 of Simpson's *This View of Life* and ch. 6 of Dobzhansky's *The Biology of Ultimate Concern.* See also the teleological vitalism of Edmund W. Sinnott, an American biologist of Sir Julian Huxley's generation, in his books, *The Biology of the Spirit* (New York, 1955), and *Matter, Mind and Man: The Biology of Human Nature* (New York, 1957).

10. C. H. Waddington, *The Ethical Animal* [1960] (Chicago, 1967), pp. 5, 59, and 65.

11. Waddington, "The Human Animal," in Huxley, ed., *The Humanist Frame,* p. 73.

12. *The Ethical Animal,* p. 15.

13. Ibid., pp. 186–87.

14. J. D. Bernal, *The Social Function of Science* [1939] (Cambridge, Mass., 1967), p. 155.

15. Ibid., p. xix.

16. Bernal, *Science in History,* third edition (New York, 1965), p. 978. See also Bernal, *The Freedom of Necessity* (London, 1949).

17. Alfred Korzybski, *Manhood of Humanity: The Science and Art of Human Engineering* (New York, 1921), pp. 59 and 92.

18. Ibid., p. 192.

19. See Buckminster Fuller, *Operating Manual for Spaceship Earth* (Carbondale, Ill., 1969); and C. P. Snow, "The Future of Man," *The Nation,* CLXXXVII (September 13, 1958), 124–25.

20. John Rader Platt, *The Step to Man* (New York, 1966), pp. 168, 183, 200, and 203.

21. Arthur C. Clarke, *The Promise of Space,* New York, 1968, p. 313.

16. *The Social Sciences*

1. Stanley Casson, *Progress and Catastrophe: An Anatomy of Human Adventure* (New York, 1937).

2. Robert Briffault, *Rational Evolution* (New York, 1930), p. 6.

3. V. Gordon Childe, *Man Makes Himself,* revised edition (New York, 1951), pp. 9, 19, and 11.

4. Childe, *Progress and Archaeology* (London, 1944), pp. 114–15.

5. *Man Makes Himself,* p. 187; and Childe, *What Happened in History* [1942], revised edition (Harmondsworth, Middlesex, 1954), p. 282.

6. Bronislaw Malinowski, *Freedom and Civilization* (New York, 1944), p. 4.

7. Margaret Mead, *New Lives for Old* (New York, 1956), pp. 5–6 and 458.

See also her *Continuities in Cultural Evolution* (New Haven, Conn., 1964), which discusses the mechanisms of cultural evolution and rejoices in "the miracle of man's progress on earth." The more man learns of his cosmos, she adds, the more his own place in it inspires awe (p. 291).

8. See Robert Redfield, *A Village That Chose Progress: Chan Kom Revisited* (Chicago, 1950).

9. Redfield, *The Primitive World and Its Transformations* (Ithaca, N.Y., 1953), pp. 145–46 and 159.

10. Ibid., pp. 163 and 165. See also M. F. Ashley Montagu, *The Direction of Human Development* (New York, 1955).

11. Jean Fourastié, *The Causes of Wealth,* tr. and ed. Theodore Caplow (Glencoe, Ill., 1960), pp. 205 and 229; and Fourastié and Claude Vimont, *Histoire de Demain* (Paris, 1956), p. 127. Cf. Robert Heilbroner, *The Future as History* (New York, 1960), especially pp. 204–209; and Kenneth E. Boulding, *The Meaning of the Twentieth Century* (New York, 1964).

12. See Franz Oppenheimer, *The State: Its History and Development Viewed Sociologically,* tr. John M. Gitterman (Indianapolis, 1914); and Oppenheimer, *System der Soziologie* (Jena, 1922–64).

13. See Karl Mannheim, *Essays in the Sociology of Knowledge,* tr. and ed. Paul Kecskemeti (New York, 1952), pp. 116–22.

14. Mannheim, *Man and Society in an Age of Reconstruction* (London, 1940), p. 152.

15. For Mannheim's final thoughts on planning and on the role of religion, see his *Diagnosis of Our Time* (New York, 1943); and *Freedom, Power and Democratic Planning* (New York, 1950)

16. Morris Ginsberg, *The Idea of Progress: A Revaluation* (London, 1953); Ginsberg, "A Humanist View of Progress," in Sir Julian Huxley, ed., *The Humanist Frame* (New York, 1961), pp. 111–28; and Ginsberg, "Moral Progress: A Reappraisal," in A. J. Ayer, ed., *The Humanist Outlook* (London, 1968), pp. 129–44. The influence of Ginsberg is apparent in Leslie Sklair's recently published doctoral thesis at the University of London, *The Sociology of Progress* (London, 1970).

17. "A Humanist View of Progress," p. 113; and "Moral Progress: A Reappraisal," pp. 143 and 141.

18. See George A. Lundberg, *Can Science Save Us?* (New York, 1947); C. Wright Mills, *The Sociological Imagination* (New York, 1959); and Mills, *The Marxists* (New York, 1962). For a tentative concept of progress as the evolution of the prerequisites of freedom (not freedom itself), see "On the Notions of Progress, Revolution, and Freedom," by the Harvard sociologist Barrington Moore, in *Ethics,* LXXII (January 1962), 106–19.

19. Burnham Beckwith, *The Next 500 Years* (New York, 1967). For his comments on the Marquis de Condorcet, see pp. 6–8. Consult also Daniel Bell, ed., *Toward the Year 2000* (Boston, 1968); Herman Kahn and Anthony J. Wiener, *The Year 2000* (New York, 1967); John McHale, *The Future of the Future* (New York, 1969); and *The Futurist,* bimonthly journal of the World Future Society, Washington, D.C.

20. See Alfred Cobban, "The Decline of Political Theory," *Political Science Quarterly*, LXVII (September 1953), 321–37.

21. John T. Marcus, *Heaven, Hell, and History: A Survey of Man's Faith in History from Antiquity to the Present* (New York, 1967), p. 196; see also p. 265.

22. Werner Sombart, *A New Social Philosophy* [*Deutscher Sozialismus*], tr. and ed. Karl F. Geiser (Princeton, N.J., 1937), pp. 149 and 147.

23. Adolf Hitler, *Mein Kampf* [1925–27], tr. Ralph Manheim (Boston, 1943), pp. 288–89.

24. Ibid., pp. 396 and 384.

25. Walter Lippmann, *Essays in the Public Philosophy* (Boston, 1955).

26. Lippmann, *An Inquiry into the Principles of the Good Society* (Boston, 1937), pp. xiii and 19–20.

27. Ibid., p. 389. See also R. V. Sampson, *Progress in the Age of Reason* (Cambridge, Mass., 1956), ch. 11.

28. Léon Blum, *For All Mankind*, tr. W. Pickles (New York, 1946), pp. 185 and 181.

29. G. D. H. Cole, *The Simple Case for Socialism* (London, 1935), pp. 18, 47, and 56.

30. Cole, *A History of Socialist Thought* (London, 1953–60), V:321–22.

31. Harold J. Laski, *The Rise of European Liberalism* (London, 1936), pp. 18–19.

32. Laski, *Reflections on the Revolution of Our Time* (New York, 1943), pp. 93 and 411; and Laski, *Faith, Reason, and Civilization* (New York, 1944), pp. 183–84 and 187.

33. Michael Harrington, *The Accidental Century* (New York, 1965), pp. 39 and 42.

34. See, for example, Isaac Deutscher, ed., *The Age of Permanent Revolution: A Trotsky Anthology* (New York, 1964), pp. 360–65.

35. N. I. Konrad, "Notes on the Meaning of History," translated from *Vestnik Istorii Mirovoi Kul'tury* (1961) in *Soviet Studies in History*, I (Summer 1962), 3–23. See also Constantin Borgeanu, "Sur l'Actualité de l'idée de progrès," in *Akten des XIV. Internationalen Kongresses für Philosophie* (Vienna, 1969), IV:584–90. Borgeanu is a young Rumanian philosopher who has written several studies of the idea of progress from a Marxist perspective.

36. N. I. Boldyrev, "To Help Create the New Soviet Man," translated from *Sovetskaia Pedagogika* (1962), in Harry G. Shaffer, ed., *The Soviet System in Theory and Practice: Selected Western and Soviet Views* (New York, 1965), p. 294; and G. H. Shakhnazarov, ed., *Social Sciences*, ibid., p. 296.

37. Ivan Efremov, *Andromeda: A Space-age Tale*, tr. G. Hanna (Moscow [1959?]).

38. See Maurice Cornforth, *Historical Materialism*, revised edition (London, 1962), pp. 26–28 and 73–76; Howard Selsam, *What Is Philosophy?: A Marxist Introduction* (New York, 1938), chs. 3 and 5; and Selsam, *Ethics and Progress* (New York, 1965).

39. See especially the anthology edited by Erich Fromm, *Socialist Humanism* (Garden City, N.Y., 1965), which includes thirteen essays from various anti-Stalinist socialist scholars in Poland, Czechoslovakia, and Yugoslavia.

40. Andrei D. Sakharov, *Progress, Coexistence and Intellectual Freedom,* tr. *The New York Times* (New York, 1968).

41. Harvey Cox, "Afterword," in Daniel Callahan, ed., *The Secular City Debate* (New York, 1966), pp. 200–203.

42. Ernst Bloch, "Man and Citizen According to Marx," tr. Norbert Guterman, in Fromm, ed., *Socialist Humanism,* pp. 225–26.

43. Bloch, *Das Prinzip Hoffnung* (Frankfurt am Main, 1959), III:1628. See also Bloch, "Differentiations in the Concept of Progress" and "Theses" in *A Philosophy of the Future,* tr. John Cumming (New York, 1970), pp. 112–44, where he applauds the Western idea of progress but rejects the traditional unilinear model of progress in favor of a "polyphonic" model incorporating the achievements of the non-Western peoples.

17. The Meta-Psychology of Progress

1. Gerald Heard, *The Ascent of Humanity: An Essay on the Evolution of Civilization from Group Consciousness through Individuality to Super-Consciousness* (New York, 1929), p. 6. See also Heard, *The Human Venture* (New York, 1955); and Heard, *The Five Ages of Man: The Psychology of Human History* (New York, 1963).

2. B. F. Skinner in Skinner and Carl Rogers, "Some Issues Concerning the Control of Human Behavior," *Science,* CXXIV (November 30, 1956), reprinted in Robert Perrucci and Marc Pilisuk, eds., *The Triple Revolution: Social Problems in Depth* (Boston, 1968), pp. 339 and 354. See also Robert Boguslaw, *The New Utopians: A Study of System Design and Social Change* (Englewood Cliffs, N.J., 1965), especially pp. 114–18; and George Kateb, *Utopia and Its Enemies* (New York, 1963), especially ch. 6.

3. Abraham Kaplan, "Freud and Modern Philosophy," in Benjamin Nelson, ed., *Freud and the Twentieth Century* (New York, 1957), pp. 228–29.

4. Sigmund Freud, *The Future of an Illusion* [1927], tr. W. D. Robson-Scott and James Strachey (Garden City, N.Y., 1964), p. 4; and Freud, *Civilization and Its Discontents* [1930], tr. James Strachey (New York, 1962), pp. 81 and 62.

5. *Civilization and Its Discontents,* pp. 23 and 58. See also *The Future of an Illusion,* pp. 4–5.

6. *The Future of an Illusion,* pp. 87–88 and 90; see also pp. 8 and 72–73; and *Civilization and Its Discontents,* pp. 62 and 88.

7. Philip Rieff, *Freud: The Mind of the Moralist* (Garden City, N.Y., 1961), p. 210.

8. *The Future of an Illusion,* pp. 81 and 87.

9. Ibid., pp. 81–82.

10. Erich Fromm, *Beyond the Chains of Illusion: My Encounter with Marx and Freud* (New York, 1962), p. 26.

11. Fromm, *The Sane Society* (New York, 1955), p. 27 and fn 1.

12. Ibid., p. 354; and Fromm, *You Shall Be as Gods: A Radical Interpretation of the Old Testament and Its Tradition* (New York, 1966).

13. *Beyond the Chains of Illusion*, p. 179; *The Sane Society*, p. 355; and Fromm, *Escape from Freedom* (New York, 1941), p. 264.

14. *The Sane Society*, pp. 352 and 362–63. See also Abraham H. Maslow, *Toward a Psychology of Being* (New York, 1962).

15. See Herbert Marcuse, "Epilogue: Critique of Neo-Freudian Revisionism," in *Eros and Civilization: A Philosophical Inquiry into Freud* [1955] (New York, 1962), pp. 217–51 passim. Fromm attacks Marcuse in *The Revolution of Hope: Toward a Humanized Technology* (New York, 1968), pp. 8–9, fn 3.

16. *Eros and Civilization*, p. 92.

17. Ibid., p. 18.

18. Ibid., pp. 197–99.

19. Marcuse, *One-Dimensional Man: Studies in the Ideology of Advanced Industrial Society* (Boston, 1964), p. 257.

INDEX

Extended treatment of a thinker's ideas of progress or anti-progress is indicated by page references set in **boldface.**

Racism: in the United States, 122-25; fears of racial failure, 171-73; and Adolf Hitler, 322-23

Radhakrishnan, Sarvepalli, 102

Ranke, Leopold von, 106, 177, **178**, 187

Ratzenhofer, Gustav, 43, 45

Rauschenbusch, Walter, 95, **100-102**, 104

Redfield, Robert, **316**

Reiser, Oliver L., 282 fn, 308-309 fn

Relativism, 147, 174-76; in historical thought, 176-88; in anthropology, 188-94; in sociology, 194-99; in the comparative study of civilizations, 199-203. *See also* Historicism

Religion: of humanity, 14, 26, 33, 88; collapse of faith during *la belle époque*, 25-26; 20th-century scepticism and longing, 147. *See also* Christianity; Theology

Renouvier, Charles, 105

Rickert, Heinrich, 178, 194

Rickman, H. P., 179

Rieff, Philip, 338

Ritschl, Albrecht, **82-83**, 180

Robertson, J. M., 26

Robinson, James Harvey, 122, 294

Robinson, John A. T., 258, 261

Rockefeller, Nelson A., 241

Roepke, Wilhelm, 233

Romanticism, 150-52, 164

Rosenberg, Alfred, 322

Rostow, W. W., 317

Rousseau, Jean-Jacques, 19, 139

Rudhyar, Dane, 309 fn

Russell, Bertrand, 72, 240, 267, **268-71**, 274

Sabatier, Auguste, 89 fn, 94

Sade, D. A. F., comte ["marquis"] de, **168-69**

Sahlins, Marshall D., 314

Saint-Pierre, Charles Irénée Castel, abbé de, 17, 18, 19, 336

Saint-Simon, Claude Henri de Rouvroy, comte de, 19

Sakharov, Andrei D., 331

Salomon, Albert, 234

Salomon, Roger B., 171

Sampson, R. V., 8, 17

Sartre, Jean-Paul, 205, 218, **219**, 221

Savage, Minot J., 96

Schäffle, Albert, 39

Schelling, F. W. J. von, 19, 56

Schillebeeckx, Edward, 266

Schiller, F. C. S., 80 fn

Schleiermacher, Friedrich, 82

Schopenhauer, Arthur, **158-59**, 162, 163

Schweitzer, Albert, 95, 207, 240, 243, 244, **254-57**, 258

Science: in the 17th century, 15-16, 17-18; turns to natural history, 16; influence during *la belle époque,* 29; as agency of progress, 30; and ideas of progress, chs. 4 and 15

Science fiction: and counter-utopism, 229-31; and ideas of progress, 307-308; in the Soviet Union, 330-31

Seidenberg, Roderick, **232-33**

Service, Elman R., 314

Shaw, Bernard, **74-76**, 149, 240

Sheckley, Robert, 230

Shestov, Leon, 212

Shklar, Judith N., 233, 235

Simmel, Georg, 178

Simon, W. M., 34

Simpson, George Gaylord, 301, 302

Skinner, B. F., **335-36**

Small, Albion, 43

Smith, Ronald Gregor, 261 fn

Smuts, J. C., **73-74**, 240, 274

Snow, C. P., 307

Socialism: revolutionary *vs.* evolutionary, 126; and ideas of progress, 126-42, 323, 325-33; and the denial of progress, 234-35. *See also* Marx, Karl; Marxism

Sociology: as discovery of the laws of progress, 19, 26, 38-54, 318-21; and relativism, 194-99

Sombart, Werner, **322**

Somervell, D. C., 199

Sorel, Georges, 138

Sorokin, Pitirim A., 185, 199, **201-202**

Spencer, Herbert, 18, 20, 95; and emergence of social sciences, 29, 38; definition of progress, 39, 63; criticized by L. T. Hobhouse, 51; as *bête noire* of anti-positivist philosophy, 56; on the eternal return, 62

Spengler, Oswald, 158, **163-64**, 228, 269; favorable reception of *The Decline of the West,* 146; and Brooks Adams, 165, 166-67; and relativism, 185; and